Food Security in the High North

This book explores the challenges facing food security, sustainability, sovereignty, and supply chains in the Arctic, with a specific focus on Indigenous Peoples.

Offering multidisciplinary insights with a particular focus on populations in the European High North region, the book highlights the importance of accessible and sustainable traditional foods for the dietary needs of local and Indigenous Peoples. It focuses on foods and natural products that are unique to this region and considers how they play a significant role towards food security and sovereignty. The book captures the tremendous complexity facing populations here as they strive to maintain sustainable food systems – both subsistent and commercial – and regain sovereignty over traditional food production policies. A range of issues are explored from food contamination risks, due to increasing human activities in the region, such as mining, to changing livelihoods and gender roles in the maintenance of traditional food security and sovereignty. The book also considers processing methods that combine indigenous and traditional knowledge to convert the traditional foods, which are harvested or hunted, into local foods.

This book offers a broader understanding of food security and sovereignty, and will be of interest to academics, scholars, and policy makers working in food studies, geography and environmental studies, agricultural studies, sociology, anthropology, political science, health studies, and biology.

Kamrul Hossain is a research professor and director of the Northern Institute for Environmental and Minority Law at the University of Lapland. He led several research projects with a focus on human rights and human security in the Arctic, and has widely published in these disciplines.

Lena Maria Nilsson is an experienced nutritional epidemiologist with a research focus on traditional Sami food as a determinant of health and on food security in the Arctic. Since 2019, she is the vice director at, Várdduo, the Centre for Sami Research at Umeå University, Sweden.

Thora Martina Herrmann is a cultural geographer with expertise in action-research projects in polar regions on place-based Indigenous knowledge and identity and the social-cultural dimensions of human-environment interactions. She works in First Nation, Inuit, Mapuce, and Sámi contexts.

Routledge Research in Polar Regions

Series Editor: Timothy Heleniak, Nordregio International Research Centre, Sweden

The Routledge series in Polar Regions seeks to include research and policy debates about trends and events taking place in two important world regions: the Arctic and Antarctic. Previously neglected periphery regions, with climate change, resource development and shifting geopolitics, these regions are becoming increasingly crucial to happenings outside these regions. At the same time, the economies, societies and natural environments of the Arctic are undergoing rapid change. This series seeks to draw upon fieldwork, satellite observations, archival studies and other research methods which inform about crucial developments in the Polar regions. It is interdisciplinary, drawing on the work from the social sciences and humanities, bringing together cutting-edge research in the Polar regions with the policy implications.

Resources, Social and Cultural Sustainabilities in the Arctic
Edited by Monica Tennberg, Hanna Lempinen, and Susanna Pirnes

Arctic Sustainability, Key Methodologies and Knowledge Domains
A Synthesis of Knowledge I
Edited by Jessica K. Graybill and Andrey N. Petrov

Food Security in the High North
Contemporary Challenges Across the Circumpolar Region
Edited by Kamrul Hossain, Lena Maria Nilsson, and Thora Martina Herrmann

For more information about this series, please visit: www.routledge.com/Routledge-Research-in-Polar-Regions/book-series/RRPS

Food Security in the High North

Contemporary Challenges Across the Circumpolar Region

Edited by
Kamrul Hossain,
Lena Maria Nilsson, and
Thora Martina Herrmann

Routledge
Taylor & Francis Group
LONDON AND NEW YORK

First published 2021
by Routledge
2 Park Square, Milton Park, Abingdon, Oxon OX14 4RN

and by Routledge
605 Third Avenue, New York, NY 10017

First issued in paperback 2022

Routledge is an imprint of the Taylor & Francis Group, an informa business

Publisher's Note
The publisher has gone to great lengths to ensure the quality of this reprint but points out that some imperfections in the original copies may be apparent.

British Library Cataloguing-in-Publication Data
A catalogue record for this book is available from the British Library

Library of Congress Cataloging-in-Publication Data
A catalog record has been requested for this book

ISBN: 978-1-138-37006-7 (hbk)
ISBN: 978-0-367-52426-5 (pbk)
ISBN: 978-1-003-05775-8 (ebk)

DOI: 10.4324/9781003057758

Typeset in Times New Roman
by codeMantra

Contents

Figures

Tables

Contributors

Sergei Andronov, Ph.D. (Medicine), is a senior researcher in Arctic Research Scientific Centre in Western Siberia. His scientific interests are focused on adaptation of respiratory organs, prevention of chronic non-specific lung diseases, and changes in the traditional diet of the tundra residents of YNAO (Yamal-Nenets Autonomous Okrug).

Peter Bjerregaard, MD, Dr. Med. Sci., was born in 1947. He is a professor of Arctic health since 1996. He worked as a district medical officer in Northern Greenland in 1978–1980 and then for four years in Kenya until his research career started in 1989. His main research interests are the social epidemiology of diet, mental health, and risk factors for cardiovascular disease among the Inuit in Greenland. He also spends time advising the Greenland Government on health issues. He is former president of the Circumpolar Health Research Network, the International Union for Circumpolar Health, and the Danish-Greenlandic Society for Circumpolar Health. He was a scientific editor of *International Journal of Circumpolar Health* in 1996–2012. Peter has published scientific articles, reports, and books mostly about the health of the Inuit in Danish, Greenlandic, and English languages.

Sylvie Blangy is a researcher at the French National Research Center (CNRS) hosted at the CEFE (Centre for Functional Ecology and Evolutionary) in Montpellier, France. Sylvie coordinates the French Research Group on Participatory Action Research and Citizen Sciences (https://websie.cefe.cnrs.fr/gdrparcs/). She was the co-director of the OHMI Nunavik (International Human-Environment Observing Network) in northern Québec (http://ohmi-nunavik.in2p3.fr/) from 2012 to 2019. Sylvie has been working since 2006 in the Arctic and sub-Arctic regions, in collaboration with the Cree, Inuit and Saami, communities to develop participatory-action research projects in aboriginal contexts. The goal is to address their concerns such as the social and cultural impacts of industrial development, or the role of tourism in sustaining their culture, language and lifestyle. She is exploring new ways to link local expertise and scientific knowledge (http://websie.cefe.cnrs.fr/tuktu/).

Elena Bogdanova, Ph.D. (Economics), Ph.D. (Phil.), is an associate professor in Northern Arctic Federal University. She is the leader of national research projects amd a participant in the expeditions to the High North. She has done over 90 works on regional, health economics, and indigenous peoples' human rights in the Arctic.

Corinna Casi, MA in Philosophy, is a doctoral candidate in environmental ethics at the Faculty of Social Sciences, University of Helsinki, Finland. She is also a member of HELSUS (Helsinki Institute of Sustainability Science) and a doctoral researcher (University of Lapland, Finland) in the legal and ethical team of the ValuBioMat project, funded by the Academy of Finland (2019–2022). Her doctoral research focuses on non-economic value of nature such as ecological and aesthetic values as well as on Indigenous ethics and traditional ecological knowledge. In September 2019, she won the Vonne Lund Award for her article 'Sami Food Practices and Traditional Ecological Knowledge' during EURSAFE 2019 Conference, at the University of Tampere, Finland. She taught at Helsinki Summer School in 2015–2019. In the fall of 2017 and 2018, she was a guest lecturer at the University of Palermo, Italy. In November 2014, she was a guest lecturer at the University of Latvia, Riga for a month; in 2013, she was a visiting lecturer at the University of Iceland, Reykjavik.

Véronique Coxam is the head of the Nutrition Skeleton and Metabolism Team and deputy head of the Human Nutrition at INRAe. She develops translational research regarding the nutritional prevention of chronic disease (especially musculoskeletal diseases).

Tine Curtis, MA, PhD, is the head of the Centre for Health Promotion and Disease Prevention at Local Government Denmark and research director of the municipality of Aalborg. She is also an adjunct professor at both the University of Southern Denmark and the University of Aalborg. Her work focuses on the implementation of evidence-based knowledge in local health promotion and disease prevention initiatives as well as on practice and research cooperation in the development of new knowledge. Previously, Tine worked at the National Institute of Public Health in Denmark as a researcher at the Centre for Public Health in Greenland and later as head of Centre for Intervention Research. Her research on public health in Greenland includes health services and social epidemiology.

Bruce C. Forbes, a research professor at the Arctic Centre of the University of Lapland, has a background in applied ecology and geography in northern high latitudes, with special emphasis on permafrost regions. His experience is circumpolar, encompassing studies of rapid land use and climate change in Alaska, the Canadian High Arctic, various regions of northern Russia, and northernmost Fennoscandia. His approach is strongly interdisciplinary and participatory, aiming for the co-production of knowledge, particularly concerning local and regional stakeholder-driven research questions. He has conducted fieldwork annually in the Arctic for over 35 years.

Hinnerk Gnutzmann is a post-doctoral researcher at the Leibniz University in Hannover, Germany. He is an applied economist, working primarily in industrial organisation and international trade. Hinnerk's current research in the project 'FINEXCA' focuses on competition in the fertiliser industry and the impact of fertiliser prices on agricultural commodities. He holds a PhD in economics from the European University Institute in Florence, an MPhil from Utrecht University, and a BA from Cambridge University.

Assi Harkoma is a junior researcher at the Northern Institute for Environmental and Minority Law, Arctic Centre/University of Lapland. Assi's background is in political sciences and international law (minor). She has a Master's degree from the Faculty of Social Sciences of the University of Lapland. In her research, Assi focused on human rights policy and international human rights law as it applies to Indigenous peoples, and she has done community-based participatory research in several research projects.

Thora Martina Herrmann, DPhil (Geography) studies the socio-cultural dimensions of human-environment interactions, focusing on the impacts of socio-environmental and climate changes on Indigenous societies in polar regions. Her expertise is also in action research-creation projects in polar regions working with First Nations and Inuit in Canada and with Sami in Northern Europe on place-based Indigenous knowledge and identity using visual art-based methodologies, such as filmmaking and photovoice.

Kamrul Hossain is a research professor and the director of the Northern Institute for Environmental and Minority Law (NIEM) at the Arctic Centre in the University of Lapland. He is an adjunct professor of international law, and has also served as the (acting) professor of public international law at the University of Lapland. Dr. Hossain has extensively published in the field of international law and human rights law that apply to the Arctic. He has been leading many national and international research projects. He served as a special editor for several significant international journals including the *Yearbook of Polar Law*. He regularly teaches at the University of Lapland and periodically at other foreign universities. He was a visiting research fellow at a number of foreign institutions including the University of Technology Sydney, University of Toronto, Scott Polar Research Institute of the University of Cambridge, and the Muroran Institute of Technology in Japan.

Zhanna Kasparyan, PhD (Economics), is academic secretary of Barents Centre of the Humanities Kola Science Centre RAS. She graduated from the St. Petersburg State Academy of Veterinary Medicine in 1994. In 2010, she was successfully graduated from the Kola branch of Petrozavodsk State University with a degree in 'state and municipal management'. At the same time, she defended her PhD thesis 'Improving regional socio-economic policy towards people with the special needs'. She has participated in a number of international research projects. Her scientific interests include regional economy, health and well-being in the Arctic, mathematical methods for economic analysis, and state regulation of socio-economic processes.

Ruslan Kochkin is the head of the Experimental Biological Laboratory at Arctic Research Scientific Centre in Western Siberia. He studies traditional food, bioresources of the Arctic, and human adaptation to the climatic factors in the Arctic region.

Annie Lamalice holds a PhD in Geography and Ecology of Health from the University of Montreal and the University of Montpellier through which she has studied the transformation of the Nunavik food system (Canadian Arctic).

Christina V.L. Larsen is a sociologist by training (University of Copenhagen, 2006) and completed her PhD in public health (University of Southern Denmark, 2014). She has lived and worked in the Arctic with her family since 2006. She is the research director of the Centre for Public Health in Greenland at the National Institute of Public Health and responsible for the research-based consultancy for the Greenland Ministry of Health. Dr Larsen is also a Fulbright Scholar and part of the second Fulbright Arctic Initiative cohort (2018–19). She has been the president of the Circumpolar Health Research Network since 2015. Dr Larsen has been involved in Arctic Council Mental Health Projects for a decade and has represented Greenland in the Arctic Human Health Expert Group since 2015.

Geraldine Laurendeau is a multidisciplinary artist from Montreal (Canada). She has a background in Fine Arts, Landscape Architecture and Environmental Studies. She's also a trained ethnologist and collaborates with First Nations, museums and research institutions as an independent curator, designer and consultant on projects related to environment, land planning, heritage, culture and biodiversity conservation, food security and health.

Andrey Lobanov, PhD (Medicine), is a professor at Arctic Research Scientific Centre in Western Siberia. His scientific activity is focused on the issues of health and social well-being of the Indigenous population of YNAO, adaptation, and traditional nutrition in the Arctic.

Pavel Lomov has a Master's degree from Petrozavodsk State University (Kola Branch), Apatity, Russia in 2016. In 2011, he received his PhD in computer science from Institute for Systems Analysis of Russian Academy of Sciences (ISA RAS), Moscow, Russia. Now he works as a senior researcher at the Institute of Informatics and Mathematical Modelling, Apatity, Russia. His research interests are ontology engineering, ontology learning, ontology visualisation, semantic integration of information, domain modelling, and human-machine interaction.

Philip A. Loring is an associate professor, Arrell Chair in Food, Policy and Society at the University of Guelph. He is ecological anthropologist with interests in food security and sovereignty, community sustainability, and environmental change. In his research he collaborates with First Nations in Alaska, and Western and Arctic Canada on projects related to conflict, climate change adaptation, and social justice in coastal community food systems. He draws from a diverse set of tools and methods, including ethnography, film, and social network analysis.

Marion Macé graduated with a Master's degree in environmental assessment from Montreal's Concordia University in 2018. Since then, she has been involved in various northern gardening initiatives in Nunavik (Northern Quebec) and worked as the Greenhouse manager in Inuvik, Northwest Territories. In November 2019, she began her current role coordinating the Pirursiivik: Greenhouse and Social Arts project in Inukjuak (Nunavik) which aims to develop a year-round greenhouse, to promote healthy eating habits and to contribute to food security.

Moeko Minagawa is an Attaché Temporaire d'Enseignement et de Recherche (ATER) at Toulouse University 2, Jean-Jaurès. She belongs to the Centre d'Etudes Japonaises and also to the Centre de Recherches Sociologiques et Politiques de Paris (LabTop). Her current research interests are in health/ food issues including human security and ethics of care approaches.

Johana Evelyn Montalvan Castilla is a social anthropologist, an archaeologist, and a researcher in the field of human and societal security. Since 2012, she has been focusing on the study of a diverse array of societal development issues, particularly occurring in Peru and Norway. Her main research interests are connected to the importance of traditional knowledge for today's society. The main focus of her current research lies also on the impact that digitalisation and information and communication technologies have on society, with a focus in the Arctic and Norway. She currently lectures at UiT, the Arctic university of Norway, in topics connected to societal security and safety and has been a guest lecturer and researcher at several other universities and institutions.

Ildikó Asztalos Morell is an associate professor in rural development at the Swedish University of Agricultural Sciences. She has published 50 chapters and articles in gender, rural, postsocialist transition, and marginality studies. She co-edited *Gendering Postsocialism* (2018) and *Gender Regimes, Citizen Participation, and Rural Restructuring* (2008).

Lena Maria Nilsson is research coordinator at the Arctic Research Centre at Umeå University. She has a PhD in public health (2012), with a thesis focusing on traditional Sami lifestyle factors as determinants of public health. The main part of her research is based on data from the Northern Sweden Health and Disease Study (NSHDS) cohort. She is also involved in research on food and water security in the circumpolar North. So far, this project has resulted in one report, one chapter in the Arctic Resilience Interim Report 2013, and five peer-reviewed papers. Nilsson has also initiated an epidemiological follow-up study based on unique historical survey data on Sami and non-Sami schoolchildren, monitored in northernmost Sweden between 1929 and 1931. Since January 2019, Nilsson is appointed co-Director at Vaartoe, the Centre for Sami Research at Umeå University, Sweden. She is also the secretary of the Nordic Society for Circumpolar Health (2012–ongoing), and a member of the steering group of Neon, the Nordic Nutrition Epidemiological Network

(2012–ongoing). So far, Nilsson has published 63 peer-reviewed papers and 6 book chapters (H index: 23, i10 index: 43). Based on her broad expertise, Nilsson is often invited to speak to both scientific and public audiences. A list of publications (both peer-reviewed and other ones) is available at: https://www.umu.se/en/staff/lena-nilsson/?flik=publikationer" https://www.umu.se/en/staff/lena-nilsson/?flik=publikationer.

Ingelise Olesen is research coordinator at the Centre for Public Health in Greenland at the National Institute of Public Health. She was born in Narsaq, Greenland, in 1961 and has lived most of her adult life and worked in Nuuk, Greenland. Authorised as midwife in 1992, she worked at the hospitals in Narsaq and Nuuk. She holds a Master's degree in health education and health promotion from the University of Aarhus, Denmark (2015). Since 1999, she has coordinated the Greenland Health Surveys. She has conducted qualitative research on food insecurity and patient experience. As an Indigenous person, she has interest and expertise in cultural aspects of health and well-being in Greenland.

Anne Ouma holds a PhD in environmental and human geography from the University of Umeå Sweden and an MSc in environmental science and technology from IHE UNESCO Water Institute, Delft Holland. Previously Ouma worked, inter alia at the United Nations Development Programme (UNDP), and at the Swedish International Development Cooperation (Sida) on issues related to biodiversity conservation and Sustainable human development. She has experience in providing expertise to development projects on economic, environmental, social determinants of sustainable human development for over a decade. Her current research focus and interests are connected to the role of traditional/indigenous knowledge within contemporary health and well-being. Ouma has published and serves as associate editor for some journals, while she teaches and is periodically a visiting research fellow at the University of Dar es Salaam and at other foreign universities and institutions. She has lived and worked in the Arctic with her family for more than a decade and is lead researcher for action research-based consultancies for Region Vasterbotten County Health Ministry on Health Governance partnerships with Kenyan counties. Currently, Ouma is affiliated to the Arctic Centre at Umea University and Centre for Sami Studies as a Senior Researcher.

Andrei Popov, Ph.D. (Medicine), is a leading researcher at Arctic Research Scientific Centre in Western Siberia. His scientific activity is focused on the issues of health and social well-being of the Indigenous population of YNAO and the impact of climatic factors on its health.

Noor Jahan Punam is a junior researcher at NIEM, Arctic Centre and she is pursuing her doctoral studies at the Faculty of Law, University of Lapland.

Her research focuses on accommodating Indigenous peoples' traditional knowledge in international climate change law with a special reference to the European High Arctic, where she combines conceptual frameworks with Indigenous traditional knowledge for the promotion of Indigenous peoples' resilience against climate change. She received her Master's degree in comparative and international law from University of Eastern Finland. She is specialised in environmental and climate change law and did European Union law as her minor studies. She completed a postgraduate diploma in law in Northumbria University in the UK. Prior to that, she completed her Bachelor's degree in law in the University of London.

Christer Pursiainen is a professor of societal safety and environment at the Arctic University of Norway (UiT). He has previously worked, inter alia, in the European Commission, Council of the Baltic Sea States, Nordregio – the Nordic Centre for Spatial Development, University of Helsinki, RECEP – Russian Centre for Economic Policy, and the Finnish Institute of International Affairs. Pursiainen has more than 100 scholarly publications in such fields as societal security, crisis management, critical infrastructure, political and cultural history, and international politics.

Dele Raheem holds a Doctor of Food Sciences from the University of Helsinki. Dele's research interest is in food bio-processing, preservation, and crosscutting issues related to food security. He has gained extensive research and industrial experience in the past three decades. He also obtained a Master's degree in education from the University of Greenwich, London, UK. He has published and co-authored 1 book, 7 book chapters, and 20 articles in peer-reviewed journals. Currently, he is affiliated to the Northern Institute for Environmental and Minority Law, Arctic Centre at the University of Lapland as a senior researcher. He is a professional member of the Finnish Association of Academic Agronomists and the Food Climate Research Network based in Oxford, UK.

Jose L. Martinez Lastra holds a BSc. degree in Electrical Engineering and a M.Sc. and a Dr.Sc. degrees in Automation Engineering. He joined Tampere University of Technology (Finland) in 1997, and became Full Professor in 2006. Since 2019, Dr. Lastra serves as Director of the CINTECX: Research Center on Industrial Technologies and Energy (Spain) and He leads the FAST-Lab. at Tampere University. His research interest is in applying ICT to the field of automation. Dr. Lastra serves as an Associate Editor of the IEEE Transactions on Industrial Informatics, and he was Technical Editor of the IEEE/ASME Transactions on Mechatronics (2015-16).

Dr. Sc. (Tech.) **Borja Ramis Ferrer** received his doctoral degree in the sphere of Factory Automation and Industrial Informatics. His research interests include the deployment of knowledge-based and cyber-physical systems in factory automation. Dr. Ramis has strong knowledge on the state-of-the-art technologies and mega-trends in the Industrial domain, including

Industrie 4.0 and the 2nd wave of the digital revolution. In this context, he has (co)authored and published more than 50 scientific papers.

Maxim Shishaev is the chief scientist at Institute for Informatics and Mathematical Modelling of Technological Processes of the Kola Science Center RAS. He is also a professor at Murmansk Arctic State University. He was born in 1970. He has studied robototechnique at St. Petersburg Polytechnic University (1993). Doctor of Technical Sciences (2010), Professor of Russian Academy of Sciences (2015). Prof. Shishaev has authored over 170 research papers in the areas of predictive modelling, networking, knowledge-based information systems, and information support of regional development.

Victòria Soldevila Lafon is a lecturer of applied economics at Universitat Rovira i Virgili (URV). She obtained her PhD in economics at the University of Barcelona in 2008. Her thesis analyses the impact of environmental regulations on pig farming in Catalonia. She holds a Master's degree in international comparative rural policies studies in Universitat Autònoma de Barcelona. Before joining URV, Victoria taught at the University of Barcelona, l'École Supérieure de Commerce de Toulouse (ESEC) in Barcelona, and Universidad Nacional Autónoma de Nicaragua (UNAM) in Managua. She has been involved as faculty in several International Comparative Rural Studies Summer Schools (University of Lapland, 2019; Tuskegee University, 2018; among others). Her research interests include agri-food value chains, rural development, food waste, and food security. She was involved in evaluations of LEADER programme in Spain. She has published in Land Use Policy.

Piotr Śpiewanowski (PhD, European University Institute, 2014) is an assistant professor at the Institute of Economics, Polish Academy of Sciences. He has devoted the last four years of his research life to fertiliser economics and the implications of market structure, supply disruptions, and energy prices for prices of nitrogen, potassium, and phosphorus fertilisers. Each of the nutrients received at least one academic publication. Besides, his research interests lie in climate policy and competitiveness of trade-exposed, energy-intensive industries and the aviation market. He has published articles in academic journals including *American Journal of Agricultural Economics* and *Nephrology Dialysis Transplantation.*

Shirley Thompson is an associate professor at the Natural Resources Institute, University of Manitoba. Dr Thompson is the principal investigator of the Mino Bimaadiziwin partnership, which is building capacity and houses in Northern Manitoba through partnerships for community-led post-secondary education in Garden Hill and Wasagamack First Nations. Also, an Indigenous Food Sovereignty community college program was planned for Brokenhead First Nation to start in May 2020 but disrupted by COVID-19. For more information check: http://ecohealthcircle.com/.

Preface

Arctic communities are facing environmental, technological, and economic changes that put the traditional food supply system at risk. It has been suggested that access to sufficient, safe, and nutritious foods is declining within Arctic communities. In this book, we explore some of the complex challenges facing food sustainability, supply chain, and food safety and sovereignty. The overarching conceptual framework we employ is that of food security as part of human security. While the Arctic, in general, is the primary region focused on in this book, we also put a particular emphasis on the European High North (EHN), as this region has been less explored as far as food security challenges are concerned.

This book is the outcome of the project entitled 'Food (in)Security in the Arctic: Contribution of Traditional and Local Food to Promote Food Security with Particular Reference to the European High North'. This project, funded by the Nordic Council of Ministers under its Arctic Cooperation Programme administered by Nordregio, started in mid-2017 and is hosted at the Northern Institute for Environmental and Minority Law at the Arctic Centre of the University of Lapland, Finland. Approximately 20 scholars from a diversity of disciplines, from a range of countries, and whose research concerns food security, either directly or indirectly, participated in this three-year project. While we note with regret that it was not possible to include all the contributions in this book, we did include most of them, and especially those in which the authors specifically addressed relevant overarching questions as follows: Why is traditional food important for enjoying greater human rights and human security? How do climate and socio-environmental changes in the Arctic affect traditional food systems, creating consequences related to the enjoyment of the most fundamental human rights from the viewpoint of food security? Who are the main vulnerable groups, and how do they become the victims of human rights violations in relation to food security and sovereignty? How is food security interlinked with safety issues, creating implications for human health? And how are vulnerable groups, such as women, the elderly, and Indigenous peoples, negatively affected due to increasing food insecurity in terms of a lack of physical/economic access and a lack of delivery of traditional foods?

These questions are systematically analysed in individual chapters, taking into consideration the importance and uniqueness of the food systems in the region.

During the last three years, we held several meetings – including three workshops held in Stockholm (Sweden), Umeå (Sweden), and Rovaniemi (Finland) – during which each of the book's chapters was presented about and discussed.

We are particularly grateful to the Council of Baltic Sea States (CBSS), Umeå University and its Arctic Research Centre, and the Arctic Centre of the University of Lapland for cooperating with us and providing us with the venues for organising scientific workshops and other meetings. The chapters included in this volume have gone through several rounds of external and internal review processes. We included external scholars who had expertise in the related fields as blind reviewers. We are grateful to all the reviewers for their hard work commenting on the manuscripts. We also convey our sincere thanks to all the contributing authors involved in this book project and extend special thanks to Assi Harkoma, who assisted us during the editing process, and to Marcin Dymet, who helped us coordinate the language-editing process. We finally acknowledge the generous grant awarded by Nordregio under its Arctic Cooperation Programme, without which this book would not have been produced.

February 2020
Kamrul Hossain
Lena Maria Nilsson
Thora Martina Herrmann

Introduction

Conceptualising food (in)security in the High North

Kamrul Hossain, Lena Maria Nilsson and Thora Martina Herrmann

In recent years, food insecurity has become a crucial issue in the circumpolar Arctic region. An investigation into why this issue has become so prevalent requires the exploration of a wide range of factors that affect the foods and food systems of the region. These factors include both natural phenomena as well as geopolitical, socio-economic, and cultural aspects. Global climate change and its disproportionate impacts on the region contribute to the rapid transformation of the region in terms of its environment, economy and geophysical, political and socio-cultural characteristics. At times, this transformation arguably offers new opportunities in relation to economic globalisation, but it also significantly alters the natural functioning of the region's ecosystems. The overall implications of climate change place heavy pressure on the areas of environment, land use, and natural resource management. Hence, it also brings about adverse consequences for, among other things, traditional food supply chains, and traditional food systems at large. Moreover, the combined effect of technological advancement and economic globalisation puts pressure on traditional food supplies, especially in the Nordic part of the circumpolar Arctic, given that traditional foods have become relatively less available and that the supply of imported foods is becoming increasingly available in many parts of the region, and especially in its European parts. This situation results in adverse implications on the availability of safe and nutritious foods for many communities in the Arctic.

This book is the result of a three-year research-network project aiming to address food security challenges and the promotion of food security, of which traditional and local foods are integral factors, both at the regional and local levels. Therefore, the focal points of this book are to address complex challenges concerning food sustainability, the supply chain, and food safety and sovereignty from the viewpoint of human food security. While the Arctic, in general, is the referent region, this book particularly highlights the conceptualisation of food insecurity in relation to the Nordic part of the Arctic region – referred to as the *European High North* (EHN). Figure 0.1 shows a map that highlights the EHN. Given that the circumpolar Arctic region is not homogenous, and that there are regional variations in terms of demography, political system, economy, and livelihood

DOI: 10.4324/9781003057758

practices, an exploration of food insecurity requires an understanding in of the different geophysical settings within the region itself. Food insecurity in the North American Arctic, in Greenland or in the Russian Arctic, for example, is not the same as it is in the Arctic region of the Nordic countries. Therefore, framing a concrete definition of *food security* that is applicable across the Arctic region is complex. However, the Rome Declaration on World Food Security and the Plan of Action of the World Food Summit has offered a general understanding of it: 'Food security exists when all people, at all times, have physical and economic access to sufficient, safe and nutritious food to meet their dietary needs and food preferences for an active and healthy life' (FAO, 1996, Action 1).

In the glossary of the State of Food Insecurity in the World 2001, the Food and Agricultural Organization (FAO) of the United Nations refines the definition of *food security* that we refer to in our discussions in this book, as follows: *food security* is a 'situation that exists when all people, at all times, have physical, social and economic access to sufficient, safe and nutritious food that meets their dietary needs and food preferences for an active and healthy life' (FAO, 2001, Glossary, 9 para.).

This conceptualisation analyses food security by referring to four pillars: availability, access, utilisation and stability or sustainability. *Availability* often refers to the physical presence of a food supply, measured by level of production, stock level, and net trade, among other things. *Access* relates to ensuring food security at the household level; crucial here are the affordability of food at an individual level (i.e., economic access) and the physical access ensured by stability in the market infrastructure, prices, and the political state of affairs in a given context. *Utilisation* is one of the most important aspects of food security, which co-exists with safety concerns in food practices. Food security is jeopardised in food systems where safety and nutritional and dietary needs are compromised. As such, within utilisation, it is necessary to ensure the health value of food; food products must contain sufficient energy and nutrients, food preparation must be done appropriately and diversity in diet must be available. The fourth pillar – *stability* or *sustainability* – indicates the presence of a resilient infrastructure to ensure that the other three pillars remain stable so that people's ability to access and utilise food remains sustainable over time.

When we discuss the whole circumpolar region, we generally refer to the broader conceptualisation of *food security* stated above; however, we also make specific reference to the EHN region, where all four of these components are not equally significant in food security discourse. There are certain specific issues that result in food security challenges in the EHN region, both from the viewpoint of the above conceptualisation and from other issues that affect regional food security systems, such as challenges pertaining to the exercise of food sovereignty.

The EHN is an undefined geographic area that is generally understood to refer to the northern parts of Finland, Sweden, and Norway, as well as some north-western parts of Russia. However, we make specific reference to the Sápmi region while addressing various issues in connection to food security. Sápmi is the traditional area of the Sami people, the Indigenous People of the EHN. There are no clear borders defined for Sápmi, so what is shown in the map in Figure 0.1 is an approximation based on cultural presence and recognition acknowledged in modern times.

Within the Sápmi region, except for Russia's Kola Peninsula, despite the presence of wide areas used for traditional livelihood activities and sparse human settlements – sometimes perceived as wilderness to outsiders – the physical infrastructures across the region are well networked. The road and transport facilities provide easy access to almost all areas in the region and connect various remote human settlements. Basic services, including access to all essential food products from local supermarkets, are largely available. Therefore, the components of food security, such as availability, access and stability or sustainability, are not of major concern in the region. Foods are generally available in local supermarkets, as they are imported and easily transported even to remote locations in the region. On the one hand, the presence of well-connected infrastructure provides easy physical access, while on the other hand, in terms of affordability, the prevailing welfare model in these countries (i.e., Finland, Norway, and Sweden) offers basic support through social security schemes, allowing individuals to have the economic means to access food. In the aspect of stability, as it relates to the supply of and access to adequate food, the region is relatively well equipped with a food supply, providing even greater sustainability in terms of food security. Do these conditions, however, make the region food secure? To some extent, the answer is 'yes', if *food security* only refers to the availability of and access to food for the purpose of basic consumption. However, as we highlighted above and expand on in many chapters in this book, the

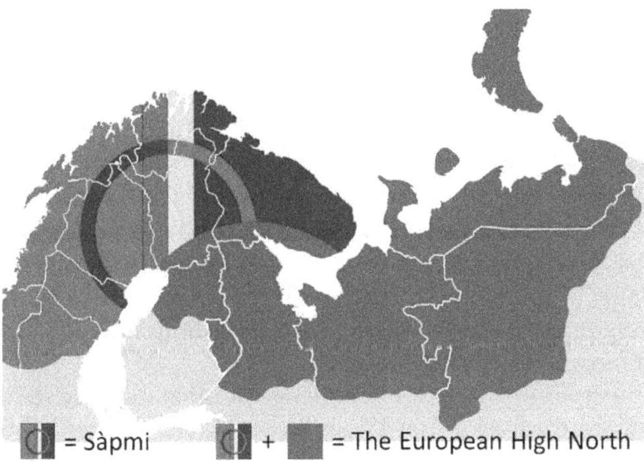

Figure 0.1 The Sápmi region of the EHN.
Author of the map: Lena Maria Nilsson

conceptual understanding of *food security* as defined by the FAO and in relation to other aspects relevant to food practices – for example, spirituality, sociocultural, and psychological wellbeing, political participation, and control over food practices (by relevant actors) – we clearly see significant missing points that make the region food insecure.

The main missing point from the viewpoint of the FAO definition here is so-called *utilisation*, which refers to the availability of sufficient nutritious foods containing the required health value. The dietary needs vary in different locations on earth and are defined by the particular characteristics that a region possesses. Locally or traditionally available foods harvested in and collected from the natural environment provide natural substances for maintaining healthy living conditions. In the EHN, naturally available foods, such as various kinds of berries, provide ingredients that help meet the dietary needs of those in the area, but these resources have largely been abandoned. This is caused by many factors, including the lack of human capital to collect these resources and the non-commercialised processes involved in their collection having a high cost attached to them when compared with store-bought foods. Traditionally, among the Sami Indigenous People, reindeer meat is part of their regular diet, both for its health value and for its cultural significance. Today, a reluctance has been observed among the Sami – and especially among the young generation of gathering, fishing, hunting, and/or reindeer herding Sami – to continue participating in their ancestral livelihood (Lawrence & Larsen, 2019). With respect to reindeer herding, this is often due to the laborious nature of the work, although the practice is less laborious nowadays as new technologies, such as snowmobiles, helicopters, and GPS tracking methods, are used more and more. Moreover, traditionally performed activities are also constrained by state-based natural resources management regimes, limiting the ability of people to be flexible in their food practices. For example, infrastructure development projects, such as the 'Arctic Corridor' project (House of Lapland, 2019), and extractive industrial activities, such as mining and oil and gas developments, place strain on the continuation of traditional food practices and threaten the safety of traditionally available foods (Sheehy et al., 2014; Nilsson & Evengård, 2015; Hossain, Raheem, & Cormier, 2018). The safety of foods has increasingly become an issue; once a food source is contaminated – due to, for example, the use of toxic materials in mining activities – it may result in various health-related risks, including contributing to obesity and diabetes (Ford, 2009). As a result, despite the high nutritional value of many traditional and local foods, they been found to be relatively less available and have, at times, been contaminated, jeopardising the overall food security of the region.

Food security is also a serious challenge in other parts of the Arctic, including remote Canadian and Greenlandic Arctic regions. A significant body of literature is available illustrating various aspects of food insecurity in the Canadian North and in Greenland, and especially amongst the

Indigenous communities (e.g., Fergurson, 2011; Council of Canadian Academies, 2014; StGermain, Galloway, & Tarasuk, 2019). Most of the literature suggests that these remote Arctic regions suffer from a lack of all four components of food security. On the one hand, the supply chains of traditionally available foods have been affected the consequences of climate change and other stressors related to environmental changes; on the other hand, imported foods in these regions are either relatively less available or their costs are overwhelmingly high, affecting affordability. Moreover, the melting of the Arctic's coastal and marine areas has been exacerbated by increased human activities, contaminating marine food resources and resulting in serious consequences to human health. Across this region, Indigenous food systems are heavily affected. Some research has suggested that Indigenous Peoples in the North American North will remain food insecure unless a collective will towards dismantling inequities is taken up, creating room for Indigenous Peoples to solve the problem on their own terms (Loring, 2017).

Against this background, this book explores traditional food production challenges, risks pertaining to food contamination from ongoing and increasing human activities (e.g., mining), gender roles in the maintenance of traditional food systems within local and Indigenous communities, and changing livelihoods affecting food patterns. It also explores the processing methods that combine Indigenous and traditional knowledge, through which an exercise in food sovereignty in the regional context can be manifested. While addressing these issues, we adopted a multi-disciplinary and inter-disciplinary framework, wherein various disciplines, such as food sciences, agriculture, policy studies, human rights, health sciences, biology, and political sciences, interact.

We particularly address contemporary challenges facing the region in relation to food security as it relates to the notion of human security. The concept of *human security* was popularised during the early 1990s as an alternative to what traditional security stands for (e.g., military-oriented aspects of security). Within this framework, our understanding of security has been broadened and deepened, incorporating various concerns that affect individuals and communities both in sub-state and sub-regional contexts. The Human Development Report (HDR) of 1994, endorsed by the United Nations Development Programme (UNDP), popularised the concept, in which food security represents one of the seven components identified (UNDP, 1994). The other six components are health, environmental, economic, personal, community, and political security. The concept of human security, as demonstrated in the HDR, suggests a rather broad formulation highlighting a number of freedoms, the absence of which causes obstacles for individuals and communities to flourish at a sustainable level. Freedoms from fear, want, and indignity are the three substantial and interconnected aspirations that individuals and human communities require to maintain a dignified life. Although food has been placed as only one of the seven components of human security, it also reinforces – or is interconnected with – other components. For example, a lack of safe and nutritious foods that contain proper dietary value causes health risks, affecting health security. Similarly, cultural aspects, such as participation in or connection to practices related to traditional and local foods pertain to the maintenance of community identity, especially for Indigenous groups; a lack of availability to traditional or local foods thus enhances the threat to community security. Further, food security and environmental security are interconnected, given that any harm to or contamination of the natural environment may very well cause food contamination too, especially for foods that are locally harvested and produced. As such, food security – as it relates to human dignity and ensuring freedoms from fear and want – provides a powerful framework to promote human wellbeing not only in terms of physical and psychological contexts but also in strengthening a community identity. In this way, political aspirations are strongly linked to food security, as they relate to establishing control over the maintenance and continuation of a food system. We therefore also refer to food sovereignty, as it is linked to community security, with particular reference to local and Indigenous Peoples of the circumpolar Arctic.

We define *food sovereignty* as a situation in which local and Indigenous Peoples are in control of the processes leading to food security in their home areas, as expressed by the Final Declaration of the World Forum on Food Sovereignty in Havana, Cuba in 2001:

> Food security is the peoples' right to define their own policies and strategies for the sustainable production, distribution and consumption of food that guarantees the right to food for the entire population, on the basis of small and medium-sized production, respecting their own cultures and the diversity of peasant, fishing and indigenous forms of agricultural production, marketing and management of rural areas, in which women play a fundamental role.
>
> (World Forum on Food Sovereignty, 2001)

An illustration of this definition would be a situation in which reindeer herders are in control of the full utilisation of traditional grazing areas and are able to adjust, in all situations, their practice according to changes in weather and foraging conditions. It could also refer to a situation in which local people maintain traditional knowledge about how to utilise the local wild plant, berry, fish, and game stock for local food production.

This book also uses various other concepts, such as food resilience and traditional and local knowledge in food practices and in traditional food systems. We define *food resilience* as 'capacity over time of a food system and its units at multiple levels, to provide sufficient, appropriate and accessible food to all, in the face of various and even unforeseen disturbances' (Tendall et al., 2015, p. 19). The notion of food resilience does not, however, include any references to traditional knowledge. The concept of food sovereignty can therefore also be applied in terms of the global market contribution to the food system of the EHN, as well as food banks provided by authorities to Indigenous Peoples (e.g., as is done in the Canadian Arctic). *Traditional local knowledge* generally refers to knowledge conveyed narratively or through practical learning from one generation to the next for such a long period of time that no one knows if the knowledge is based on evidence or, if it is, what this evidence consists of or how it has arisen. In contrast, we refer to *traditional Sami knowledge* – in Sami, *Árbediehtu* – as ecological skills, knowledge, and values inherited from previous generations in the Sami society, such as reindeer husbandry, hunting, fishing and gathering food, and skills related to mastering the Arctic landscape, weather, climate, wildlife, and cultural needs. In several chapters in this volume, we further illustrate Sami traditional knowledge with respect to the promotion of food security by integrating traditional or local foods. By *traditional food*, we largely refer to food items that have been normalised in the EHN for such a long period of time that no one any longer remembers the time that existed before the food item was normalised. Traditional food mainly consists of food items with a direct link to natural resources and local food production, such as lingonberries (*Vaccinium vitis-idaea*), but commodities such as coffee and sugar may also be included if trade traditions extend far back in time.

This book consists of three parts. The first part analyses issues related to Indigenous Peoples, livelihood practices, and traditional knowledge in the context of food production, consumption, and diversity. This part includes five chapters. In the first chapter, Castilla explores the value of stockfish for strengthening the local food system and the role of stockfish in enhancing local food security. She highlights the importance of the transmission of the traditional ecological knowledge (TEK) involved in the production and preparation of stockfish by conducting original ethnographic research on the Lofoten Islands. Her findings suggest that, according to the local population, stockfish is a significant factor in their food security and in the resilience of the local food system. Based on her research findings, however, she suggests that interests in stockfish are decreasing among the younger generations. Therefore, she argues for a need to develop strategies aiming at strengthening the transmission of TEK related to stockfish, which eventually would lead to the promotion of food security among the local and Indigenous population. In Chapter 2, Harkoma and Forbes investigate the potential of Indigenous knowledge-based traditional pasture management and a rotational grazing system. According to them, such a process is capable of securing reindeer herding as a subsistence base for the production of healthy and culturally appropriate food as part of traditional food systems. In this context, they highlight food sovereignty for Indigenous Sami people by putting a special emphasis on its six pillars, including food and its value orientation in a localised context where food systems are to be controlled by the people who possess local and traditional knowledge for its sustainable production.

In the third chapter, Minagawa highlights the food insecurity of reindeer herders after the 1986 Chernobyl nuclear accident. She focusses on how different voices and measures were sought in response to health risks associated with the aftermath of the disaster, to sustain Sami reindeer herders' livelihoods. She concludes that, in decision making, economic needs were prioritised above health risks, which has had consequences on food security. Bogdanova et al., in Chapter 4, discuss traditional nutrition as a part of the traditional culture of the nomadic Indigenous population in the Arctic zone of Western Siberia. They argue that traditional nutrition effectively prevents health problems and enables people to adapt to the harsh climate conditions of the High North. They further analyse the implementation of Russian national policies on traditional reindeer herding, which is affected by issues of food security and food sovereignty. Thereafter, Bjerregaard, in Chapter 5, explores the food preferences of the Inuit in Greenland, the role of traditional Greenlandic food (*kalaalimernit*), and the prevalence and social variation of food insecurity. He argues that *kalaalimernit* is preferred to imported food, but that imported food is also popular due to its lower cost, its ease of preparation and a wish for variation. In this regard, he argues that food insecurity is associated with socio-economic conditions and is reported much more often in remote East Greenland than it is in West Greenland, even after adjustment for the social and economic variables presented in the chapter. To conclude, he suggests is that a low level of consumption of *kalaalimernit* is generally by choice and not due to a lack of access or money.

Part II of the book presents multi-disciplinary perspectives on food (in) security, highlighting human rights, ethics, and environmental justice, regimes, and crisis management in relation to traditional food and food systems. In Chapter 6, Hossain and Punam analyse traditional food systems, specifically those in the EHA, that include activities such as berry picking, hunting, reindeer herding, fishing, and processing and consuming these local foods. They focus on the interplay between human rights and the traditional food systems in the EHA. Their conclusion suggests that local people, including Indigenous communities, play a key role in traditional food systems and that the maintenance of a stable system is vital for both Indigenous and non-Indigenous people. They further examine how a human rights framework can integrate traditional food systems, thereby showing how a stable food system promotes food security and food sovereignty as aspects of human rights. Casi, in Chapter 7, investigates the relationship between traditional food systems and the identity of Sami Indigenous communities. She creates a theoretical framework drawing from three non-Indigenous fields of study: food ethics, food security, and food sovereignty. Whereas food ethics discourse is used to outline the significance of traditional food in Sami culture, referring mainly to the accessibility and availability of adequate and sufficient food for a healthy and active life, food sovereignty is used to highlight the relationship between food and issues of inequality and injustice and the importance of strengthening local food production. According to her, in order to promote the empowerment of local communities, it is essential to decolonise food security and indigenise food sovereignty discourse. In Chapter 8, Shishaev et al. discuss issues related to food security management in the western part of the Russian Arctic, taking information support tasks into account. They examine key actors, indicators, threats, and existing instruments for food security management in the Russian Arctic zone, paying attention to existing relevant data sources, particularly those concerning local food, and to the food security problems of the Indigenous and rooted populations. The discussion of the applied ontology in this chapter determines how to assess knowledge of food security at the individual level and the role of local food in food security in the Russian Arctic. According to the authors, the penetration of digital technologies coupled with modern intellectual information processing capabilities offers new opportunities for the effective monitoring and management of food security. The chapter, therefore, offers findings concerning the use of digital technologies to build effective information support systems for food security management in the Russian Arctic.

Pursiainen, in Chapter 9, considers three possible scenarios in relation to food crisis management: radiological fallout and contamination, animal disease epidemics and the impact of oil spills on fisheries. Focussing mostly on the EHN, and especially on Finland, he asks whether crisis management systems are currently in place that could handle these types of crises. Utilising the crisis management cycle, the chapter reviews in some detail issues such as risk assessment, prevention, preparedness, response, recovery, and post-crisis learning. He concludes that sectorial-level crisis management systems are rather well developed, while a more holistic perspective is missing. In the following chapter, Soldevila Lafon outlines the challenges of food security using the food regimes approach. The concept of a food regime allows an analysis of the role of agriculture and food in different stages of the global capitalist economy. She confines her discussion to three historical food regimes: the First Food Regime (lasting from 1870 to 1914), the Second Food Regime (lasting from the end of World War II to the end of the twentieth century) and the Third Food Regime (which, according to the author, is currently emerging). Recent agrarian and food crises have highlighted the weaknesses of the Third Food Regime in addressing problems of food insecurity, even in the Global North. Given that the Arctic is especially vulnerable, as recent adverse external factors (e.g., climate change) are threatening food security in the region, she analyses the Third Food Regime and the challenges to food security it presents. According to her, the dynamics of the Third Food Regime and external factors reinforce each other and undermine food security strategies in the Arctic, which need particular attention.

The last part of this book, Part III, consists of four chapters highlighting possible future strategies to build resilient food regimes in order to enhance food security and sovereignty. This part is built on works that offer innovative, inspiring, transformative, and future-forward strategies for a healthy, equitable, resilient, and culturally diverse future of food security that is shaped by the people, communities, and their institutions in the circumpolar Arctic. Nilsson, in Chapter 11, provides reflections on Swedish food strategies from a Sami and an Arctic perspective. According to her, a food-secure area depends, to a large extent, on food imports and national and international trade. In Sweden, the level of food sovereignty is low and, after the nation's entrance into the European Union (EU) in 1995, is rapidly declining. She stresses that a future sustainable food strategy for Sápmi/Northern Sweden should be outlined and agreed on by the Swedish government and the Swedish Sami Parliament. It should be product-oriented, ecological, in balance with nature and based primarily on plants and animals adapted to an Arctic or sub-Arctic boreal climate zone. The strategy should also specify caloric values and nutritional content, include local and traditional perspectives, include strategies for waste reduction, promote knowledge sharing from different knowledge systems and strengthen relationships among all inhabitants of the area. In Chapter 12, Raheem et al. highlight the gaps that have been measured by the technological development – the Internet of Food. The authors examine how food security gaps in the EHN can be mitigated by collating data to leapfrog local foods into the digital era via the Internet. They argue that to achieve sustainability, changes are required in food supply chains and the entire food system. Consumers need information to make informed choices about what to eat; they need to know where food comes from, the conditions under which it grew and the food's nutritional profile. The food industry has been slow to take advantage of the Internet. According to the authors, with increasing interest in redistributed manufacturing, circumpolar regions such as the EHN will need to adapt to digitisation. The Internet of Food is an emerging trend that will make food traceable, transparent and trustworthy and empower consumers with more personalised food that caters precisely to their individual food, diet, and health choices.

Next, Gnutzmann and Śpiewanowski investigate the supply of fertilisers and the role of Arctic deposits in connection to the enhancement of food security, in Chapter 13. They emphasise that fertiliser supply, in which Arctic resources play a prominent role, is a key element of global food security. The chapter discusses reserves and production levels as well as the effect of fertiliser on global food prices. According to the authors, the geological characteristics and low cadmium content of the Arctic make Arctic fertiliser particularly valuable, especially as EU food safety regulations tighten. In Chapter 14, Herrmann et al. show how community-led Initiatives contribute to shaping the future of food security and sovereignty in Canada. Based on a number of case studies from Inuit and First Nations communities, the authors identify an array of community-based food initiatives that, according to them, promote better access to and knowledge of subsistence activities that encourage local food production, and help reduce inequalities in access to a healthy diet. They examine how these initiatives have improved access to affordable, nutritious food in remote communities while at the same time improving food sovereignty and wellbeing and preventing diet-related chronic diseases. Finally, in the last chapter, Ouma sheds light on traditional food knowledge by comparing north–south perspectives, discussing the views, interests, intuitions, stories, reflections, and experiences shared by participants in a talking circle. The talking circle was held in 2011 during the People in Places Conference, which was hosted by the Coastal Community–University Research Alliance (CURA), a six-year project at Dalhousie University, Halifax. The Coastal CURA focusses on the inclusion of Indigenous communities in integrated resource management processes and explores the use of local knowledge to address food security. Ouma analyses key messages that raise important issues about the relationship of the biocultural diversity web with food security and governance. According to her, embracing Indigenous cultural practices related to food systems, food security and governance could help engage and inform mainstream official policies in the challenges of food (in)security, both in the Arctic and beyond.

References

Council of Canadian Academies. (2014). *Aboriginal food security in Northern Canada: An assessment of the state of knowledge.* Ottawa, ON: The Expert Panel on the State of Knowledge of Food Security in Northern Canada.

Fergurson, H. (2011). Inuit food (in)security in Canada: Assessing the implications and effectiveness of policy. *Queen's Policy Review, 2*(2), 54–79. Retrieved from https://www.queensu.ca/sps/qpr/sites/webpublish.queensu.ca.qprwww/files/files/16%20Inuit%20food%20insecurity.pdf.

Food and Agricultural Organization (FAO). 1996. Rome Declaration on World Food Security and World Food Summit Plan of Action. World Food Summit 13-17 November 1996. Rome. Retrieved January 15, 2020, from http://www.fao.org/3/w3613e/w3613e00.htm.

Food and Agricultural Organization (FAO). (2001). *The state of food insecurity in the world 2001.* Rome, Italy: Author. Retrieved January 15, 2020, from http://www.fao.org/3/y1500e/y1500e06.htm#P0_0.

Ford, J. D. (2009). Vulnerability of Inuit food systems to food insecurity as a consequence of climate change: A case study from Igloolik, Nunavut. *Regional Environmental Change, 9*(2), 83–100. doi:10.1007/s10113-008-0060-x.

Hossain, K., Raheem, D., & Cormier, S. (2018). *Food security governance in the Arctic-Barents region.* Basingstoke, UK: Springer.

House of Lapland. (2019). *The Arctic railway – Risks and opportunities.* Rovaniemi, Finland: Author. Retrieved January 15, 2020, from https://www.lapland.fi/uploads/2019/10/c9834acc-arctic-railway.pdf.

Lawrence, R., & Larsen, R. K. (2019). Fighting to be herd: Impacts of the proposed Boliden copper mine in Laver, Älvsbyn, Sweden for the Semisjaur Njarg Sami reindeer herding community. Stockholm, Sweden: Stockholm Environment Institute (SEI). Retrieved January 10, 2020, from https://www.sei.org/wp-content/uploads/2019/04/sei-report-fighting-to-be-herd-300419.pdf.

Loring, P. (2017, April). Food (in)security and food sovereignty in the North. *Northern Public Affairs*, 56–59. Retrieved January 10, 2020, from http://www.northernpublicaffairs.ca/index/volume-5-issue-1/food-insecurity-and-food-sovereignty-in-the-north/.

Nilsson, L. M., & Evengård, B. (2015). Food security or food sovereignty: What is the main issue in the Arctic? *The New Arctic*, 213–223. doi:10.1007/978-3-319-17602-4_16.

Sheehy, T., Kolahdooz, F., Schaefer, S. E., Douglas, D. N., Corriveau, A., & Sharma, S. (2014). Traditional food patterns are associated with better diet quality and improved dietary adequacy in Aboriginal peoples in the Northwest Territories, Canada. *Journal of Human Nutrition and Dietetics*, *28*(3), 262–271. doi:10.1111/jhn.12243.

St-Germain, A. F., Galloway, T., & Tarasuk, V. (2019). Food insecurity in Nunavut following the introduction of Nutrition North Canada. *Canadian Medical Association Journal*, *191*(20), E552–E558. doi:10.1503/cmaj.181617.

Tendall, D., Joerin, J., Kopainsky, B., Edwards, P., Shreck, A., Le, Q., … Six, J. (2015). Food system resilience: Defining the concept. *Global Food Security*, *6*, 17–23. doi:10.1016/j.gfs.2015.08.001.

United Nations Development Programme. (1994). *Human Development Report 1994*. New York, NY: Oxford University Press. Retrieved from http://hdr.undp.org/sites/default/files/reports/255/hdr_1994_en_complete_nostats.pdf.

World Forum on Food Sovereignty. (2001, September). *Final declaration of the World Forum on Food Sovereignty*. Havana, Cuba: Author. Retrieved January 10, 2020, from http://www.ukabc.org/havanadeclaration.pdf.

PART I

Food security, traditional knowledge and livelihoods

1 'The role of stockfish in local food security

Traditional knowledge, transmission and change in Lofoten, Norway'

Johana Evelyn Montalvan Castilla

Introduction

Lofoten is an archipelago located in Nordland county in northern Norway. National and foreign travellers see it as exotic, and it is a remote yet popular destination. Lofoten is characterised by high, steep mountains and dramatic, abrupt peaks, untouched portions of land, beautiful beaches and bays, and the rich, open ocean surrounding the islands.

Despite its location in the Arctic Circle, the archipelago has relatively mild temperatures and a subpolar-type oceanic climate. This temperature anomaly is caused by the Gulf Stream. The North Atlantic Current and the Norwegian Current, both extensions of the Gulf Stream, contribute to higher temperatures at this high latitude.

Arctic societies have flourished for thousands of years along the coast; their livelihoods have always been strongly connected to the sea. In Lofoten, the earliest archaeological sites evidencing human settlements are about 5,500 years old (D'Anjou et al. 2012). The presence of the Atlantic cod (*Gadus morhua*) has been of pivotal importance for the subsistence and development of early and modern societies in this region.

Drying is one of the oldest food preservation methods known to mankind. To secure nutritious food supplies throughout the year, northern Norwegians learned to preserve fish, particularly Atlantic cod, by hanging it until it dried. The end result is a product known as stockfish. It can have a shelf life of more than five years.

Although much has been written about stockfish, fisheries, and the economic importance of fish for the country (i.e., Hovelsrud et al. 2010; Perdikaris 1996; Sønvisen et al. 2011; Sørheim 1997; Standal and Sønvisen 2015; Standal et al. 2016), few current or contemporary studies address changes in the transmission of traditional knowledge in an Arctic marine context.

This chapter explores the importance of stockfish today, especially its significance for food security in Lofoten and the Arctic region. It particularly focuses on the transmission of traditional knowledge involving stockfish.

DOI: 10.4324/9781003057758-1

The potential of stockfish to enhance food security and the resilience of the local food system is highlighted as well.

Resilience refers to the degree of a system's ability to withstand a shock and remain stable, to promptly adapt to new situations, or to function unhampered by catastrophic changes (Pingali et al. 2005). Household production of stockfish provides a long-lasting source of nutritious food, contributing to the resilience of the local food system.

In the Arctic, foods sold in stores are often imported from ecosystems located far away. Consequently, their availability and prices can be affected by changes in ecosystems and markets elsewhere in the world (Nilsson et al. 2013a, 114). To examine the resilience of Arctic food systems, the multiple variables and current changes that affect Arctic food security must be understood (Nilsson et al. 2013a). This chapter seeks to contribute by shedding light on this topic.

The chapter uses historical and archaeological sources along with the findings of ethnographic research conducted in May and July 2018 in the Lofoten archipelago. The first part of the chapter addresses the particularities of the region's climate, environment, and geography. These particularities create the conditions that allow the production of stockfish. Next, the process of hanging and drying codfish to make stockfish is explained. The longstanding historical importance of stockfish to Norway is also explored.

After evaluating and contextualising the importance of stockfish to the region's food security in the past, this chapter examines the changes surrounding the transmission of traditional ecological knowledge (TEK) and the present concerns of local inhabitants about its transmission. Original fieldwork findings are summarised in this section. The findings from the interviews, conversations, participant observation, and other fieldwork conducted in Lofoten are discussed. The implications of these findings are also explored, particularly in relation to the potential of stockfish to enhance or strengthen food security and the resilience of the local food system.

Methodological considerations

This study was conducted in the town of Svolvær on the island of Svinøya and in the villages of Kabelvåg and Henningsvær in Lofoten. These areas belong to the Vågan municipality in Nordland county. Svolvær is the administrative centre of the municipality.

These locations were chosen for their proximity to each other and geographical accessibility and because they have been traditional northern Norwegian fishing villages for thousands of years (D'Anjou et al. 2012). Svolvær has 4,630 inhabitants (SSB 2017); Kabelvåg and Henningsvær have 1,801 and 457 inhabitants, respectively (Vågan kommuneplan 2017).

Social science research methods, including participant observation and other ethnographic methods, were used. In participant observation, the researcher spends time with the local community and takes part in at least some activities in the community (Becker and Geer 1957). In this ethnographic method, data are collected by observing the behaviour and experiences of the participants in particular social contexts. This method also considers participants' own interpretations of their experiences and behaviours, as these are often shaped by the constraints of their society (Dewan 2018). As such, rather than 'studying' people, ethnography means 'learning' from the people we encounter (Spradley 2016, 3).

Qualitative data collection methods, such as semi-structured interviews and formal and informal conversations, were employed. Central themes and arguments that illustrate the perceived importance of stockfish were identified from the responses. Among these, several points were highlighted: the availability of stockfish; its financial benefits and the security it offers (at the local and national levels); its nutritional content and other health-related aspects; its role in decreasing dependence on imported foods; the lifespan of stockfish; and its cultural, social, and historical dimensions. In total, about 15 people (male and female) aged 16–85 participated. Some of the participants were locals that the researcher met during the first field trip and their friends and relatives. They were contacted via snowball sampling. Other participants were local people of diverse backgrounds who willingly initiated contact. These participants learned about the study from an article in the local newspaper or through social media.

Informal conversations with locals and travellers also took place on the main square or in the open marketplace in Svolvær and other public places. All semi-structured interviews were audio-recorded. When I was invited to continue the conversation at the homes of local families, only written notes were taken. The study followed the guidelines of the Norwegian Data Protection Centre (NSD 2018). All information was anonymised. Personal or sensitive data about the participants were not recorded.

The questions in the semi-structured interviews addressed the transition of traditional, local knowledge within families and the local community. The importance or lack of such knowledge and changes in its transmission were addressed. Further, the questions allowed the participants to express their opinions about food security in the region and the role stockfish plays in it.

Stockfish

Stockfish is cod fish (particularly Atlantic cod) that has been dried solely by exposure to cold wind and air. Unlike other fish-preserving methods, no salt or other additive is used. The production of stockfish is common in the Lofoten and Vesterålen islands; it requires very specific climatological conditions which are perfectly met in these two parts of Norway.

Traditionally, cod are fished during the winter and then gutted and beheaded. The bodies are then hung by the seashore on traditional wooden

Figure 1.1 Stockfish hanging from the racks in Svinøya, Lofoten. Photo by the author.

racks known as *hjell*. The fish are usually hung to dry from January to early April. They are placed at particularly windy locations to air dry gradually. Very cold nights freeze the fish, while daytime temperatures thaw them slightly (Perdikaris 1999). These small but continuous temperature variations are all that is needed to produce high-quality stockfish. Once dried, stockfish has a storage life of several years (Figure 1.1).

In Norway, cod fisheries are seasonal and rely on cod spawning areas that extend from Troms to Møre (Sunnana 1992, in Perdikaris 1999). Cod traditionally migrates from the southern part of the Barents Sea to the Norwegian coast in the winter and early spring. The spawning grounds are very extensive and close to the shore (Nakken 1993). This enables fishermen to harvest a great deal of fish.

The special climatological conditions of Lofoten deserve some attention. The climate of Lofoten is abnormally mild for its Arctic latitude due to the warm streams from the North Atlantic Drift (Ryvarden 1991), also known as the Gulf Stream. The archipelago experiences strong winds, particularly in late autumn and winter. Snow and sleet are not uncommon in winter, when cod are fished. The cold winds and temperature variations enable the production of stockfish (Figures 1.2 and 1.3).

'Stockfish built this nation!': stockfish through history

As the Norse King Sagas of Snorre Sturlason (1226, in Aak et al. 2011) mentions, fish-drying techniques in northern Norway date back more than 1,000 years ago. Archaeological research in the Norwegian Arctic indicates that intensive dried fish production extends back into the Iron Age (ca. 500 BC), giving cured cod a substantial role before the historical stockfish trade that began around 1200 AD (Perdikaris 1999).

Figure 1.2 Fish hanging and drying from racks in Svinøya, Lofoten. Photo by the author.

Figure 1.3 Stockfish production in Svinøya, Lofoten. Photo by the author.

Jenssen (2012, 8) asserts that without cod, the Vikings would not have survived. Lacking salt, they learned to dry cod. It soon became an important source of food and trading item. Dried fish were taken on sea journeys. As a nutrient-rich food with an unlimited expiration date, codfish enabled the Vikings to survive long trips to Greenland, Europe, and the United States of America, which they reached 500 years before Columbus' arrival.

Preserving fish has long been important, not only among northern Norwegians, but also among Sea Sami populations, especially because of local variations in the use of natural resources (Berg 2014). Commercial fisheries began using traditional hanging and drying methods around 1150–1200 AD (Bertelsen 1992); at that time, Lofoten and Vesterålen provided stockfish to urban and rural populations.

During the fourteenth century, the Hanseatic League monopolised the distribution of stockfish, which was mostly exported from Lofoten to northern Norway. The Hanseatic League was a German trading company that operated in the North Sea and Baltic Sea regions. They established an office in Bergen between 1360 and 1761.

While massive amounts of stockfish, cod liver oil, hide, butter, fur, and tallow were exported from the North, grain was imported from the Baltic countries. Norway thus received goods such as flour, malt, salt, beer, and hemp, along with fabrics, hardware, and glass (Trebbi 1996).

(Insert picture: *Hanseatic League headquarters and storage rooms are still well preserved in the city of Bergen on the western coast of Norway.*)

Stockfish provided a vital source of revenue for the king and church, and became an essential object of regular tax policies in the kingdom of Norway (Bjørgo 1986, 42). It was cod that gave Norway the economic power to establish churches and cathedrals, to consolidate the monarchy, and to establish bureaucracy, government, and a parliament (Jenssen 2012). The city of Bergen would not have flourished without stockfish. As many study participants expressed it: 'Cod and stockfish built this nation!'

Today, most people have modern refrigerators and freezers that can preserve fish and other foods for months. However, many people in Lofoten still produce their own stockfish. Many participants in this study (or their friends or relatives) produce stockfish for household consumption. This seems to indicate that many Lofoten households still preserve fish in traditional ways by hanging it to dry. It is certainly not uncommon to see the fish left to dry on balconies or in other ventilated areas in local houses. Some also buy stockfish from local sellers or producers. The practice and knowledge involved in producing, keeping, preparing, and then consuming stockfish deserves further exploration.

Exploring traditional ecological knowledge

TEK includes multiple bodies of knowledge accumulated over many generations (Drew 2005; Ruddle 1993). It results from the interactions between people and the natural world; it is accrued through trial and error and transmitted to future generations orally or through practical experience (Drew 2005; Ohmagari and Berkes 1997). Thus, TEK constitutes a way of knowing, and as such, it shares some characteristics with Western science. For instance, both are based on collections of observations (Berkes et al. 2000). TEK and Western science do have differences as well. Anthropologist Claude Levi-Strauss argues that they represent two different ways of acquiring knowledge about the universe. In his words, 'the physical world is approached from opposite ends in the two cases: one is supremely concrete, the other supremely abstract' (Levi-Strauss 1962, in Berkes et al. 2000).

Despite the validity and importance of TEK, it should be noted that, today, due to changing environmental and socio-economic conditions, not all traditional practices around the world are ecologically wise or adaptive (Berkes et al. 2000). It is beyond the scope of this chapter to discuss maladaptive TEK practices in other regions or cultures, but they should be acknowledged.

In the literature on ecological knowledge and coastal societies, another important term, local ecological knowledge (LEK), is often used. LEK is the knowledge that a particular group of individuals have about their local ecosystems. Unlike TEK, LEK does not necessarily involve cultural or historical continuity regarding the use of resources (Olsson and Folke 2001) or transmitted practices. The production of Lofoten's stockfish entails the cultural transmission of preservation practices that have been done for many generations and can be traced back to Viking times. Thus, the practice itself involves the continuation of historical and cultural values.

TEK about food preservation techniques has long played an important role in food security for the peoples of the Arctic (Bergman et al. 2004; Perdiakaris 1999). Later, in the seventeenth and eighteenth centuries, as Sweden expanded its imperial ambitions, it turned to the North – not just for furs, but also for meat and stockfish to feed its growing armies (Lundmark 1989; Wheelersburg and Kvist 1996, in Green 2016, 83).

Another example of the importance of TEK in the Arctic regions can be found in Norwegian explorer Roald Amundsen's 1911 expedition; he became the first man to reach the South Pole. He took *pemmican*, a highly nutritious food made by drying or smoking meat, to the South Pole. Along with salted fish, this was one of the main foods for the expedition crew. Amundsen's observations and integration of traditional adaptation strategies and foods of Arctic indigenous groups allowed him and his team to finish their expedition in good health.

It can be argued that traditional preservation techniques emerged from necessity. Just like northern Norwegians settled in Lofoten, Arctic Indigenous groups, such as the Sami, lived in Lofoten as well (NRK 2017). However, little is known about their time there or how they lived. It is known that the Sami settled in other parts of northern Norway and developed different subsistence strategies. For instance, they hung and dried fish that they subsequently stored. This helped them survive in environments that experienced strong seasonal fluctuations that affected productivity and the availability of resources.

The transmission of TEK in Lofoten

In the Lofoten archipelago, the transmission of TEK about the practice of drying cod has not stopped. The fish are dried mainly during the months of January to early April. Some people produce stockfish for household consumption; other individuals and companies produce it to sell it. This practice has continued, presumably almost unchanged, for over a thousand years. The unsalted fish are hung without any additives; merely the power of nature and the right environmental conditions allow it to dry gradually. Several interviewees asserted confidently that this food preservation practice 'continues to be the same one we learned from our ancestors'. However, it does seem that several small changes have occurred as the knowledge was transmitted and the newer generations received it and put it in practice (Figures 1.4 and 1.5).

Figure 1.4 Stockfish produced at home by a woman in Henningsvær. Photo by the author.

Figure 1.5 A closer look at fish left to dry. Once stockfish is produced, it will be used for household consumption or sold to travellers. Photo by the author.

Today, nearly everything can be bought from supermarkets, and food is usually stored in refrigerators and freezers for later consumption. People are highly dependent on imported foods. They also depend on electricity and digital services that monitor temperatures and function of freezers in supermarkets and in institutions such as hospitals.

For a society's or culture's TEK to be passed on to newer generations, younger people must have a continuing need for or interest in putting such knowledge into practice. Therefore, this study sought to explore whether members of the younger generations were interested in TEK about stockfish, including fishing, producing, preparing, and consuming stockfish.

Only three study participants reported that they sometimes fished for cod to produce stockfish. When they did, it was a social activity often involving friends and relatives of their own age or older. They also reported that their children were 'not that interested' in joining these activities and preferred to do other things. Other participants said they bought fish from local fishermen to produce stockfish themselves.

Producing stockfish entails gutting the fish and removing the heads. The fish are then hung on balconies or in other open areas. The cold winds and temperature fluctuations then handle the drying process. Both male and female participants produced stockfish in this way at home or, in a few cases, in small designated areas at properties they owned. Only one woman reported that her oldest daughter gladly joined her in producing stockfish for sale and for consumption by the family. Some male participants reported that a son joined them in cleaning and hanging the fish. One man with one daughter and two sons (aged 10, 15, and 20, respectively) said, 'I must pass on this tradition to my kids as well'. He also reported that his children were 'picky and difficult' and less interested than in his own generation or previous ones in fishing or in drying and cooking fish.

Despite several reports of children's reduced participation, several people mentioned that their children had 'observed the process itself' on repeated occasions without taking part. This may indicate that TEK is still transmitted to younger generations, albeit in a more limited way.

The preparation of stockfish dishes in most households involves adults rather than children and is a very social occasion. Close friends and relatives are often invited to help as some of the dishes are time-consuming to prepare. Some children and teenagers occasionally observe or participate in this process.

Aside from one family that ate a primarily vegetarian diet that rarely includes fish, most participants reported eating stockfish every month. Families' annual consumption of stockfish ranges from around 15 to up to 30 times a year. *Klippfisk* (bacalao), *buknafisk* (semi-dried fish), and stockfish are some of the preferred types of dried fish. Many families also produce stockfish snacks.

Some interviewees strongly believed that their teenage children or grandchildren and their friends would see the value in producing, preparing, and eating stockfish when they reached adulthood. One added:

> When one grows up or becomes an adult, then one starts to see the value in cooking food that takes more time to be cooked. For example, traditional food made for holidays or special occasions. You put more of your soul in that, you work a bit extra hard and continue the preparations and cooking during the weekend.

Similarly, a local man from Kabelvåg commented:

> I believe the younger generations will appreciate stockfish more when they become older. It is difficult to get them to consume stockfish. It must be finished soaking and ready. This means it must be frozen. It can then easily take a day before it is dethawed. Then we have the same problem again. I hope they will see the value and joy in the old traditions. It is important that [those traditions] do not die.

Another man who hung and dried fish with friends added, 'We do this so that we can have food we can consume all throughout the autumn and winter.' He reported that unlike his children, nieces, and nephews, the children of one of his friends sometimes helped clean, hang, and cook the fish. Like several other participants, he pointed out that today, it is difficult to recruit people to fish, produce, or cook stockfish. According to these participants, many modern Norwegians prefer office jobs to traditional occupations. A local chef noted, 'I see it is difficult to recruit people to work as fishermen, farmers, chefs, or cooks. These days, everybody wants to sit in an office.'

Although many locals produce stockfish and eat it throughout the year, many also buy stockfish from companies and local producers. Some participants pointed out that today, it is mostly Eastern Europeans (particularly Polish men) rather than locals who work in the fish landing and processing facilities. This was corroborated during a field visit. Since cleaning and hanging fish for stockfish production is a seasonal job, it is not appealing to most Norwegians, who prefer jobs that offer steady year-round incomes. Therefore, several study participants expressed concern that foreign workers were not being taught the traditional way of hanging fish from the racks, which could lead to a decrease in the quality of stockfish.

This concern is illustrated in three comments by participants. One said, 'The fish must be cleaned and hung with its back facing south-west and its abdomen facing north. If this is not taught to the Polish and other foreign workers, the stockfish will be low quality'. Another commented, 'Trying different ways of hanging the fish has taught us that the "old ways", the ways of our ancestors, are the best way to hang the fish'. A third participant expressed it more dramatically: 'Very few Norwegians know the history of stockfish; even fewer know how it should be hung. This knowledge is almost disappearing'.

Most study participants were adults who had children, nieces, or nephews; some even had grandchildren. One concern that came up often in the conversations was decreasing interest in eating, cooking, and producing stockfish. Most participants said that young people and young adults seemed to prefer fast food these days. It is primarily adults who take the initiative to produce and cook stockfish. As one put it, 'They prefer foods that can just be reheated or microwaved and eaten because they take less effort and time (...); such foods are not always the healthiest option'.

Another participant pointed out that the food and advertising industries had strongly influenced food preferences, saying, 'The big capital invested in advertisement campaigns also dictates a trend of consumption. TV and media, in different ways, encourage fast food and associate it with being cool, popular, good, trustworthy, and casual'. Another woman remarked, 'In supermarkets, convenience food is abundant. Frozen heat-and-eat food is available and shelved at eye level. There are many sweets and unhealthy snacks...but what about stockfish and dried fish products? How much high-profile placement and advertisement do they get?'

The comments above illustrate that food preferences and traditions are not static. Time and societal changes can significantly impact food preferences, and this can in turn decrease interest in a specific food item and perceptions of its value. However, the fact that older generations said they valued and appreciated stockfish and therefore continued to produce it for household consumption deserves attention. The following section considers why locals still consider stockfish important and the production and consumption of stockfish a valuable tradition worth continuing.

Stockfish and its importance for food security

According to the Food and Agriculture Organization (FAO 1999), to be food secure, individuals and communities must be able to reliably access food. Moreover, sufficient nutritious food must be available, and that food must be of an acceptable quality to meet dietary needs. In short, food security is having physical and economic access to adequate quantities of safe, nutritious food.

In light of the findings of this study regarding (particularly) younger locals' decreased interest and participation in the TEK and practices surrounding the production of stockfish, an essential question remains. In today's society, is TEK about stockfish still important, and if so, how or why?

This study's findings strongly point to the potential role of stockfish as a contributor to food security in Lofoten. In all 15 open interviews, participants indicated that stockfish was important in today's society, and different reasons were given to support this argument. The bar chart (ranging from 0% to 100%) in Figure 1.6 summarises the reasons given. These answers will soon be further explored in the context of food security.

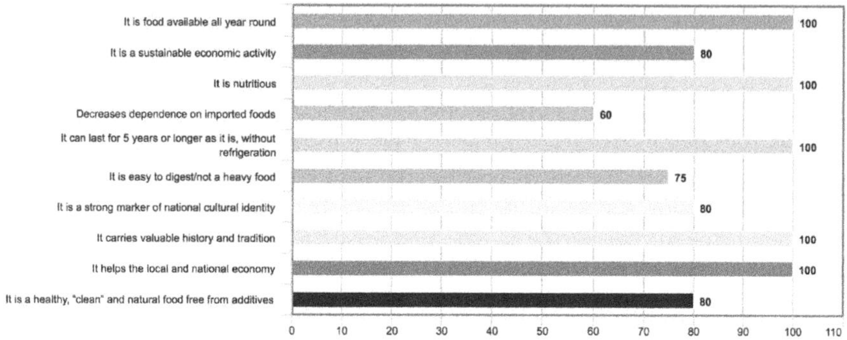

Figure 1.6 Bar chart illustrating the importance of stockfish, according to local inhabitants.

Availability and access are very important aspects of food security. Once dried and ready for consumption, stockfish can be eaten in all seasons and easily stored at home. Stockfish is currently exported to different countries. Among these, Italy represents the biggest market. However, the fact that stockfish can be produced at a local, household level and can provide a year-round source of protein and nutrients means that it can enhance food security in Lofoten.

The production of stockfish is also considered a sustainable economic activity. Strong national regulations apply to commercial cod fisheries. They regulate how much fish can be fished, during what seasons, and in what areas. However, there are also some threats to the sustainability of the marine ecosystem. Many locals mentioned that they participate in local organisations that monitor the protection of the marine ecosystem of the Lofoten archipelago and the islands of Senja and Vesterålen. Some primary concerns relate to the possible impacts of oil extraction, ship pollution, unsustainable tourism, and perceived climate change.

A local civil organisation, 'Peoples' Action for an Oil-Free Lofoten, Vesterålen, and Senja' (Norwegian: *Folkeaksjonen oljefritt Lofoten, Vesterålen og Senja*), brings many local inhabitants, activists, and even politicians together every year. Their gatherings disseminate information and organise collective action to fight state or private corporations seeking access to these Arctic regions to search for oil. All interviewees in this study expressed concerns about the possible threats oil extraction poses to the region's food security. They emphasised that the oceanic ecosystem is delicately balanced, and should an oil spill occur, many fish would die or simply not even be born. After a fish's eggs have been laid, the natural process of hatching would be interrupted by oil or other pollutants in the water. Without cod, stockfish and other fish products could no longer be produced.

Traditional food is an essential element of culture and an important source of nutrients (Nilsson et al. 2013, 818). Many locals highlighted the nutritional content of stockfish along with its other healthy attributes. Stockfish is one of the richest existing protein sources; just 100 grams of stockfish contain 79 grams of pure protein. It also contains many important B vitamins (such as thiamin, riboflavin, and niacin), minerals (such as iron and calcium), and omega fatty acids (Norwegian Seafood Export Council 2018).

Stockfish also reduces the region's dependence on imported foods. One kilo of dried stockfish is equivalent to approximately five kilos of fresh fish (Norwegian Seafood Export Council 2018). One participant said: 'Norway imports many items from all over the world. And they take it for granted that this will continue indefinitely'. Such expressions are supported by statistics from the industry and food sectors. Today, Norway produces only 40% of the food that is consumed in the country (Agricultural Cooperatives in Norway 2018). The nation is thus highly dependent on imported foods. Some of the most important imported food staples are meats, cheeses, and vegetables. Rainfall was lower than usual in Norway and much of Europe in 2018, which affected production and the prices and availability of certain foods. Since stockfish is highly nutritious and easy to store, it could help enhance local or regional resilience if the usual flow of imported foods were ever interrupted.

Another advantage of stockfish is its shelf life; it can be preserved for years without much effort. In the homes of some local inhabitants, dried fish hangs from the veranda, or new and old stockfish is stored in a designated room. When it is needed, the stockfish is soaked or watered for several days (usually an entire week) and then cooked in different ways. After a week of watering or soaking, the fish has grown several times larger, recovering its original size. The ability to inexpensively produce and store a highly nutritious food that can last for years significantly improves the food security of individual households. Although stockfish is usually said to last for two to five years, three study participants said they had eaten ten-year-old stockfish that still tasted good and retained its nutrients.

Furthermore, stockfish does not always have to be soaked before consumption; it can also be eaten as-is after drying. Many participants said they use a hammer to break it into small pieces to eat as snacks. It is also easy to digest. Some participants consider this important, commenting that many processed foods or foods high in starch or sugar had undesirable effects on their digestion. Therefore, many companies sell stockfish as a snack in Norwegian supermarkets (Figures 1.7 and 1.8).

There is also a social and cultural dimension and value to traditional food and food supplies in local and Indigenous communities (Hossain 2017; Nilsson et al. 2012, 2013). Food and ancestral knowledge about food help build cultural and personal identity, as they are part of a cultural heritage (Royal Commission on Aboriginal Peoples 1996, 194, in Ford 2008). Participants often stressed that they and their ancestors grew up eating stockfish and that 'it was part of their culture'. Several mentioned that they brought stockfish with them when they left Lofoten for a visit or to relocate and introduced it to new friends and acquaintances.

Figure 1.7 Stockfish snacks in Norwegian supermarkets. Photo by the author.

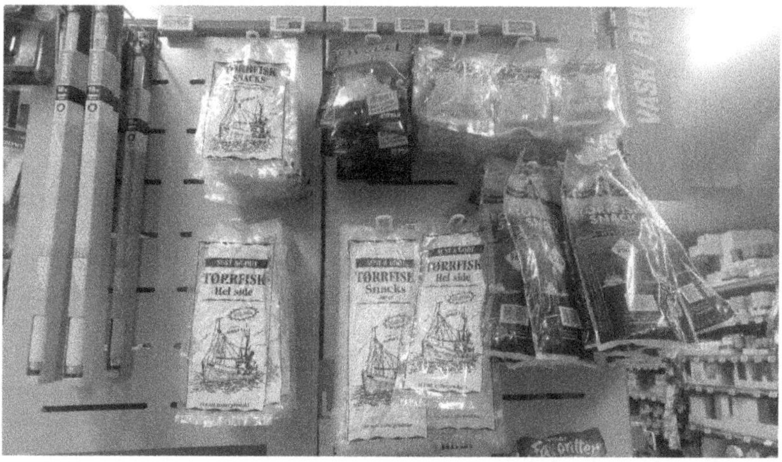

Figure 1.8 Stockfish snacks made from different kinds of cod. Photo by the author.

In the main square, I initiated conversations with national and foreign travellers. Norwegians from the South or the West Coast, for example, bought many kilos of stockfish and other dried fish products such as *buknafisk* (half-dried fish) or *bacalao* (dried salted fish) to take home. Some had a parent who had emigrated from northern Norway to other parts of the country and that they 'taught them about their culture and how good stockfish was'. As a result, they grew up eating stockfish and other dried fish products. Similarly, one Norwegian man who bought stockfish in Lofoten was married to a Philippine woman. He said that teaching her about his culture and northern Norway 'meant teaching her about stockfish and enjoying stockfish as a snack and in meals'. A young woman who originally came from Lofoten had moved to southern Norway and returned to Lofoten on vacation. She and her husband bought many kilos of stockfish to take home. The husband said that he had learned to appreciate stockfish a few years before when they started dating; she and her family had invited him to eat stockfish at their home.

Locals also mentioned the value of the history and tradition surrounding stockfish. As one said:

> When we know the history and meaning of stockfish, we appreciate it more. At the same time, it seems that society has changed so much. The values around food have changed. People want things that can be made fast, like pizza or unhealthy foods. I would call this 'a new life philosophy' built on the idea that everything should be fast, because one must work, maybe one is too tired (...), but other people still appreciate other values, such as stockfish and good food. There are foodies; there are people who are interested in the history and nutrition of our traditional food. This is important for the security of the region.

Another participant pointed out that 'the most important value in it [stockfish] is the history and tradition it carries. It is important to take care of that; big parts of Norway are built on stockfish'.

Statements like these stress the importance of the historical knowledge embedded in the traditions surrounding stockfish. As mentioned earlier in this chapter, stockfish has played an important role throughout Norway's history, providing high revenues as an export and fostering economic development. Concomitantly, this fostered urban, cultural, and social development that has benefited Norwegian society for centuries.

All participants stressed that stockfish still contributes substantially to the local and national economy. It is a source of income for the many people involved in its local and national production and sale. Beyond that, at the industry level, in 2017 alone, Norway exported stockfish for 645 million kroner (Norges Sjømatsråd 2018). Stockfish is imported primarily to Italy, Nigeria, the UK, and the USA.

Some concerns were also raised about the sale of stockfish, specifically the stockfish bought from Iceland and sold to tourists and travellers in Lofoten. Since electricity is cheaper in Iceland, stockfish is produced faster there by drying the fish artificially. A local commented, 'Some of the stockfish sold at some of the shops or kiosks in Lofoten features the name "Lofoten," but if you read the back of the package, you'll see it's fish from Iceland'. Another added, 'People are being misinformed. Fish dried artificially in Iceland is sold and advertised as though it were produced locally'.

Finally, local participants said they valued locally produced stockfish because it is a healthy, 'clean', natural food free from additives. Some expressed concerns about genetically modified foods or unhealthy convenience foods. Participants said they feel comfortable eating local stockfish as they know it is healthy and does not contain any harmful additives.

Discussion

The opinions, experiences, and perspectives of local inhabitants on the importance of stockfish and TEK highlight the contemporary social, economic, and cultural importance of this traditional food. Currently, stockfish represents an important source of income for the region and for the sectors and the people involved in the fishing, production, and sale of the end product. However, beyond this, the stockfish in individual households may have the potential to prevent food insecurity. Its nutritional properties and long shelf life and the fact that it can also be consumed without additional preparation enable it to contribute to the region's resilience and decrease its vulnerability.

The household production of stockfish can foster resilience. As previously discussed, currently, about 60% of the food consumed in Norway is imported. In the event of a crisis or short-term event preventing access to the imported foods that the region and country depend on, stockfish could help households maintain food security. It could also increase food security in the event of an electrical or digital failure or a security breach affecting the operation of freezers or refrigerators at supermarkets, a problem that could lead to a food shortage. As recently as 2019, it was reported that the freezers belonging to some supermarkets, food stores, and hospitals in different countries could easily be accessed online, an indication of their potential vulnerability (Brombach 2019).

Storms and extreme weather are also a concern to several municipalities in northern Norway because they may threaten electricity supply, block roads, and even isolate communities (Hovelsrud et al. 2010, 36). This can cause food security concerns that include serious disruptions of food supplies, particularly due to infrastructure failures or challenges in distributing food (Hovelsrud et al. 2010; Nilsson et al. 2013).

Currently, northern Norway is considered a food secure area, because 'all people, at all times, have physical and economic access to sufficient safe and nutritious food to meet their dietary needs and food preferences for a healthy and active life' (FAO 1996, in Pinstrup-Andersen 2009, 5). However, this region depends to a large extent on imported food and on national and international trade (for another example of this in the European North, see Chapter 11).

In a food emergency, local and traditional foods such as stockfish may substantially decrease the dangers of dependency and reduce hunger while providing adequate nutrition. Intervention strategies that aim to increase the resilience of the local food system should therefore include TEK. Reinforcing TEK and its transmission, particularly TEK about stockfish, could play a key role in avoiding food crises.

Globally, TEK about food and natural resources is under great threat (see Chapter 15 of this volume). The findings of this study seem to indicate that there are changes in the transmission of TEK. Newer generations appear to be less engaged or interested in participating in the production of stockfish and preparation of stockfish dishes than previous ones.

Sixty percent of this study's participants expressed concerns over the perceived 'loss' of the local and national history surrounding stockfish. Based on personal experience, many said that youth and younger generations knew very little about the role of stockfish in Norway's history. Many commented that young people preferred convenience foods to traditional ones such as stockfish, and that they believed that perhaps only upon reaching adulthood would the new generation understand the value in stockfish and stockfish meals. The same could be said about its potential to increase the society's resilience.

Concerns about the transmission of knowledge related to the production and preparation of stockfish lead us to reflect on how this knowledge should be preserved. Several study participants suggested that the history and importance of stockfish should be taught in schools, along with how to prepare it. They believe this would be a good way to interest children and youth in continuing the tradition of preserving fish and ensuring that their households have access to local, nutritious food all year. Undoubtedly, multiple strategies could be used to strengthen the TEK about stockfish acquisition, production, preparation, and consumption. Further studies in the region, especially studies including younger participants, are therefore suggested. Such research could be used to plan strategies that include promoting stockfish to enhance the resilience of the local or regional food system and overall societal resilience.

Acknowledgements

I kindly thank the local inhabitants who participated in the study and generously shared their knowledge, time, and opinions and warmly welcomed me into their homes and workplaces. Signe Sønvisen, colleagues from food security workshops in Stockholm and Umeå, and an anonymous reviewer kindly offered valuable insight and feedback for this chapter.

References

Aak, A., T. Birkemoe, and G.K. Knudsen. 2011. Efficient mass trapping: Catching the pest, Calliphora vicina (Diptera, Calliphoridae), of Norwegian stockfish production. *Journal of Chemical Ecology* 37(9): 924.

Agricultural Cooperatives in Norway. 2018. Norway is completely dependent on imported food. Retrieved 2 January 2019 from https://www.landbruk.no/internasjonalt/i-dag-er-norge-helt-avhengig-av-importert-mat/

Becker, H.S. and B. Geer. 1957. Participant observation and interviewing: A comparison. *Human Organization* 16(3): 28–32.

Berg, E. 2014. *Sami traditions: Márkomeannu's contribution to the revitalization of Sami food traditions.* M.phil. Faculty of Humanities, Social Sciences and Education. University of Tromsø.

Bergman, I., L. Östlund, and O. Zackrisson. 2004. The use of plants as regular food in ancient subarctic economies: A case study based on Sami use of Scots pine inner bark. *Arctic Anthropology* 41(1): 1–13.

Berkes, F., Colding, J., & Folke, C. 2000. Rediscovery of traditional ecological knowledge as adaptive management. *Ecological applications, 10*(5), 1251–1262.

Bertelsen, R. 1992. An archaeological perspective on the Medieval north-south connection. In *Medieval Europe 1992: exchange and trade pre-printed papers*. York: Medieval Europe, 177–184.

Bjørgo, N. 1986. Kontakten mellom Bergen og Nord-Norge i melomalderen. [The contact between Bergen and northern Norway in middle ages]. In I. Oye (ed.), *Kjøpstad og Rikssentrum*. Bergen: Bryggen Museum, 41–53.

Brombach, H. 2019. Kjøleanlegg hos en rekke matbutikker og sykehus ligger åpent på nett. Retrieved from https://www.digi.no/artikler/kjoleanlegg-hos-en-rekke-matbutikker-og-sykehus-ligger-apent-pa-nett/458119.

D'Anjou, R., R. Bradley, N. Balascio, and D. Finkelstein. 2012. Climate impacts on human settlement and agricultural activities in northern Norway revealed through sediment biogeochemistry. *Proceedings of the National Academy of Sciences* 109: 20332–20337.

Dewan, M. 2018. Understanding ethnography: An 'exotic' ethnographer's perspective. In: Mura P., Khoo-Lattimore C. (eds.), *Asian qualitative research in tourism*. Singapore: Springer, 185–203.

Drew, J.A. 2005. Use of traditional ecological knowledge in marine conservation. *Conservation Biology* 19(4): 1286–1293.

FAO 1999. *The state of food insecurity in the world*. Report. Retrieved 1 October 2017 from http:// www.fao.org/docrep/007/x3114e/x3114e00.htm.

Ford, J. 2008. Vulnerability of Inuit food systems to food insecurity as a consequence of climate change: A case study from Igloolik, Nunavut. *Regional Environmental Change* 9: 83–100.

Green, A. 2016. *Tastes of sovereignty: An ethnography of Sami food movements in arctic Sweden*. PhD dissertation in Applied Anthropology. Oregon State University.

Haaland, A. and Å Svihis. Kyst og havlandet. Kyst-Norge i et historisk perspektiv 2009. Publisher: ABM-utvikling; Kystverket; Riksantikvaren; Fiskeridirektoratet.

Hossain, K. 2017. *Security – a shared concept? Are the Sami legitimate actors in the securitisation move?* Third International Conference on Human Security. University of Belgrade – Faculty of Security Studies Human Research Center. Retrieved 10 March 2018 from http://conference.human- security.info/wp-content/uploads/2017/07/HS-Conference-Proceedings-2017.pdf.

Hovelsrud, G.K., H. Dannevig, J. West, and H. Amundsen. 2010. Adaptation in fisheries and municipalities: Three communities in northern Norway. In G. K. Hovelsrud & B. Smit (Eds.), *Community adaptation and vulnerability in arctic regions*. Dordrecht: Springer, 23–62.

Jenssen, F.A. 2012. *Torsk: Fisken som skapte Norge*. Oslo: Kagge Forlag AS.

Nakken, O. 1993. Causes of trends and fluctuations in the Arcto-Norwegian cod stock. *ICES 1993 Cod and Climate Symposium* No. 12.

Nilsson, A., L.M. Nilsson, A. Quinlan, and B. Evengård. 2013a. Food security in the Arctic: Preliminary reflections from a resilience perspective. *Arctic Resilience Interim Report 2013*. Stockholm: Stockholm Environmental Institute and Stockholm Resilience Centre, 113–116.

Nilsson, L., G. Destouni, J. Berner, A. Dudarev, G. Mulvad, J. Odland, A. Parkinson, C. Tikhonov, A. Rautio, and B. Evengård. 2013b. *A call for urgent monitoring of food and water security based on relevant indicators for the Arctic*. The Netherlands: Springer.

Nilsson, L., A. Winkvist, M. Brustad, J. Jansson, I. Johansson, P. Lenner, B. Lindahl, and B. Van Guelpen. 2012. A traditional Sami diet score as a determinant of mortality in a general northern Swedish population. *International Journal of Circumpolar Health* 71(1), 18537. Retrieved from www.tandfonline.com/doi/pdf/10.3402/ijch.v71i0.18537.

Norges Sjømatsråd. 2018. A new record for white fish. Retrieved 21 January 2019 from https://seafood.no/aktuelt/nyheter/nok-et-rekordar-for-hvitfisk/.

Norwegian Seafood Export Council. 2018. Tørrfisk brosjyre. Brochure on stockfish, N-9291.

NRK 2017. Our roots are more important for us than what we would like to admit. Retrieved 1 January 2019 from https://www.nrk.no/sapmi/xl/_-rottene-vare-er-viktigere-for-oss-enn-vi-liker-a-innromme-1.13701498.

NSD – Norwegian Data Protection Centre. 2018. Guidelines on the protection of personal information and data management. Retrieved 15 March 2018 from http://www.nsd.uib.no/personvernombud/en/notify/index.html.

Ohmagari, K. and F. Berkes. 1997. Transmission of Indigenous knowledge and bush skills among the Western James Bay Cree women of Subarctic Canada. *Human Ecology* 25: 197–222.

Olsson, P. and C. Folke. 2001. Local ecological knowledge and institutional dynamics for ecosystem management: A study of Lake Racken watershed, Sweden. *Ecosystems* 4(2): 85–104.

Perdikaris, S. 1996. Scaly heads and tales: Detecting commercialization in early fisheries. In *Archaeofauna. Ichthyoarchaeology and the Archaeological record.* Proceedings of the 8th meeting of the ICAZ Fish Remains Working Group. Madrid: A. Morales (ed.), 5 (1996): 21–33.

Perdikaris, S. 1999. From chiefly provisioning to commercial fishery: Long-term economic change in arctic Norway. *World Archaeology* 30(3): 388–402.

Pingali, P., L. Alinovi, and J. Sutton. 2005. Food security in complex emergencies: Enhancing food system resilience. *Disasters* 29: S5–S24.

Pinstrup-Andersen, P. 2009. Food security: Definition and measurement. *Food Security* 1(1): 5–7.

Ryvarden, L. 1991. *Lofoten og Vesterålen. Reise og Kulturguide.* Oslo: Nortrabooks.

Ruddle, K. (1993). *The transmission of traditional ecological knowledge. Traditional ecological knowledge: Concepts and cases.* Ottawa, Canadian Museum of Nature and IDRC, 17–31.

Sønvisen, S. A., Johnsen, J. P., & Vik, J. 2011. The Norwegian coastal employment system: what it was and what it is. *Maritime Studies (MAST)* 10 (1): 31-56.

Sørheim, H. 1997. The origin of commercial fisheries and the trade of stockfish in the northern part of western Norway. In G. de Boe and F. Verhaeghe (eds.), *Exchange and trade in Medieval Europe: Papers of the Medieval Europe Brugge 1997 Conference*, 3, 107–117.

Spradley, J.P. 2016. *Participant observation.* Long Grove, Illinois: Waveland Press.

Standal, D. and S.A. Sønvisen. 2015. Gear liberalization in the Northeast Arctic cod fisheries – implications for sustainability, efficiency and legitimacy. *Marine Policy* 53: 141–148.

Standal, D., S.A. Sønvisen, and F. Asche. 2016. Fishing in deep waters: The development of a deep-sea fishing coastal fleet in Norway. *Marine Policy* 63: 1–7.

Trebbi, M. 1996. The Hanseatic Museum and Schøtstuene. Bergen: Grieg Medialog AS.

Vågan municipality. 2017. Vågan kommuneplan 2017–2029. Retrieved 20 October 2018 from https://www.nfk.no/byromseminar-nordland/_f/p35/i4d840902-16f8-47fa-991b-22ddf2de352d/kommuneplanens-arealdel-med-kystsonen-2016-2028-formannskapets-forslag-22-mai-2017.pdf.

2 Traditional reindeer rangeland management and a (human) rights-based approach to food sovereignty

Assi Harkoma and Bruce C. Forbes

Introduction

Rangeland conditions and shrinking operational space in Finland

The reduction in extent of viable reindeer pastures is creating challenges and constraints (see Forbes 2006; Kitti et al. 2006; Magga et al. 2009; Johnsen 2014; Kumpula et al. 2014) to food sovereignty all over Sapmi. For Finnish reindeer herders, the situation is especially severe. In Finland, the extent and productivity of lichen-dominated winter pastures in particular have decreased and, since World War II, lichen cover has decreased in fell areas ('Avoimuutta poropuheisiin' 2016), where reindeer herding is an important source of livelihood for the Sami. Such of reindeer pastures could jeopardise the ability of the Sami to practice traditional reindeer herding and produce reindeer meat in traditional ways. A combination of factors affect the quality and quantity of pasture vegetation, including trampling of lichens during the snow-free season, climate change, altered or degraded pasture lands (Forbes 2006; Kumpula et al. 2014), regulatory challenges (Kitti et al. 2006; Magga et al. 2009), other uses of the land, and increased human activities with numerous economic interest (Pape and Löffler 2012; Kumpula et al. 2014; Saamelaiskäräjät 2016). The situation has led to a heated public discussion.

In this discussion, reindeer herding itself is often presented as the greatest threat to nature in the fell areas of Finnish Lapland (Ruukki 2016; Järvinen 2017a, 2017c). Sami reindeer herders are accused of damaging the pastures – destroying biodiversity, including rare plants (Kivipelto 2016; Mainio 2016; Valikainen 2017; Järvinen 2017c, 2018a, 2018b; Kyrölahti 2019; 'Naali ei ole ainoa uhattu' 2019), increasing erosion (Ruukki 2016), and negatively impacting ecosystems (Valikainen 2017; Järvinen 2017a) – by herding reindeer in Sami herding areas. According to the public discourse, the problem is 'overgrazing' ('Avoimuutta poropuheisiin' 2016). Sami herders are held responsible because they have grown their herds and exposed the plants to intensive, long-term grazing, and associated trampling. As a result, available pastures cannot provide enough winter fodder for the animals because the

DOI: 10.4324/9781003057758-2

lichens do not have enough time to regenerate (Järvinen 2017a). Supplementary feeding is now needed in winter to keep the reindeer alive (Mainio 2016); this allows herders to further increase herd sizes ('Avoimuutta poropuheisiin' 2016; Ruukki 2016; Vehmas 2016b). Thus, the media-driven discourse characterises this form of livelihood as 'intensive livestock farming' (Ruukki 2016; Vehmas 2016b), which is ethically, economically, and ecologically unsustainable (Ruukki 2016; Järvinen 2017b). Reindeer are 'killing the unique nature in the fell areas' (Vehmas 2016b) and reindeer herders are 'committing an environmental crime' (Valikainen 2017).

While these public discussions have created a narrative focused on the 'overgrazing' of ground lichen, the strongly negative tone of the public discourse to date has been driven by non-experts in rangeland ecology and management. More balanced scientific analyses have been published over the past decade or so (Forbes 2006; Helle and Jaakkola 2008; Forbes and Kumpula 2009; Forbes 2010; Kumpula et al. 2014). According to these scientific overviews of Finnish reindeer rangelands and the increase in supplemental feeding, the underlying causes are far more complex and nuanced than portrayed in the media.

The narrative regarding 'overgrazing' becomes problematic when it disseminates misleading information about the reasons underlying its narrow focus on the decline of formerly lichen-dominated winter rangelands. The public discussions comparing the conditions of lichen pastures in the Finnish–Norwegian border zone provide a perfect example of such an oversimplified narrative. Several newspapers have published articles and opinion pieces about oversized reindeer herds demolishing the lichen on the Finnish side, supporting these claims with images of the lichen pastures around the border (Ruukki 2016; Järvinen 2017a). Yet rangeland researchers do not deny the fact that declining pasture conditions are a significant problem in the fell areas (Forbes 2006; Vehmas 2016b; 'Lapin luonto tuhoutuu poronhoidon vuoksi väittää tietokirjailija' 2018). There are clearly visible differences in the amount of ground lichen in Enontekiö (Finland) and Kautokeino (Norway), but the number of reindeer is approximately the same (2.5–3.5 reindeer per square mile on both sides of the border) (Oksanen et al. 2016). Therefore, the size and density of the respective herds do not fully explain the quality and quantity of Finnish fell lichen pastures (Oksanen et al. 2016). However, there are significant differences in seasonal grazing and, especially, summer trampling pressure on the Norwegian and Finnish sides of the border (Kumpula 2006, 172). The lichen in Finland has decreased since the 1970s because the reindeer 'double graze' the same pastures throughout the year (Forbes 2006, 19). The reindeer consume ground lichens from beneath the snow cover in winter and trample the dry lichen in summer (Forbes 2006, 19; Helle and Jaakkola 2008; Lavia 2016; Oksanen et al. 2016; Vehmas 2016b). In Norway, reindeer are managed differently and only have access to the lichen rangelands near the border during the winter. In Norway, Sami herders have continued to practice traditional herding migration – reindeer are herded from coastal summer pastures to inland winter pastures near the Finnish border (Kumpula 2006, 167; Lavia 2016; Oksanen et al. 2016; Vehmas 2016b).

The reasons for declining ground lichen include the unavailability of seasonably appropriate grazing areas, the continuing loss of pasture to a variety of land uses, and hotter, drier summers during which lichens cannot photosynthesize (Kumpula et al. 2014). Reindeer management in Finland has experienced significant changes in the past five decades (Kumpula 2006, 167; Helle and Jaakkola 2008; Kumpula et al. 2014). In the past, the Sami practiced nomadic reindeer herding in large, connected, undisturbed grazing areas. Finnish Sami reindeer migrated all the way to the Arctic Ocean and back to Finland (Kumpula 2006, 167). The creation of nation-states – Norway, Sweden, Finland, and Russia – and closed borders in the 1950s put an end to nomadic herding. Reindeer herders could no longer practice long-distance migrations like those used in the reindeer herding areas of Sweden and Norway (Salvesen 1995, 110–111; Kumpula 2006, 167; Forbes and Kumpula 2009; Reinert et al. 2009, 19–21; Forbes 2010; Vehmas 2016a). Over time, factors such as competing land use, reindeer management policies, and regulatory challenges reduced and fragmented the grazing areas even further, forcing reindeer herders to practice their livelihood in much smaller and more limited areas (Muller-Wille et al. 2006; Forbes 2010; Vehmas 2016a; Miettunen 2017). For example, fences between the reindeer cooperatives restrict reindeer from moving between winter and summer pastures ('Poronhoitajat haluavat saada äänensä kuuluviin alansa päätöksenteossa' 2005; Kumpula 2006, 167; Vehmas 2016a). The ongoing loss of grazing lands since the 1950s has left reindeer herders less capable of using lands and natural resources and practicing their livelihood in a sustainable way. It is against Sami herders' self-interest to exhaust the rangelands upon which their traditional livelihood and culture depend (Vehmas 2016b). Therefore, 'overgrazing' is an overly simplistic explanation for the declining pasture conditions (Sorjanen 2016); the real reason lies in the ever-shrinking operational space for contemporary reindeer rangeland management ('Poronhoitajat haluavat saada äänensä kuuluviin alansa päätöksenteossa' 2005; Kumpula et al. 2014; 'Lapin luonto tuhoutuu poronhoidon vuoksi väittää tietokirjailija' 2018).

Reindeer reduction policies versus traditional pasture management

The government is responsible for Finnish policies on reindeer husbandry. The Natural Resources Institute of Finland (Luke) (see Luonnonvarakeskus) monitors the dynamics of reindeer rangeland by conducting regular inventories of ground and arboreal lichen and of green sources of forage in the reindeer management area. Inventories were conducted in 1995–1996, 2005–2008, and 2017–2019. In the northernmost reindeer cooperatives or districts (20), the inventory comprises taking measurements from 625 specific test sites in the lichen pastures. Thirteen of the northernmost cooperatives are in the Finnish portion of Sapmi (Paliskuntain yhdistys 2014, 9). Based on these pasture inventories (see Kumpula et al. 2019) and other relevant information, the Finnish Ministry of Agriculture and Forestry (*Maa- ja metsätalousministeriö*) regulates the total reindeer population. The Ministry determines the largest permissible numbers of living reindeer for each district based on the alleged capacity of winter pastures (Luke 2016). In doing so, the Ministry sets parameters that guide reindeer management policy for the next ten years. This has led to a political discussion about whether and how reindeer management policy should evolve in the twenty-first century.

Policies aimed at reindeer population control have long been the main governmental tool wielded to prevent the rangeland degradation perennially attributed to 'overgrazing'. In practice, these policies mean that the Ministry reduces reindeer populations in the northernmost cooperatives, which are in the Finnish parts of Sapmi, and the Sami are ordered to cull their herds. The discourse focused on 'overgrazing' often refers to reindeer reduction policy, and many people openly support mandated slaughter of reindeer to regulate 'carrying capacity' (Mainio 2016; Ruukki 2016; Sorjanen 2016). The public discussion puts pressure on the Ministry to reduce the total number of reindeer.

Reindeer reduction, however, is problematic. The maximum permitted number of reindeer is different for each herding cooperative (Paliskuntain yhdistys 2014, 11, Paliskuntain yhdistys a). For example, in one of the northernmost cooperatives in the Käsivarsi district, the maximum number of reindeer is 10,000 (Paliskuntain yhdistys b). Reindeer herders are then forced to compete for suitable rangeland and respective herd sizes. The government pays reindeer owners with at least 80 reindeer a 'living support' or subsidy (Valtioneuvoston asetus 2019/2020). If a herd is already small and the herder is struggling to gain support and sustain themselves and their family, reducing the number of reindeer can tip the scales, making reindeer herding unprofitable and forcing the herder to give up their traditional livelihood. The Sami argue that reindeer reduction policies could threaten their livelihood and the maintenance of Sami culture and identity. Furthermore, such policies infringe on Indigenous rights. In Norway there has been public discussion about the negative effects of reindeer reduction policies (see Henley 2017; Martyn-Hemphill 2017).

Therefore, Finnish Sami have suggested an alternative reindeer management policy. According to the Sami, reindeer reduction is not the only tool for maintaining viable rangeland conditions, nor the only way to restore lichen-rich pastures. The sustainable solution lies in pasture management and rotational grazing systems based on Indigenous knowledge (Saamelaiskäräjien lausunto 2016). Scholars, the Sami Parliament, and Sami representatives of reindeer husbandry agree that the most effective way to restore the lichen pastures is to revitalise seasonal rotation among pastures (Mainio 2016; Sorjanen 2016). In practice, this means strict separation of winter and summer pastures (Sorjanen 2016). Revitalisation of the rotational grazing system will not be an easy task. It will require appropriate participatory modes of research that combine different ways of knowing about traditional grazing lands, collaborative decision-making process and adjustments to reindeer herding legislation.

To tackle this question, this chapter will look at food sovereignty in the context of reindeer-herding Sami and their struggles related to rangeland conditions and shrinking operation space in Finnish Lapland. First, it will determine what food sovereignty actually means, especially for Indigenous Peoples. It then examines and discusses how pasture management and rotational grazing systems based on Indigenous knowledge can enhance food sovereignty. Moreover, it examines how (human) rights-based approaches can strengthen food sovereignty. It concludes by providing recommendations to enhance food sovereignty and improve food security in Sami communities.

Food sovereignty

Food sovereignty is a well-discussed topic and a reference point for discourses in the field of food security. It has become an important theoretical concept; it is 'an ever-expanding series of ideas and principles' and the subject of policy frameworks, and it is included in governmental programs. It is also a key driver behind political change in food movements (Grey and Patel 2014, 431–432; see León 2007). It is commonly defined as:

> … the Right of peoples, communities, and countries to define their own agricultural, labour, fishing, food and land policies, which are ecologically, socially, economically and culturally appropriate to their unique circumstances.
>
> It includes the true right to food and to produce food, which means that all people have the right to safe, nutritious and culturally appropriate food and to food producing resources and the ability to sustain themselves and their societies.
>
> (Food sovereignty: A right for all, 2002)

The concept of food sovereignty is related to concepts such as food security and the right to food. Food security – adopted by FAO member states – and food sovereignty consider different aspects of food systems. Food security, which focuses mainly on the supply and availability of food (Food security 2006; Rudolph and McLachlan 2013; Martens et al. 2016, 20), is relatively neutral in terms of power relations (Gordillo and Jerónimo 2013, 7). On the contrary, food sovereignty recognises 'the power inherent in food systems' (Martens et al. 2016, 20). It includes: 'a set of principles that protect the policy space for peoples and countries to define their agricultural and food policies, and their models of production and food consumption patterns' (Windfuhr and Jonsén 2005, 11). Food sovereignty recognises the importance of how food is produced (Windfuhr and Jonsén 2005, 11; Gordillo and Jerónimo 2013, 7) and the asymmetric power relations and structures involved in food production (Gordillo and Jerónimo 2013, 7). It also assesses food-related challenges and threats. In fact, food sovereignty can be seen as a prerequisite for genuine food security (Via Campesina 1996, 1). The right to food, on the other hand, can be used as a tool to achieve food sovereignty. It asserts every individual's right to adequate food (Windfuhr and Jonsén 2005, 11, 17).

The concept of food sovereignty developed among farmers in the 1980s and early 1990s as a reaction to globalization of the economy and agriculture (Wittman et al. 2010, 1; see Glibo and Pascual Jr. 2005). Changes in national and international agricultural policies have led to multiple crises affecting food, such as rural poverty, hunger, and malnutrition (Windfuhr and Jonsén 2005, 1, 3). As an alternative to 'decades of destructive economic policies based on the globalisation of a neoliberal, industrial, capital-intensive and corporate-led model of agriculture,' food sovereignty was introduced to give control over food markets, environments, land, and rural culture back to local communities and the people who produce and consume food (Wittman et al. 2010, 2). Over the years, a global social network of non-governmental organisations (NGOs), civil society organisations (CSOs), social movements, conferences, forums, and declarations have supported the idea of food sovereignty and contributed to the development of the food sovereignty policy framework (Windfuhr and Jonsén 2005, xi, 1, 11; see Anderson 2018).

Food sovereignty includes six basic principles – the six pillars of food sovereignty. They were drafted during the World Food Summit (WFS) in 1996 by transnational agrarian movement, La Vía Campesina, and further developed at the 2007 Nyéléni Forum for Food Sovereignty (see Mali 2007). According to these pillars, food sovereignty (1) focuses on food for people, (2) values food providers, (3) localises food systems, (4) places control at the local level, (5) promotes knowledge and skills, such as Indigenous knowledge, and (6) works with and protects nature (Gordillo and Jerónimo 2013, 3–4).

Food sovereignty and Indigenous Peoples

Indigenous Peoples are part of this food movement and have been instrumental in putting food sovereignty at the centre of policies on food production, agriculture, and rural development (Wittman et al. 2010, 2). In fact, food sovereignty is closely connected to Indigenous movements and the advancement of Indigenous Peoples' rights (Shawn 2008; Morrison 2011; Desmarais and Wittman 2014; Grey and Patel 2015) because Indigenous Peoples face more food-related challenges and insecurities than other groups (Ledrou and Gervais, 2005; Kuhnlein and Burlingame 2013, 5; Tarasuk et al. 2014; Martens et al. 2016, 20; First Nations Information Governance Centre, 2012; UN Permanent Forum on Indigenous Issues 2012).

For this reason, Indigenous Peoples have developed their own form of food sovereignty (Grey and Pater 2015, 436), Indigenous food sovereignty (IFS), and related movements that address food-related challenges (Martens et al. 2016, see also the Indigenous Food System Network).

The definition of Indigenous Peoples, also known as first peoples, aboriginal peoples and native peoples, varies in different parts of the world. There are no universal criteria defining them as a single group. Rather, Indigenous Peoples have 'the right to define themselves according to their culture' (Kuhnlein et al. 2009, 3; see also Kuokkanen 2019). In the context of food sovereignty, Indigenous Peoples can be referred to as peoples who 'liv[e] in their rural homelands [and] depend on traditional food systems rooted in the historical continuity in their regions, where food is harvested with traditional knowledge from the natural environment, and prepared and served in local cultural settings' (Kuhnlein and Burlingame 2013, 6).

For Indigenous Peoples, food sovereignty is centrally the ability to make their own decisions about their respective food systems (Grey and Patel 2015, 431). It holds that Indigenous Peoples have the inherent right to define their traditional food systems and food-related policies and strategies for sustainable production – that is, that they have control over their traditional food systems (Grey and Patel 2015, 433).

Indigenous Peoples' traditional food systems have certain qualities that affect decisions about those systems. Traditional food systems are 'composed of items from the local, natural environment that are culturally acceptable' (Kuhnlein and Receveur 1996, 418; see Woodley et al. 2009). They are intrinsically connected to traditional lands, waters, and territories, and are based on Indigenous knowledge (Kuhnlein et al. 2009, 3). Indigenous knowledge systems, cultures, and identities are maintained through active participation with traditional lands and food systems (Indigenous Food System Network). Therefore, traditional food systems contribute to the collective health and wellbeing of the individual and the community (Neufeld and Richmond 2017, 94) – 'not only physical but also emotional, mental and spiritual aspects of heath, healing and protection from disease' (Kuhnlein et al. 2009, 3).

For Indigenous Peoples to be able to make their own decisions regarding food systems, traditional food systems must often be revitalised. This is because of the impacts of colonisation on Indigenous food sovereignty and traditional food systems. Colonial policies and practices have caused 'a drastic decline in the health and integrity of Indigenous cultures, ecosystems, social structures, and knowledge systems', which are integral to the production of healthy Indigenous foods (Indigenous Food System Network). The revitalisation of traditional food systems is connected to a larger Indigenous cultural, social, and political resurgence. In fact, food sovereignty can be seen as 'the continuation of anti-colonial struggles in ostensibly postcolonial context' (Grey and Patel 2015, 433). Furthermore, the revitalisation of Indigenous food systems requires Indigenous knowledge.

Yet, all too often, Indigenous food sovereignty is compromised, reducing food security in Indigenous territories. Often the problem lies in Indigenous Peoples' lack of political power to make their own decisions about traditional food systems. Clearly, the challenges related to food sovereignty require policy reforms and need to be addressed with and by Indigenous communities.

Indigenous knowledge-based traditional pasture management as a promoter of food sovereignty

This section examines how Indigenous knowledge-based traditional pasture management can potentially enhance Sami food sovereignty. It focuses on Sami efforts to (re)claim control over their traditional food systems by proposing an alternative policy to reindeer reduction. This alternative involves traditional pasture management and rotational grazing systems. The analysis will relate the current situation of reindeer-herding Sami to the six pillars of food sovereignty (described in the "Food sovereignty" section).

Pillar I: Focusing on food for people

According to the first pillar, food sovereignty 'focuses on food for people.' People have a right to adequate supplies of healthy, culturally appropriate food. This is true for all individuals, peoples, and communities, including those who are marginalised. The aim is to satisfy people's need for food (Sélingué 2007, 76).

Traditional food systems allow Indigenous Peoples to produce and consume healthy, culturally appropriate food, and therefore satisfy their need for food. These food systems rely on subsistence activities and productive resources. Subsistence activities might include reindeer herding, fishing, hunting, and berry picking (Helander 2008). Land, water, and species such as reindeer, fish, game and berries, as well as other natural resources, are productive resources (Windfuhr and Jonsén 2005, 14). For the Sami, traditional reindeer herding is an important subsistence activity, and they have exploited reindeer and lichen pastures located in their traditional lands as productive resources for centuries, if not millennia (Müller-Wille et al. 2006; Forbes and Kumpula 2009).

Food sovereignty 'puts people's need for food at the centre of the policies', including policies affecting food. The Ministry's expected or potential policy of reindeer reduction as a solution to declining pasture conditions causes food insecurity for the Sami because it threatens a subsistence activity – traditional reindeer herding. Traditional reindeer herding was productive in the past thanks to traditional pasture management and a seasonal rotational grazing system (Kumpula 2006, 167). Reapplying these methods to reindeer husbandry and incorporating them in reindeer management policies could secure a subsistence base for the production of heathy, culturally appropriate food through traditional food systems, thereby enhancing food sovereignty.

Pillar II: Valuing food providers

According to the second pillar, food sovereignty 'supports sustainable livelihoods'. This means that food sovereignty supports the contributions of food providers, such as Indigenous Peoples, who practice sustainable livelihoods (Sélingué 2007, 76). It is important that Indigenous food providers can define their own food systems and produce their own food using sustainable, ecologically sound methods – traditional food systems. These systems include the sustainable practice of subsistence activities and the use of productive resources. In Finland the current system does not meet the requirements of ecological reindeer husbandry (Oksanen et al. 2016) because it does not allow enough operational space for seasonal pasture rotation. Practicing reindeer husbandry using traditional pasture management would support the efforts of Sami reindeer herders to keep traditional, sustainable reindeer herding practices alive, and guarantee profitability and the continuity of the traditional livelihood in the future.

Food sovereignty also 'respects the work of all food providers'. It requires rejecting 'those policies, actions, and programmes that undervalue them, threaten their livelihoods and eliminate them' (Sélingué 2007, 76). Indigenous food sovereignty emphasises that Indigenous Peoples' right to food cannot be constrained by colonial laws, policies, or institutions (Indigenous Food System Network). In principle, the Sami Parliament (Saamelaiskäräjät 2016) supports the national government's conservation strategies and policies, but it disagrees with the Finnish government about how lichen pastures should be restored and biodiversity goals implemented. The Parliament rejects the planned reindeer reduction policy because this policy focuses on reindeer herding and ignores other factors that rangeland experts have documented as contributing to the long-term deterioration of pasture conditions (Helle and Jaakkola 2008; Kumpula et al. 2014). The reindeer reduction policy proposed by the government would directly and negatively impact traditional reindeer herding (Saamelaiskäräjät 2016, 3–6). Traditional pasture management offers an alternative to the reindeer reduction policy, which undervalues Sami reindeer herders and threatens their traditional livelihood. Therefore, traditional pasture management enhances food sovereignty.

Pillar III: Localising food systems

According to the third pillar, food sovereignty 'localises food systems.' This involves placing food providers and consumers at the centre of decision-making on food issues (Sélingué 2007, 76). Decisions about policies and rules that govern food are often made at the international, regional, and national level. Food sovereignty means that the decision-making processes should highlight the perceptions and needs of local food providers and consumers and therefore reinforce local food systems (Withfurth and Jonsén 2005). Otherwise, international, regional, and national policies and rules that govern food can form different kinds of structures that delocalise food systems.

Reindeer herders are obliged to follow the policies and rules that apply to reindeer herding in the nation-state where they reside. Finland has many such policies and rules. Since World War II, reindeer herding, designated as an economic enterprise, has been subjected to increasingly strict agricultural norms (Paine 1994; Laakso 2002; Forbes 2006). Since Finland joined the EU in 1995, reindeer husbandry has been considered a form of agriculture and regulated according to industrial livestock standards (Forbes 2006; Saamelaiskäräjät 2016, 3).

The Reindeer Husbandry Act (848/1990) regulates the system based on reindeer herding cooperatives. Finland's reindeer management area is divided into 54 local reindeer herding cooperatives. Every reindeer herder belongs to a cooperative, and each cooperative is responsible for the animals in one specific, limited area (Paliskuntain yhdistys). The boundaries of reindeer herding cooperatives are not based on so called *'siida' system* (a.k.a. traditional village system or reindeer herding groups) (see Reinert et al. 2009, 17, 19; Sara 2009). According to the Sami Parliament the Reindeer Husbandry Act (848/1990) does not recognise traditional Sami reindeer herding nor its significance to Sami culture and language (Saamelaiskäräjät 2016, 3). Furthermore, reindeer herders feel that they have no influence over these top-down regulations (Forbes 2006).

Traditional pasture management localises the food system. The Sami Parliament states that policies and rules governing reindeer husbandry should be based on the perceptions and needs of reindeer herders and on Sami culture (Saamelaiskäräjät 2016, 3). In Finland, the state is required to consult the Sami on decisions affecting them (Allard 2018, 31–32, see Finland's Strategy for the Arctic Region 2013, 22). Section 9 of the Sami Parliament Act (974/1995) states that the authorities must negotiate with the Sami Parliament in all far-reaching and important measures which may directly and specifically affect the status of the Sami as Indigenous People. These consultation rights include policies related to Sami livelihoods (Allard 2018, 31–32). Therefore, the Sami Parliament has demanded that the government engage them in the decision-making process and consider reindeer herders' perceptions and needs in reindeer husbandry policies. This would mean instituting traditional pastures management and rotational grazing systems (Saamelaiskäräjät 2016, 3). Traditional pasture management is a bottom-up approach that places Sami herding culture and the needs of reindeer herders at the centre of the decisions on reindeer management policy; this would enhance Sami food sovereignty.

Pillar IV: Putting control in the hands of local food providers

According to the fourth pillar, food sovereignty 'places control over territory, land, grazing, water, seeds, livestock and fish populations on local food providers' (Sélingué 2007, 76). When Indigenous food providers own, control, and/or have access to territories, lands, and natural resources, they are more able to practice sustainable local food production. Locally controlled, traditional food systems and subsistence activities enable Indigenous Peoples to use, manage, and share territories, lands, and natural resources in environmentally sustainable ways, which conserves biodiversity.

Traditional pasture management places control of grazing lands and reindeer in the hands of local reindeer herders in the Sami homeland. According to Sami reindeer herders, even more important than nutrition is access to good quality pasture lands, characterised by 'grazing peace' (Kitti et al. 2006). However, access to pastures is compromised for several reasons. Competitive land use, infrastructure, the exploitation of natural resources, other human activities, etc. prevent reindeer from accessing traditional grazing areas. These activities are disturbing the peace in many pastures; as a result, these pastures cannot be used for reindeer herding during certain times of the year (see Kumpula et al. 2007, 2014; Anttonen et al. 2011; Hast and Jokinen 2016; Sandström et al. 2016). The Sami Parliament has pointed out that competitive land use etc. has made the traditional rotation between winter and summer pastures increasingly difficult (Saamelaiskäräjät 2016, 6). As a result, Sami herders cannot regulate the use of lichen and other natural resources in those areas. Traditional pasture management returns control over their lands, territories, natural resources, biodiversity, and reindeer back to the Sami reindeer herders, enhancing food sovereignty.

Food sovereignty 'recognises the need to inhabit and share territories.' This means that, although Indigenous territories often cross geopolitical borders, Indigenous Peoples have a right to inhabit and use their traditional lands and territories. This pillar also promotes positive interactions between Indigenous food providers from different regions and territories and from different sectors, preventing conflicts and promoting their resolution when they occur (Sélingué 2007, 76). Traditional pasture management is based on the same principle. It requires large, connected, undisturbed grazing lands and territories; reindeer naturally migrate across these lands. Traditional grazing lands extend well beyond the borders of nation-states and reindeer cooperatives. Traditionally, the North Calotte has been considered a shared pasture area where Sami reindeer herders worked together to govern reindeer herding in Fennoscandia. Reindeer herding in accordance with the *siida* system is an integral part of traditional Sami pasture management (Reinert et al. 2009, 33; Mazzullo 2010). Reviving rotational grazing systems would require comprehensive long-term planning; such a system could not be limited by reindeer cooperatives (Kumpula et al. 2019). Traditional pasture management recognises the Sami's need to inhabit and share different regions and territories and therefore enhances food sovereignty.

Food sovereignty also 'rejects the privatisation of natural resources' (Sélingué 2007, 76). The long-term plans to preserve and secure the availability of winter pastures often include rejecting the privatisation of these natural resources. For example, it is recommended that plans for land use in the northernmost reindeer cooperatives should spare existing grazing lands, where reindeer can still graze peacefully, and minimise other land use in those areas. Old-growth forests outside nature reserve areas should also be spared in areas that are significant for reindeer herding (Kumpula et al. 2019, 69–70). Therefore, traditional pasture management rejects the privatisation of natural resources and enhances food sovereignty.

Pillar V: Building on Indigenous knowledge and skills

The fifth pillar states that food sovereignty 'builds on traditional knowledge'. This means developing and utilising the skills and local knowledge of Indigenous food providers and establishing local organisations that conserve, develop, and manage localised food production (Sélingué 2007, 76). According to the Sami Parliament, the Finnish government currently does not use Indigenous knowledge effectively in planning and developing reindeer husbandry (Saamelaiskäräjät 2016, 3).

Traditional pasture management and rotational grazing system are based on Indigenous knowledge. This knowledge is required to properly manage different herd sizes within and among different geographic areas and seasonal pastures, which offer widely disparate grazing conditions. Sami reindeer herders use Indigenous knowledge to control the size and composition of herds in proportion to available pasture (Bjorklund 2004, 125–126) and to maintain sustainable levels of grazing. Therefore, the Sami Parliament insists that Sami reindeer herders' knowledge of pasture management – holistic knowledge about the nature and condition of pastures and an understanding of how different factors affect pasture conditions – be used to develop and manage reindeer herding in the northernmost reindeer cooperatives (Saamelaiskäräjät 2016, 3). Basing reindeer husbandry on Sami reindeer herders' knowledge of traditional pasture management will help conserve, develop, and manage local food production through traditional food systems, enhancing food sovereignty.

Food sovereignty also 'uses research to support and pass this knowledge to future generations'. It is essential to develop appropriately linked rangeland research and management systems to support the use of skills and knowledge of Indigenous food providers. This will support localised food production and pass on Indigenous knowledge to future generations (Sélingué 2007, 76). According to the Sami Parliament, previous state- and university-driven research on pasture conditions, lichen coverage, and sustainable grazing levels has not considered Indigenous knowledge or the views of reindeer herders on the condition of grazing lands (Saamelaiskäräjät 2016, 3).

Previous state- and university-driven research on pasture conditions and lichen coverage has been far too narrow in scope. Bioeconomic assessments of pastures have focused on measuring vegetation – lichen biomass, coverage of ground lichen, and changes in coverage. Studies have identified a 75% decline in lichen coverage and a 60% decrease in lichen biomass in the reindeer pastures of the northernmost Lapland ('Avoimuutta poropuheisiin' 2016). These studies were considered objective and their results reliable measurements of changes in vegetation and pasture conditions. However, while bioeconomic assessments encompass ecological and economic factors, social and cultural factors are still too often ignored in state- and university-driven research. Reindeer herding is part of Sami culture. Reindeer herders' perceptions and observations of pasture 'quality' extend well beyond vegetation characteristics (Kitti et al. 2006; Forbes and Kumpula 2009). The valuable Indigenous knowledge of these practitioners and local researchers needs to be acknowledged and included in future research (Brunet et al. 2014; Riseth 2014). University-affiliated academics who evaluate vegetation changes may lack the in-depth experience and practical knowledge of pasture conditions that Sami herders have. Future research must address issues of reindeer management in the local sociocultural context (Forbes 2006, 16).

Therefore, appropriate participatory modes of research that apply Indigenous knowledge combine different ways of knowing about traditional grazing lands (Forbes 2006; Roturier and Roué 2009; Brunet et al. 2014; Riseth 2014) and contribute to the revival of rotational grazing systems are essential. The next step is recognising Sami herders as researchers and enhancing locally driven and indigenous-led research projects. Recognising and implementing Indigenous knowledge of pasture conditions in scientific research is vital because this will provide a foundation for government planning of reindeer husbandry policy over ten-year periods. If reindeer herders, university-affiliated researchers, and policymakers collaborate and revitalise the rotational grazing system, young generations of herders can learn how to manage pastures according to Sami traditions. Indigenous knowledge will then be passed on to future generations, further enhancing food sovereignty.

Pillar VI: Working with nature

The sixth pillar of food sovereignty calls for optimising the contributions of ecosystems, improving resilience, and rejecting destructive production methods. Food sovereignty uses the 'contributions of nature in diverse, low external input [...] ecological production [...] methods that maximise the contributions of ecosystem and improve resilience and adaption, especially in the face of climate change' (Sélingué 2007, 76).

The narrative on 'overgrazing' often presents the reindeer themselves as the greatest threat to nature and endangered species in Finnish fell areas

(Mainio 2016; Kivipelto 2016; Ruukki 2016; Järvinen 2017a; Valikainen 2017) but disregards the fact that reindeer are a keystone species of fell areas (Soppela et al. 2002). Reindeer can have positive impacts on vegetation and soil and contribute to ecosystem function (Huhta 2016; Käyhkö and Horskotte 2017; Oksanen et al. 2016). Moreover, in traditional pasture management, reindeer herding is based on Indigenous knowledge about the sustainable use of natural resources (Saamelaiskäräjät, 3) and seasonal pasture rotation system allows plants such as lichen time to regenerate, creating adequate forage resources for reindeer so they no longer need supplementary feeding in fell areas.

Reindeer herding can also mitigate the impact of climate change. Grazing and trampling help prevent tall shrubs and trees from encroaching on tundra fells in a warming climate (Huhta 2016; Oksanen et al. 2016; Saamelaiskäräjät 2016, 4; Sorjanen 2016; Horstkotte et al. 2017). Intensive grazing also slows snow melt, further countering warming trends (Saamelaiskäräjät 2016, 4; Käyhkö and Horstkotte 2017). When reindeer herding is based on traditional pasture management, it is a low external input ecological production method that promotes the contributions of ecosystems and improves resilience and adaption to climate change, and it enhances food sovereignty.

An approach to food sovereignty based on (human) rights-based approach to food sovereignty

This section examines how (human) rights-based approach can strengthen claims to food sovereignty. The efforts of Indigenous Peoples to (re)claim control over their traditional food systems and/or traditional subsistence activities based on Indigenous knowledge are food sovereignty claims. Hence, it is important to recognise that Sami efforts to revitalise and (re)claim control over their traditional pasture management using rotational grazing system is in fact a food sovereignty claim. The right to food, meanwhile, is a legal concept and a human right (Food and Agriculture Organization of the United Nations 2005; Windfurh and Jonsén 2005, 19; Knuth and Vidar 2011) that can be used to enhance food sovereignty. Therefore, this section focuses on a (human) rights-based approach that can strengthen Sami claims to traditional pasture management and food sovereignty, while also promoting food security.

The right to food is protected by several international human rights instruments (Food and Agriculture Organization of the United Nations 2005, 3–7). It is defined by the UN Committee on Economic, Social and Cultural Rights (CESR) (1999) as 'the right of every man, woman, and child alone and in community with others to have physical and economic access at all times to adequate food or means for its procurement in ways consistent with human dignity' (see Windfuhr and Jonsén 2005, 19).

First, the right to food is protected under international law. The right is recognised in Article 25 of the Universal Declaration of Human Rights

(1948) and, most importantly, in Article 11 of the International Convention on Economic, Social and Cultural Rights (ICESCR), which recognises two dimensions to the right to food: the right to an adequate standard of living and the fundamental right to be free from hunger (Food and Agriculture Organization of the United Nations 2005, 3–4; Knuth 2009, 12–13). All human beings are entitled to right to food as human right, including individuals of Indigenous populations (Knuth 2009, 14).

Second, the realisation and enjoyment of the right to food depends on the effective realisation of other human rights. Particularly significant to Indigenous Peoples are the right to culture, the rights related to land, territories, resources, and the right to self-determination and right to non-discrimination, without forgetting the category of collective rights (Knuth 2019, 14–19).

Other binding and non-binding international legal instruments directly or indirectly protect Indigenous Peoples' right to food, such as the UN Declaration on the Rights of Indigenous Peoples (UNDRIP) (Food and Agriculture Organization of the United Nations 2008, 1). The UNDRIP identifies key elements of Indigenous Peoples' right to food and the realisation of that right. It delineates the right of Indigenous Peoples: 'to live in dignity, to maintain and strengthen their own institutions, cultures, and traditions, to pursue their self-determined development', to use land and natural resources and to practice subsistence activities (Food and Agriculture Organization of the United Nations 2008, 1).

The right to food, which is in itself a human right, and other relevant human rights create 'layers of protection' for Indigenous Peoples consisting of individual rights and collective rights (see Knuth 2019, 12–19). Indigenous Peoples' collective rights establish rights of peoples as groups and have additional value in comparison to individual rights because they defend groups′ collective interest. For example, certain rights are held collectively, and cultural traditions, customs and practices, such as subsistence activities, are carried out collective (Food and Agriculture Organization of the United Nations 2008, 2). Therefore, Indigenous Peoples require that not only their individual but also collective rights to food are realised (Knuth 2019, 17).

Therefore, Indigenous groups and individuals are rights holders and the right to food and other relevant human rights create human rights obligations for nation-states. States are required to fulfil human rights obligations and to domestically implement the right to food by ensuring that this right is protected by national legal systems (Knuth and Vidar 2011). Therefore, Indigenous Peoples can demand that the state fulfil these obligations to food (Windfuhr and Jonsén 2005, 19), which include: 'respecting traditional ways of living, strengthening traditional food systems, and protecting subsistence activities' such as reindeer herding (Food and Agriculture Organization of the United Nations 2008, 2), based on human rights. The right to food may be violated when (for example) in access of denial of access to land, deprivation of access to adequate and culturally acceptable food and contamination of food sources (Food and Agriculture Organization of the United Nations 2008, 2).

An alternative food sovereignty offers to existing food policies (a.k.a. food sovereignty claims) is often linked to the right to food because the two concepts aim at the same goals (Windfuhr and Jonsén 2005, 11). It is therefore useful for Indigenous Peoples to adopt a (human) rights-based approach to food-related policies. In practice, this means setting requirements for the effective implementation of alternative policies and the realisation of the (human) right to food (Windfuhr and Jonsén 2005, 11). This applies to the Sami's efforts to promote traditional pasture management as alternative to the state's long-standing reindeer reduction policy. In this case, a (human) rights-based approach and right to food provides the Sami with an additional legal argument when challenging reindeer reduction policy and the means to influence it (see Food and Agriculture Organization of the United Nations 2008, 2). It also helps them resist actions that could violate their right to food and other relevant human rights in their fight for food sovereignty.

Concluding remarks and outlook for the future

Traditional pasture management can help the Sami (re)claim control over their traditional food system. It allows the Sami to manage their reindeer and natural resources based on Indigenous knowledge autonomously and locally in their traditional land and territories. This management system would also help democratise reindeer husbandry in Finland. It supports the survival of the Sami language, of Indigenous knowledge, of Sami culture and livelihoods, and of biodiversity and nature as a whole. The (human) right-based approach can strengthen the Sami claim to food sovereignty – the claim to re-establish traditional pasture management. Traditional pasture management based on Indigenous knowledge can establish food sovereignty, and therefore food security, for the Sami in Finland.

Traditional pasture management can promote Sami food sovereignty in the following ways:

1 The use of traditional pasture management secures a subsistence base for the production of healthy, culturally appropriate food through traditional food systems.
2 Traditional pasture management supports the efforts of Sami reindeer herders to keep traditional, sustainable reindeer herding practices alive.
3 Traditional pasture management represents an alternative to the reindeer reduction policy that undervalues Sami reindeer herders and threatens their traditional livelihood.
4 Traditional pasture management places Sami culture and Sami reindeer herders and their needs at the centre of decisions on reindeer management policy.

5 Traditional pasture management returns control of Sami territories, lands, and natural resources (including reindeer) to Sami reindeer herders.
6 Traditional pasture management recognises the need for the Sami to inhabit and share different regions and territories and therefore enhances food sovereignty.
7 Reindeer husbandry conducted in accordance with Indigenous knowledge of traditional pasture management, helps conserve, develop, and manage local food production through traditional food systems.
8 When reindeer herders, researchers, and policymakers collaborate and revitalise rotational grazing systems, young people can learn how to manage pastures according to Sami traditions; this will pass Indigenous knowledge on to future generations.
9 Traditional pasture management is a low external input ecological production method that promotes the contributions of ecosystems and improves resilience in the face of climate change.

Traditional pasture management is a step towards ultimate food sovereignty, which includes 'upholding our sacred responsibility to nurture healthy, interdependent relationships with the land, plants, and animals that provide us with our food' (Indigenous Food System Network).

References

Allard, C. 2018. *The rationale for the duty to consult Indigenous peoples: Comparative reflections from Nordic and Canadian legal contexts.* Tromso: University of Norway.

Anderson, F. 2018. *Food sovereignty NOW! A guide to food sovereignty.* Brussels: European Coordination Via Campesina.

Anttonen, M., J. Kumpula, and A. Colpaert. 2011. Range selection by semi-domesticated reindeer (*Rangifer tarandus tarandus*) in relation to infrastructure and human activity in the boreal forest environment, Northern Finland. *Arctic* 64(1): 1–14.

Avoimuutta poropuheisiin. 2016. Editorial. *Lapin Kansa*, 15 September.

Bjorklund, I. 2004. Saami pastoral society in Northern Norway: The national integration of an Indigenous management system. In Anderson and Nuttall (eds.), *Cultivating arctic landscapes, knowing and managing animals in the circumpolar North.* New York: Berghahn Books; 2003, 124–135.

Brunet, N.D., G.M. Hickey, and M.M. Humphries. 2014. The evolution of local participation and the mode of knowledge production in Arctic research. *Ecology and Society* 19(2): 69.

Desmarais, A.A. and H. Wittman. 2014. Farmers, foodies and First Nations: Getting to food sovereignty in Canada. *Journal of Peasant Studies* 41(6): 1153–1173.

Finland's Strategy for the Arctic Region 2013. Governmental resolution on 23 August 2013. Prime Minister's Office Publications 16/2013.

First Nations Information Governance Centre. 2012. *First Nations regional health survey (RHS) 2008/10: National report on adults, youth and children living in First Nations communities.* Ottawa: FNIGC.

Food and Agriculture Organization of the United Nations. 2005. *Voluntary guidelines to support the progressive realization of the right to adequate food in the context of national food security.* Retrieved 15 September 2019 from http://www.fao.org/3/a-y7937e.pdf.

Food and Agriculture Organization of the United Nations. 2008. *Right to food and Indigenous peoples.* Retrieved 15 September 2019 from www.fao.org/3/a-i0728e.pdf.

Food Security. 2006. *Policy brief, 2.* Rome: FAO. Retrieved 15 September 2019 from www.fao.org/fileadmin/templates/faoitaly/documents/pdf/pdf_Food_Security_Cocept_Note.pdf.

Forbes, B.C. 2006. The challenges of modernity for reindeer management in northernmost Europe. In Forbes, Bölter, Muller-Wille, Hukkinen, Muller, Gunslay and Konstantinov (eds.), *Reindeer management in northernmost Europe: Linking practical and scientific knowledge on social-ecological systems.* Berlin: Springer, 2006, 11–26.

Forbes, B.C. 2010. Reindeer herding. In *Arctic biodiversity trends 2010: Selected indicators of change.* Akureyri, Iceland: CAFF International Secretariat, 86–88.

Forbes, B.C. and T. Kumpula. 2009. The ecological role and geography of reindeer (*Rangifer tarandus*) in northern Eurasia. *Geography Compass* 3(4): 1356–1380.

Glibo, A. and F.G. Pascual Jr. 2005. *Food sovereignty framework: Concept and historical context.* Retrieved 15 September 2019 from https://nyeleni.org/IMG/pdf/FoodSovereignityFramework.pdf.

Gordillo, G. and O.M. Jerómino. 2013. *Food security and sovereignty.* FAO. Retrieved 9 June 2020 from http://www.fao.org/3/a-ax736e.pdf.

Grey, S. and R. Pater. 2015. Food sovereignty as decolonisation: Some contributions from Indigenous movements to food system and development politics. *Agriculture and Human Values* 32(3): 431–444.

Hast, S. and M. Jokinen. 2016. Elinkeinojen yhteensovittaminen – tarkastelussa kaivostoiminta, poronhoito ja luontomatkailu. In Mononen and Suopajärvi (eds.), *Kaivos suomalaisessa yhteiskunnassa.* Rovaniemi: Lapin yliopistokustannus, 86–110.

Helander, E. 2008. Sami subsistence activities – spatial aspects and structuration. *Acta Borealia, A Nordic Journal of Circumpolar Societies* 16(2): 7–25.

Helle, T.P. and L.M. Jaakkola. 2008. Transitions in herd management of semi-domesticated reindeer in northern Finland. *Annales Zoologici Fennici* 45: 81–101.

Henley, J. 2017. Norwegian herder ordered to put down dozens of reindeer in controversial cull. *The Guardian*, 22 December.

Horstkotte, T., T.A. Utsi, Å Larsson-Blind, P. Burgess, B. Johansen, J. Käyhkö, L. Oksanen, and B.C. Forbes. 2017. Human–animal agency in reindeer management: Sami herders' perspectives on vegetation dynamics under climate change. *Ecosphere* 8(9): e01931. doi:10.1002/ecs2.1931.

Huhta, E. 2016. Kohtuullinen laidunnus on eduksi Mallan luonnonpuistolle. *Helsingin Sanomat*, 2 October.

Indigenous Food System Network. *Indigenous food sovereignty.* Retrieved 15 September 2019 from https://www.indigenousfoodsystems.org/food-sovereignty.

Järvinen, A. 2017a. Kaksi eri laidunnusta. *Lapin Kansa*, 6 September.

Järvinen, A. 2017b. Mallan luonnonpuistosta ja poroista. *Lapin Kansa*, 23 August.

Järvinen, A. 2017c. Kestämätön luontosuhde. *Lapin Kansa*, 12 August.

Järvinen, A. 2018a. Tuntipöllö kärsii ylilaidunnuksesta. *Helsingin Sanomat*, 27 April.

Järvinen, A. 2018b. Tunturikiuru ei hävinnyt ilmastonmuutokset takia vaan muista syistä. *Lapin Kansa*, 10 April.

Johnsen, K.I. 2014. The paradox of reindeer pasture management in Finnmark, Norway. *Arctic Info.* Retrieved 15 September 2019 from https://www.arcticinfo.eu/en/features/112-the-paradox-of-reindeer-pasture-management-in-finnmark-norway.

Kitti, H., N. Gunslay, and B.C. Forbes. 2006. Defining the quality of reindeer pastures: The perspective of Sami reindeer herders. In Forbes, Bölter, Muller-WIlle, Hukkinen, Muller, Gunslay, and Konstantinov (eds.), *Reindeer management in northernmost Europe: Linking practical and scientific knowledge on social-ecological systems.* Berlin: Springer, 141–165.

Kivipelto, A. 2016. *Jääleinikit katoavat porojen pötseihin Mallan luonnonpuistossa*, 28 September.

Knuth, L. 2009. *The rights to adequate food and indigenous peoples – How can the right to food benefit indigenous peoples?* Rome: FAO.

Knuth, L., and M. Vidar. 2011. *Constitutional and legal protection of the right to food around the world.* Rome: FAO.

Kuhnlein, H.V. and B. Burlingame. 2013. Why do Indigenous Peoples' food and nutrition interventions for health promotion and policy need special consideration? In Kuhnlein, Erasmus, Spigelski, and Burlingame (eds.), *Indigenous Peoples' food systems andand well-being: Interventions and policies for healthy communities.* Rome: FAO, 3–8.

Kuhnlein, H.V., B. Erasmus and D. Spigelski. 2009. *Indigenous Peoples' food systems: The many dimensions of culture, diversity and environment for nutrition and health.* Rome: FAO.

Kuhnlein, H.V. and O. Receveur. 1996. Dietary change and traditional food systems of Indigenous Peoples. *Annual Review of Nutrition* 16: 417–442.

Kumpula, J. 2006. Very high resolution remote sensing data in reindeer pasture inventory in Northern Fennoscandia. In Forbes, Bölter, Muller-Wille, Hukkinen, Muller, Gunslay, and Konstantinov (eds.), *Reindeer management in northernmost Europe: Linking practical and scientific knowledge on social-ecological systems.* Berlin: Springer, 167–186.

Kumpula, J., A. Colpaert, and M. Anttonen. 2007. Does forest harvesting and linear infrastructure change the usability value of pastureland for semi-domesticated reindeer (*Rangifer tarandus tarandus*)? *Annales Zoologici Fennici* 44(3): 161–178.

Kumpula, J., M. Kurkilahti, T. Helle, and A. Colpaert. 2014. Both reindeer management and several other land use factors explain the reduction in ground lichens (*Cladonia spp.*) in pastures grazed by semi-domesticated reindeer in Finland. *Regional Environmental Change* 14(2): 541–559.

Kumpula, J., J. Siitari, M. Kurkilahti, J. Heikkinen, and K. Oinonen. 2019. Poronhoitoalueen talvilaitumet vuosien 2016–2018 laiduninvestoinnissa: Talvilaidunten tilan muutokset ja muutosten syyt. *Luonnonvara- ja biotalouden tutkimus*, 29/2019. Helsinki: Luonnonvarakeskus.

Kuokkanen, R. 2019. *Restructuring relations: Indigenous self-determination, governance and gender.* Oxford: Oxford University Press.

Kyrölahti, A. 2019. Hävinneitä lajeja ei millään saa takaisin. *Lapin Kansa*, 9 March 2019.

Käyhkö, J. and T. Horskotte. (2017). *Globaalimuutoksen vaikutus porotalouteen Pohjois-fennoskandian alueella tundra-alueilla.* Turku: Painosalama Oy.

Laakso, A.M. 2002. *The anomalies of contemporary reindeer herding management and supporting reindeer research: The need for a new paradigm*, Pro gadu (Master's thesis). Rovaniemi: University of Lapland.

Lapin luonto tuhoutuu poronhoidon vuoksi väittää tietokirjailija: 'Porojen luontaista vaellusta ei pääse tapahtumaan.' 2018. *MTV Uutiset*, 17 September.

Lavia, R. 2016. Jutaaminen säästäisi jäkälikköjä. *Lapin Kansa*, 5 October.

Ledrou, I. and J. Gervais. 2005. Food insecurity. *Health Reports* 16(3): 47–51.

León, I. 2007. Food sovereignty: For a future without hunger. *América Latina en Movimiento*, 419. Quito: ALAI.

Luonnonvarakeskus. 2016. *Porolaitumet*. Retrieved 15 September 2019 from https://www.luke.fi/tietoa-luonnonvaroista/maatalous-ja-maaseutu/porotalous/porolaitumet/.

Magga, O.H., S.D. Mathiesen, R.W. Corell, and A. Oskal. 2009. Reindeer herding, traditional knowledge, adaptation to climate change and loss of grazing land. Alta: Norway and Association of World Reindeer Herders (WRH) in Arctic Council, Sustainable Development Working Group (SDWG).

Mainio, T. 2016. Kun Lapin poroilla loppuu ruoka, niin poromiesten on suunnattava etelään hakemaan jäkälää. *Helsingin Sanomat*, 25 September.

Martens, T., J. Cidro, M.A. Hart, and S. McLachlan. 2016. Understanding Indigenous food sovereignty through an Indigenous paradigm. *Journal of Indigenous Social Development* 5(1): 18–37.

Martyn-Hemphill, R. 2017. In Norway, fighting the culling of reindeer with a macabre display. *New York Times*, 6 December.

Mazzullo, N. 2010. More than meat on the hoof? Social significance of reindeer among Finnish Saami in a rationalised pastoralist economy. In Stammler and Takakura (eds.), *Good to eat, good to live with: Nomads and the animals in Northern Eurasia and Africa. Northeast Asian Study Series, 11.* Sendai, Japan: Centre for Northeast Asia Studies, Tohoku University, 101–119.

Miettunen, O. 2017. Porotaloudella alkaa olla väistämisen tila lopussa. *Lapin Kansa*, 6 October.

Morrison, D. 2011. Indigenous food sovereignty: A model for social learning. In Wittman, Desmarais, and Wiebe (eds.), *Food sovereignty in Canada: Creating just and sustainable food systems*. Halifax: Fernwood Publishing, 97–113.

Muller-Wille, L., J. Hukkinen, F. Muller, M. Bölter, and B.C. Forbes. 2006. Synthesis: Environmental and sociopolitical conditions for modern reindeer management in Europe's North. In Forbes, Bölter, Muller-Wille, Hukkinen, Muller, Gunslay, and Konstantinov (eds.), *Reindeer management in northernmost Europe: Linking practical and scientific knowledge on social-ecological systems*. Berlin: Springer, 365–379.

Naali ei ole ainoa uhattu. 2019. Editorial. *Lapin Kansa*, 12 March.

Neufeld, H.T. and C.A.M. Richmond. 2017. Impact of place and social spaces on traditional food systems in southwestern Ontario. *International Journal of Indigenous Health* 12(1): 93–115.

NGO/CSO Forum for Food Sovereignty. 2002. *Food sovereignty: A right for all.* Political statement of the NGO/CSO Forum for Food Sovereignty. Retrieved 15 September 2019 from https://nyeleni.org/spip.php?article125.

Oksanen, L., B.C. Forbes, J. Kumpula, T. Kumpula, J. Käyhkö, and S. Stark. 2016. *Helsingin Sanomat*, 29 September.

Paine, R. 1994. *Herds of the tundra: A portrait of Saami reindeer pastoralism.* Washington DC and London: Smithsonian Institution Press.

Paliskuntain yhdistys. 2014. *Opas poronhoidon tarkasteluun maankäyttöhankkeissa.* Rovaniemi: Pohjolan Painotuote Oy.

Paliskuntain yhdistys a. *Reindeer herding cooperatives*. Retrieved 15 September 2019 from https://paliskunnat.fi/reindeer-herders-association/cooperatives/.

Paliskuntain yhdistys b. *Käsivarsi*. Retrieved 15 September 2019 from https://paliskunnat.fi/reindeer-herders-association/cooperatives/cooperatives-info/kasivarsi/.

Pape, R. and J. Löffler. 2012. Climate change, land use conflicts, predation and ecological degradation as challenges for reindeer husbandry in Northern Europe: What do we really know after half a century of research? *Ambio* 41(5): 421–434.

Poronhoitajat haluavat saada äänensä kuuluviin alansa päätöksenteossa. 2005. Editorial. *Lapin Kansa*, 7 January.

Reinert, E.S., I. Aslaksen, I.M.G. Eira, S.D. Mathiesen, H. Reinert, and E.I. Turi. 2009. Adapting to climate change in Sámi reindeer herding: The nation-state as problem and solution. In Adger, Lorenzoni, and O'Brian (eds.), *Adapting to climate change: Thresholds, values, governance*. Cambridge: Cambridge University Press, 417–432.

Riseth, J.Å. 2014. *Sámi traditional ecological knowledge as a guide to science: Snow, ice and reindeer pasture facing climate change*. Tromso: Norut Northern Research Institute AS.

Roturier, S. and M. Roué. 2009. Of forest, snow and lichen: Sámi reindeer herders' knowledge of winter pastures in northern Sweden. *Forest Ecology and Management* 258(9): 1960–1967.

Rudolph, K. and S. McLachlan. 2013. Seeking Indigenous food sovereignty: Origins of and responses to the food crisis in northern Manitoba, Canada. *The International Journal of Justice and Sustainability* 18(9): 1079–1098.

Ruukki, J. 2016. Porojen tehotuotanto on ympäristöriski. *Helsingin Sanomat*, 19 September.

Saamelaiskäräjät 2016. *Saamelaiskäräjien lausunto luonnoksesta uhanalaisten lajien suojelun toimintaohjelmaksi*. Dnro 414/D.a.2/2016.

Salvesen, H. 1995. Sami AEdnan: Foru states – one nation? Nordic minority policy and the history of the Sami. In Tägil (ed.), *Ethnicity and nation building in the Nordic world*. Carbondale and Edwarsville: Southern Illinois University Press; 1995, 106–144.

Sandström, P., N. Cory, J. Svensson, H. Hedenås, L. Jougda, and J. Borchert. 2016. On the decline of ground lichen forests in the Swedish boreal landscape: Implications for reindeer husbandry and sustainable forest management. *Ambio* 45(4): 415–429.

Sara, Mikkel Nils. 2009. Siida and traditional Sámi reindeer herding knowledge. *Northern Review*, 30: 153–178.

Sélingué, M. 2007. *Nyéléni 2007, forum for food sovereignty*. Retrieved 15 September 2019 from https://nyeleni.org/DOWNLOADS/Nyelni_EN.pdf

Shawn, K. 2008. *Indigeneity and political theory: Sovereignty and the limits of the political*. New York: Routledge.

Soppela, P., R. Walter, B. Åhman and J. Å. Riseth (eds.). 2002. Reindeer as a keystone species in the North: Biological, cultural and socio-economic aspects. *Arctic Centre Reports 38*. Rovaniemi: Arktinen keskus.

Sorjanen, T. 2016. Laidunkierto voisi olla apu Lapin jäkäläkatoon. *Talouselämä*, 23 October.

Tarasuk, V.M., A. Mitchell, and N. Danchner. 2014. *Household food insecurity in Canada 2011*. Toronto: Research to identify policy options to reduce food insecurity (PROOF). Retrieved 15 September 2019 from https://proof.utoronto.ca/resources/proof-annual-reports/annual-report-2014/.

UN Committee on Economic, Social and Cultural Rights (CESCR). 1999. *General Comment No. 12: The Right to Adequate Food (Art. 11 of the Covenant)*. Retrieved 15 September 2019 from https://www.refworld.org/docid/4538838c11.html.

UN Permanent Forum on Indigenous Issues. 2012. *The rights of Indigenous Peoples to food and food sovereignty*. Retrieved 15 September 2019 from https://www.un.org/esa/socdev/unpfii/documents/2012/News%20and%20Media/EN%20Fact%20Sheet_Right%20to%20Food.pdf.

Valikainen, J. 2017. Jääleinikki katoaa luonnonpuistosta porojen suihin. *Lapin Kansa*, 10 August.

Valtioneuvosto. 2019. *Valtioneuvoston asetus poronhoitovuodelta 2019/2020 maksettavasta eläinkohtaisesta tuesta*. Helsinki: Valtioneuvosto.

Vehmas, M. 2016a. Ylilaiduntamista ainutlaatuisessa tunturiluonnossa vai ei: Poronhoidosta nousi äläkkä Kilpisjärvellä. *Aamulehti*, 18 September.

Vehmas, M. 2016b. Kirjoittelu on yhden miehen sotaa. *Lapin Kansa*, 16 September.

Via Campesina. 2006. *The right to produce and access land*. Retrieved 15 September 2019 from http://safsc.org.za/wp-content/uploads/2015/09/1996-Declaration-of-Food-Sovereignty.pdf.

Windfuhr, M., and J. Jonsén. 2005. *Food sovereignty, towards democracy in localised food systems*. Warwickshire: ITDG Publishing.

Wittman, H., A. Desmarais, and N. Wiebe. 2010. The origins and potential of food sovereignty. In Wittman, Desmarais, and Wiebe (eds.), *Food sovereignty: Reconnecting food, nature, and community*. andHalifax, NS: Fernwood Publishing, 1–14.

Woodley, E., E. Crowley, J. Dey de Pryck, and A. Carmen. 2009. *Cultural indicators of Indigenous Peoples' food and agro-ecological systems*. FAO and the International Indian Treaty Council (IITC).

3 Sami reindeer herders and the radioactive reindeer

Food security from different voice

Moeko Minagawa

Introduction

Sapmi is the land of the Sami people; it stretches across the northern part of Scandinavia and Russia's Kola Peninsula (Figure 3.1). The total Sami population today is estimated to be about 80,000–105,000 ('Antelet samer i Sapmi' n.d.). Of those, approximately 50,000–65,000 Sami people reside in Norway, approximately 20,000–40,000 in Sweden, approximately 8,000 in Finland, and approximately 2,000 in Russia's Kola Peninsula. There are many ways to define Sami, and there is no single universal definition of a Sami. Therefore, the Sami People are defined differently in the four countries where they live. This chapter focuses on the Sami who live in the European High North.

In Norway, the rights of Sami are provided in the Constitution and in the Sami Act (Steinlien 1989), which establishes the *Sameting* (the Sami Parliament) and provides for other matters. In this act, the Sami electoral register is defined as follows:

> All persons who make a declaration to the effect that they consider themselves to be Sami, and who either a.) have Sami as their domestic language, or b.) have or have had a parent, grandparent or great-grandparent with Sami as his or her domestic language, or c.) are the child of a person who is or has been registered in the Sami electoral register may demand to be included in a separate register of Sami electors in their municipality of residence.
>
> ('The Sámi Act' n.d.).

The Sami in Sweden were recognised as an Indigenous People group by the Swedish assembly in 1977 ('The Sami Parliament' n.d.); in 1993, the Sami in Sweden elected their own parliamentary body ('Sami in Sweden' n.d.). Those on the electoral register who speak Sami and define themselves as part of Sami society through cross-generational connections are eligible to vote for the Sami Parliament.

DOI: 10.4324/9781003057758-3

In Finland, the Sami were constitutionally recognised as Indigenous People in 1995, and the Sami parliament was established in 1996. The Sami Parliament Act in Finland defines a Sami as:

> … a person who considers himself a Sami and: 1) he himself or at least one of his parents or grandparents has learnt Sami as his first language; 2) he is a descendent of a person who has been entered in a land, taxation or population register as a mountain, forest or fishing Lapp (*sic*);[1] or 3) at least one of his parents has or could have been registered as an electoral for an election to the Sami Delegation or the Sami Parliament.
> ('Definition of a Sámi' n.d.).

Traditionally, Sami have engaged in livelihoods such as reindeer husbandry, sea and river fishing, hunting, farming, and crafts. Today, less than 4% of Sami in Norway ('Reindeer Husbandry' n.d.) and less than 20% of the Sami in Sweden (Axelsson and Sköld 2006) are actively involved in reindeer husbandry. More than 60% of Sami in Finland live outside the Sami homeland ('The Sámi in Finland' n.d.). Although most Sami today do not make their livings in these traditional industries, the food security of Sami who engage in traditional industries is worthy of attention. Since they live in close connection to nature (Axelsson and Sköld 2006), their food security is exposed to risks arising from the ecological changes caused by climate change today. For examples, climate change is impacting the efficiency of reindeer fodder. In addition, high summer temperatures are leading to increased problems with insects and aggravating reindeer's health problems. This increases the risk of less food being available for local communities (Zojer and Hossain 2017).

In addition to this, the Chernobyl nuclear disaster in 1986 caused serious damage to traditional Sami industries (Stephens 1987; Liland et al. 2009). Just after the accident, a radioactive cloud affected traditional Sami lands more than other regions due to rain and wind direction (Drozdovitch et al. 2007). Lichen and mushrooms in wild forests tend to absorb radiation. Since lichen is reindeer's main food, the reindeer have become radioactive (Strand 1994; Åhman 1999, 2007). Reindeer are important to the livelihood and food of Sami reindeer herders. However, although Cs-137 has a half-life of 30 years and more than 30 years have now passed since the disaster, Sami reindeer herders in Norway are still living under the impact of this radioactivity today (Beresford et al. 2016). Radio Free Europe/Radio Liberty reported that reindeers were still radioactive in central Norway and in Sweden (2016).

To protect food safety, the Norwegian, Swedish, and Finnish governments enacted various counter measures just after the disaster. What concrete actions have these governments taken to protect the food security of Sami reindeer herders? In those actions, how has health security been positioned in relation to food security? How has the Sami reindeer herder community then repositioned health security, and what were the implications of expanding the idea of food security?

To address these questions, this chapter is divided into three parts. First, it examines definitions of the term 'food security' provided by international organisations and considers what food security means to Sami reindeer herders. Second, it explores governments' countermeasures against the radioactive contamination of food products arising from the Chernobyl nuclear power plant accident. These government countermeasures revealed the tension between economic and health security in the area of food security. Therefore, this chapter considers what food security might mean to the Sami people from the perspective of human security. Finally, it uses an ethics of care approach to examine the unique role of Sami women in shaping this problem. Women have voiced their concerns about health. Drawing on existing literature and listening to their input, this paper traces how the idea of food security has come to encompass more meanings.

Sami reindeer herder's food security: interlinked with other kinds of securities

According to the Oxford Dictionary, 'security' means protection as well as the state of feeling safe. Protection includes 'the activities involved in protecting a country, building or person against attack, danger, etc.' (Hornby and Turnbull 2010). The term 'food security' was first used at the 1974 World Food Conference. At this conference, the term was used to refer to food supply problems such as famine, hunger, and food crises (United Nations 1975). The 1996 World Food Summit adopted this definition: 'Food security exists when all people, at all times, have physical and economic access to sufficient, safe and nutritious food to meet their dietary needs and food preferences for an active and healthy life' (FAO 1996). Meanwhile, the Food and Agricultural Organisation's (FAO) definition of food security encompasses various dimensions, including the availability, access, safety, utilisation, and stability of food.

Two years before the World Food Summit, the term 'human security' first appeared in the 1994 United Nations Development Programme (UNDP) report, the *Human Development Report*. This report revised previous practices of international development by shifting the focus of development from the security of nation states to a 'human-centred' approach. It declares the right of all individuals, especially vulnerable people, to 'freedom from fear' and 'freedom from want'. Based on the premises that development cannot be realised without peace and peace cannot be established without development, it lists seven essential dimensions of human security: economic, food, health, environmental, personal, community, and political (UNDP 1994). Amartya Sen proposed a concept of human security that reflects the 'human capabilities approach' (Sen and Ogata 2003); this idea has been applied beyond the field of development to the areas of human rights and human dignity.

The definition of food security provided by the 1996 World Food Summit follows human security. According to the FAO, 'the relationship between human security and food security is predicated on the idea of full realisation of the human right to adequate food, as fundamental human right, and one that leaves no one behind' (FAO 2016). A number of international organisations, including the UNDP, the United Nations High Commissioner for Refugees (UNHCR), United Nations Children's Fund (UNICEF), International Labor Organization (ILO), and the World Health Organization (WHO) have adopted this approach in their activities. Most of these organisations' programmes that deal with food within the human security framework primarily address the issues of famine in developing countries.

Human security is still debated actively and interpreted in many ways in academia. Debates about the relationship between human security and food security are not limited to famine; they also address problems with regard to agricultural products and food contamination. Such discussions examine food poisoning caused by natural biochemical processes and by man-induced cross-contamination, as well as the risks of radioactive materials in foods. Many debates on food security from a human security perspective followed the 2011 Fukushima Daiichi nuclear power plant accident (Bacon et al. 2014). Furthermore, several studies consider Indigenous Peoples from a human security approach that emphasises human rights (Szpak 2017; Zojer and Hossain 2017).

According to Nilsson et al., the Sami diet is difficult to define as a range of dietary habits are found in the Sami population. However, according to historical scientific literature, the Sami diet is typically characterised by a high intake of fatty fish, red meat (primarily reindeer), fat, blood and organ dishes, wild berries, and boiled, unfiltered coffee (Nilsson et al. 2012, 2). Today, Sami reindeer herders eat more of these foods than any other Sami and non-Sami in Sweden (ibid., 10). Similarly, a study on the dietary patterns of people living in the Sami regions of Norway shows that Sami-affiliated groups; a people group in which three generations speak the Sami language eat more reindeer meat than non-Sami groups (Brustad et al. 2008). Therefore, for Sami engaged in food-related traditional livelihoods, ecosystem services account for a significant portion of their diet (Zojer and Hossain 2017, 22).

Historically, the Sami paid taxes to secure their hunting rights and migrated across the borders between the then existing kingdoms until the end of the nineteenth century (Horstkotte and Käyhkö 2017). When borders between the nation states established, the circumstances of reindeer herding became different from country to country (ibid.). Today, the reindeer herding Sami migrates basically inside the national borders according to the environment and the seasons. Figure 3.1 shows the reindeer grazing grounds and the areal extent of Sapmi (ibid.).

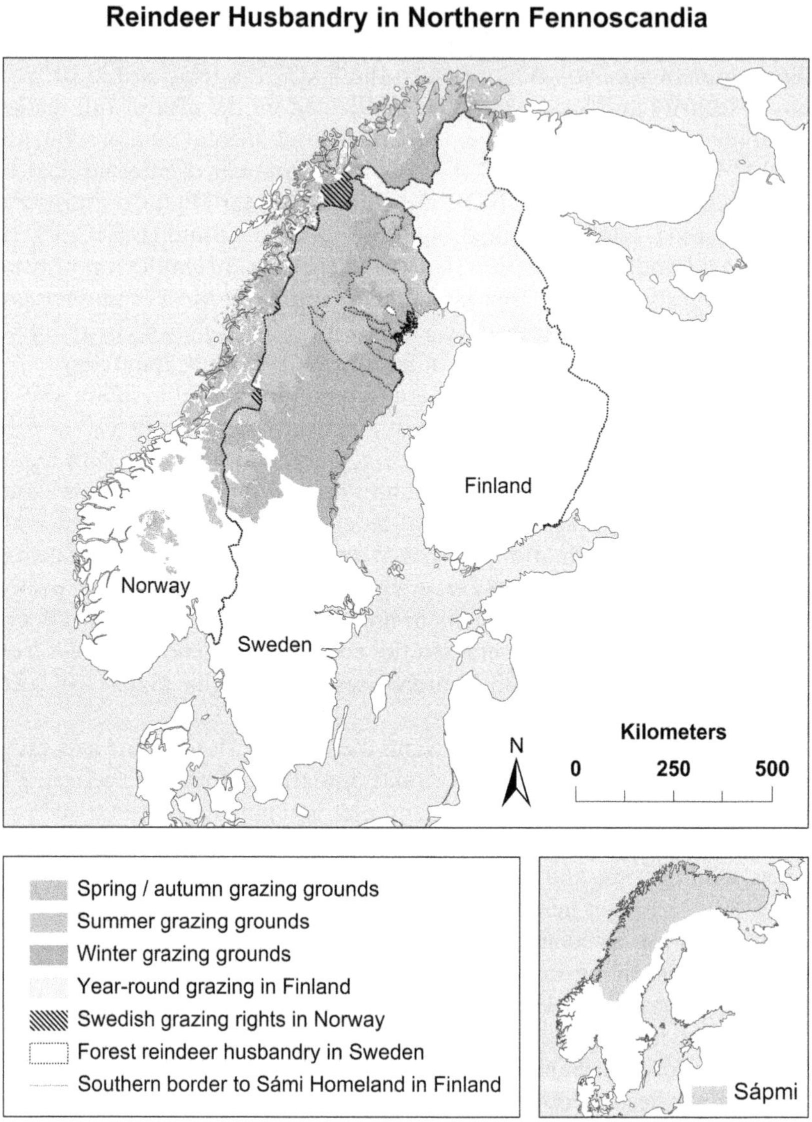

Figure 3.1 'Reindeer grazing grounds and the areal extent of Sápmi'. Created by Horstkotte and Käyhkö (2017).

In Norway, reindeer graze on pastures that account for approximately 40% of country's mainland (Roto 2015a). Six pasture territories are further divided into 89 pasture areas that may be used only by the ethnic Sami for reindeer husbandry (ibid.). Approximately 4,700 reindeer herders graze more than 250,000 domestic reindeer in Norway; most of them are based in the province of Finmark. In Sweden, reindeer are allowed to graze in at least 50% of Sweden's land. More than 4,700 reindeer owners graze approximately 250,000 reindeers ('The Sami in figures' n.d.), and there are 51 reindeer herding districts, the largest of which is in Jokkmokk. Unlike in Norway and in Sweden, in Finland, reindeer herding is not reserved for Sami; any EU citizen may own reindeer. Meanwhile, approximately 33% of the country is used for reindeer husbandry (Roto 2015b). A maximum of 203,700 reindeers are allowed in Finland, and there are approximately 4,600 reindeer herders (Mustonen and Jones 2015).

As land is indispensable for Sami's traditional livelihoods, Sami reindeer herders' food security is closely connected to the ability to use land. Therefore, the right to land is of great concern for reindeer herders (Axelsson and Sköld 2006). The ILO Convention No. 169 on Indigenous and Tribal Peoples in Independent Countries (1989) aims to protect the rights, cultures, and way of life of Indigenous Peoples. This convention is legally binding and includes the right to land. Currently, however, Norway is the only one of the three Nordic states that has ratified it. The Norwegian Constitution states that the state is responsible for creating conditions that enable the Sami people to preserve and develop their language, culture, and way of life.

Despite this, many local food sources can be exposed to potentially harmful pollution, increasing the competition over land use (Zojer and Hossain 2017, 55). In addition, climate change has forced some dietary changes and caused economic hardships, jeopardising the food security of reindeer herders (ibid., 56). Larsen and Fondahl indicated that food security is declining in the Arctic due to environmental change and globalisation (2014, 24). These risks show the relation between food insecurity and other kinds of insecurities such as environmental, economic, health, individual, and community. Indeed, Sami reindeer herders have gone through and still experience the complicated issue since before such radioactive contamination of the land.

Food insecurity following the Chernobyl nuclear accident

Even before the Chernobyl nuclear power plant accident in 1986, people in the Arctic regions were heavily exposed to the global fallout from the atmospheric bomb testing of the 1950s and 1960s (Kurttion et al. 2010). When the Chernobyl nuclear power plant accident occurred, the winds carrying the radioactive materials first moved north towards Scandinavia and later south and west through Europe (UNESCO 1996, 7, IRSN 2011). Aerial monitoring in Scandinavia and later in Western Europe illustrated the highly variable nature of the deposit (UNESCO 1996, 7–8). Some of the heaviest deposits of Cs-134 and Cs-137 were recorded in Scandinavia, Germany, Austria, and Switzerland (ibid.). The Cs-137 deposited over the Nordic countries represents more than 10% of the total amount of Cs-137 that is assumed to have been released from Chernobyl (Snihs 1996). The southern

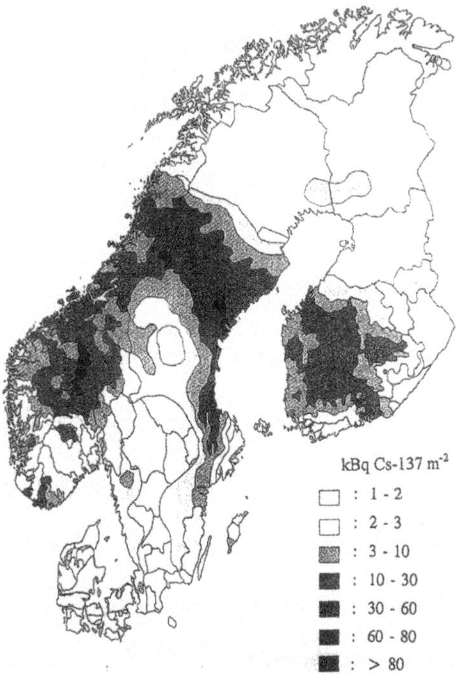

Figure 3.2 'Ground deposition of^{137}Cs, kBq m^{-2}, in Denmark, Finland, Norway and Sweden resulting from the Chernobyl accident'. Dahlgaard (1994).

half of Finland and parts of Norway and Sweden saw some of the most fallout outside the former Soviet Union; average Cs-137 levels of up to 200 kilo becquerel per square meters (kBq/m^2) were found on the ground surface (Statens strålskyddsinstitut 1986; Strand 1994; Drozdovich et al. 2007). Cs-137 was also detected in grazing animals, such as cows and sheep, that lived in contaminated areas; it was also found in reindeer that fed on contaminated lichen, which is reindeer's main source of food in winter (Hove et al. 1994; Strand 1994; Snihs 1996; Liland et al. 2009). Figure 3.2 shows the ground deposition of Cs-137, kBqm^{-2}, in Denmark, Finland, Norway, and Sweden resulting from the Chernobyl nuclear power plant accident (Dahlgaard 1994). The contamination of grazing animals was a significant problem in Norway and Sweden; as the map shows, the reindeer herding area (Figure 3.1) overlaps with the area with high Cs-137 deposits (Figure 3.2; Statens strålskyddsinstitut 1986; Blomqvist et al. 1991; Irgens et al. 1991; Strand 1994; Rekacewicz 2007).

This Cs-137 caused serious damage to Sami reindeer herders, especially those in Norway and Sweden (Statens strålskyddsinstitut 1986; Strand 1994). Cs-137 levels of up to 150,000 Bq/kg were found in reindeer from mountain areas in southern Norway from October 1986 to April 1987 (Strand et al. 1990). In Sweden, reindeer killed in highly contaminated regions after the accident showed caesium concentrations at or above the permissible limit, many higher than 10,000 Bq/kg (Statens strålskyddsinstitut 1986, 7). In Finland, reindeer breeding areas are situated in the northern part of the country, which was not seriously affected by the accident (Blomqvist et al. 1991, 19; see also Figure 3.2). However, monitoring of reindeer meat showed that a limited area in the southeast corner of the reindeer breeding area was heavily contaminated (ibid.). High levels of Cs-137 were also found in lake fish, and the Finnish authorities recommended restricting consumption of inland lake fish in the areas of highest fallout (ibid., 9–10, 18–19).

Since the end of the 1940s, the Nordic countries: Denmark, Finland, Iceland, Norway, and Sweden have collaborated on research on nuclear technology, including nuclear safety research (Marcus 1997). After the accident, in June 1987, the Nordic Committee for Nuclear Safety Research (NKS) proposed the intervention levels for radioactivity in foodstuffs as the 'Nordic model'. Similar measures were implemented by the Norwegian, Swedish, and Finnish authorities. These included banning contaminated food from circulation in food distribution channels, limiting imported foods, and investigating nationally produced foods. At the same time, the governments raised the acceptable limits of radiation levels in reindeer meat much higher than the acceptable cap set by the European Union of 600 Bp/kg (Statens strålskyddsinstitut 1986; Blomqvist 1991; Strand 1994).

However, each national government set different intervention levels for foodstuffs. In 1986, the Swedish Radiation Protection Authority (SSI) defined an acceptable radiation dose as 50 mSv in 50 years. As of May 1986, the Cs-137 limit in food products was set at under 300 Bq/kg, and distribution of food that exceeded this limit was prohibited. This should guarantee a long-term average radiation dose of less than 1 mSv/year (annual intake dose under 50,000 Bq Cs-137). As a result, that year in the slaughter season, 78% of reindeer meat was unmarketable and destroyed in Sweden (Tanoi et al. 2016, 242). The Swedish government took this action to protect the safety of its citizens by distributing as little reindeer meat as possible. The production of reindeer meat had begun to steadily increase at the beginning of 1980, but this trend changed dramatically in 1986 and 1987 ('Reindeer Husbandry in Sweden' n.d.). In June 1987, the dose limit of Cs-137 was increased to 1,500 Bq/kg (much higher than the EU standard of 600 Bq/kg); this limit was applied to 'special food' such as reindeer meat, game, lake fish, wild berries, mushrooms, and nuts. These food products are also known as 'Sami food'. Blackwell points out that the National Food Administration (SLV) promoted different radiocaesium intake limits for Sami and non-Sami, and the government did not discourage Sami from consuming meat with levels under 10,000 Bq/kg as long as the annual recommended dosage was not exceeded (Blackwell 2003). Thanks to this change, the percentage of unacceptable reindeer in the open market decreased from about 80% to about 40% (Snihs 1996, 30). In 1994, when Sweden joined the EU, the acceptable intake levels of Cs-134 and Cs-137 were reduced to 1250 Bq/kg for food products distributed in the EU market.

Norway also setting different intervention levels for reindeer meat. In May 1986, Cs-137 intake limits on food products distributed in the market were set at 370 Bq/kg for cow's milk and food products for infants and toddlers and 600 Bq/kg for all other food products. However, in that same year, the limit was pushed up to 6,000 Bq/kg for reindeer meat and game. Similarly, in July 1987, the level for wild freshwater fish and game was also increased to 6,000 Bq/kg, while the 600 Bq/kg limit continued to be applied to all other food products. In 1994, the radiocaesium intake limit was reduced to 3,000 Bq/kg, which, while lower than previously, still far exceeds the EU limit of 600 Bq/kg. Finland also raised interim intervention levels from 2,000 Bq/kg of Cs-137 in 1986 to 4,000 Bq/kg in 1987 (Blomqvist 1991, 19).

The Norwegian, Swedish, and Finnish authorities raised the intervention levels to protect the economy, culture, and lifestyle of Sami reindeer herders (Snihs 1996, 33; Bresford et al. 2016). However, this increase in limits on intake of radiocaesium also had a severe impact on Sami reindeer herders' health. After this shift, there was a significant gap in intake levels of radiocaesium from food between thc persons who were randomly selected from the Norwegian population and from the Sami population in Norway. According to the dietary surveys and whole body counting, for the former, the average effective dose equivalent during the first year after the Chernobyl accident was estimated to be 0.12–0.25 mSv and for the latter was estimated to be 1–3 mSv (Strand 1994, 7).

To determine acceptable radiocaesium intake levels, the authorities analysed the costs and benefits. They sought to keep radiation dose levels to an appropriate minimum based on economic and social costs. Increasing the limit reduced the burden on the government to provide economic compensation to Sami reindeer herders. Indeed, each Nordic country outlines the cost of this countermeasure in detail (Statens strålskyddsinstitut 1986, 7; Blomqvist 1991; Strand 1994, 6). In August 1986, the Nordic governments promised to compensate herders for reindeer they could not legally sell (Stephens 1987; Blomqvist et al. 1991, 19).

Reindeer herders also implemented several changes to minimise reindeer's internal radiation exposure. These changes included changed slaughter times, new grazing routes, feeding reindeer with caesium binders, and regular investigations (Strand et al. 1992a, 1992b; Snihs 1996, 3; Mehli et al. 2000; Beresford et al. 2016, 82). The changes were not limited to reindeer herding activities; they also involved reindeer's dietary habits (Mehli et al. 2000). For example, Norwegian authorities developed dietary advisories that provided guidelines regarding intake of the most contaminated products and another document describing how to prepare food to reduce the radiocaesium content. These were distributed to people who traditionally consume large amounts of game and fish (Skuterud and Thørring 2012).

Therefore, the Chernobyl nuclear power plant accident deeply compromised the food security of Sami reindeer herders. As mentioned before, food insecurity is interlinked with other kinds of insecurities: economic, environmental, health, individual, and community. From the perspective of human security, these governments' decisions to increase limits on intake of radiocaesium and to compensate reindeer herders secured the herders' livelihoods; economic security and all of its linked securities except for health security. Indeed, higher radiocaesium intake increases health risks. In this case, economic security and health security, although both are listed as essential dimensions of human security, become incompatible. As a consequence, here, the health risk was not prioritised.

Why was economic security valued over health security? One explanation is that health issues are uncertain. There is no clear consensus among the scientific community about the health risks posed by internal exposure to low-dose radiation; more time is needed to clarify this causal relationship. Another explanation is that, in this case, the voices calling for health security were structurally denied. This point will be explored from an ethics of care approach in the following section.

Voice of care for food security

An ethics of care approach offers a perspective that sheds light on people's everyday life and on alternative voices. Carol Gilligan pointed out the inability of the language of justice to take into account as morally pertinent the experiences and points of view of women (Laugier 2015). Gilligan highlighted the significance of caring for others, as well as practices of care that have often been undervalued because they are usually done by women (Gilligan 1993). Ethics of care recognises human vulnerability and the idea that all people are interdependent. For Joan Tronto, care can be integrated into a general ethical, social, and political approach that is not reserved for women; it is something everyone should aspire to (Tronto 1993). Furthermore, Sandra Laugier focuses on the ordinary moral problems that care poses:

> … the ethics of care draw our attention to the ordinary, to what we are unable to see, but is right before our eyes. So before being a feminine ethics, it is an ethics that gives voice an expression to humans that are undervalued precisely because they accomplish unnoticed, invisible tasks, and take care of the basic needs.
>
> (Laugier 2015, 217)

In this context, ethics of care may pay attention to the voices of Sami people that had been undervalued. To amplify these voices, this section will quote from interviews conducted by Sharon Stephens. Stephens researched the influence of the Chernobyl accident on Sami reindeer herders. Her interviews were conducted from December 1986 to January 1987 and from August to

September 1990 around Snåsa, a region in central Norway where South Sami reside and that was contaminated by radioactive materials (Stephens 1995). She interviewed a Sami herding family from Valsjöyn, Sweden who had been affected by the Chernobyl nuclear accident (Stephens 1987).

In her interview, a wife and mother in a reindeer herding family in Snåsa said:

> It seems sometimes that things have become strange and make-believe. You see with your eyes the same mountains and lakes, the same herds, but you know there is something dangerous, something invisible, that can harm your children, that you can't see or touch or smell. Your hands keep doing the work, but your head worries about the future.
>
> (Stephens 1995, 276)

Another woman, a Swedish South Sami female reindeer owner, commented, 'This is not just a matter of economics, but of who we are, how we live, how we are connected to our deer and each other. Now (in the winter of 1986–1987) we must buy everything' (Stephens ibid., 277).

These comments illustrate how their everyday lives were changed and the fear that surrounded those changes. They also express these women's concern for their children, their environment, and their future. Stephens recalls:

> [After the government increased the radiation intake limit] … a particularly painful meeting with the predominantly male planning group of the local reindeer-herding organisation. My questions – about 'the facts' of the fallout and its possible effects on the local economy, political self-determination, family life, and the future expectations of young people – were often met with silence or frustrated answers: It is too soon to know these things. (…) In contrast, women seemed to talk a great deal about Chernobyl, not just to me but to each other.
>
> (Stephens ibid., 282).

Stephens discusses the women's voices and attitudes:

> They talk about how to help the men, demoralised by their sense of the futility of their work and the uncertainties of the future, and how to help their children feel safe when their everyday food had suddenly become something dangerous and mysterious. They discussed how to get safe meat from relatives and friends up north, how to prepare store-bought foods to feed their families, and how to prepare the usual foods in different ways (for example, by repeated rinsing of the meat) in order to lower radioactivity levels. They wondered what foods they should eat themselves, especially if they were pregnant or nursing, and speculated about how to interpret changing government definitions of safe levels.
>
> (Stephens ibid.)

Stephens concluded that the women prioritised health risks associated with food over economic problems. Furthermore, these interviews demonstrate dialogue in face of dangerous circumstances; in these dialogues, women form strategies in social relationships – that is, through interdependence.

However, why did the women discuss the health risks? As implied in the quotations, Sami women are generally responsible for preparing food. According to Sarah Andrews, Sami women in Norway are generally the primary caretakers of children and they are also responsible for passing on their indigenous culture to their children (Andrews n.d.). Women also play a bigger role than men in preparing reindeer meat or making handicrafts from reindeer (Larsson 2004). Women became concerned about the health risks of radiation exposure due to practical aspects of their everyday lives and because of their roles in the private sphere.

Gradually, however, their voices became louder and began to resonate outside the private sphere. These women raised their voices to demand extensive information and comprehensive health monitoring from health agencies as well as state support for reducing becquerel counts in reindeer not just to 'acceptable' levels but to zero (Stephens 1995, 285). Stephens' work records how these women's voices led to political action. Chernobyl drove women to new forms of political involvement, bringing cultural issues into the political arena (Stephens 1995, 282). Gerd Persson, a South Sami mother and the wife of a reindeer herder, had never been involved in politics until the autumn of 1986, when she campaigned for and won a position on the Reindeer Radioactivity Commission, an influential advisory organ to the Swedish government (Stephens 1987).

A conscious Sami women's movement started to take shape as early as the 1970s (Helander and Kailo 1998, 183). In 1988, the Sami women's organisation was established. It aimed to raise women's awareness of their current roles, change social conditions, and promote gender equality (Ramnarine 2013, 177). Despite these movements, few Sami women held high management or political positions from the 1970s to the 1990s (ibid.). In 2004, Sara Larsson, a member of the Swedish Saami Parliament, commented, 'Today, reindeer herding has become modernised and male-dominated (Larsson 2004)'.

Stephens' work illustrates how Sami women succeeded in using their voices to enact political change at the end of the 1980s. Regarding ethics of care, Joan Tronto suggests that the image of care – the amorous or maternal face-to-face to which Gilligan remains attached – is too narrow to encompass all social activities that involve attentive care for others (Laugier 2015). Indeed, the needs of Sami women, encapsulated in the idea of health security, are based on the care perspective which claims health security as a universal right.

Conclusion

The Chernobyl nuclear power plant accident and the following government measures related to food safety undermined the food security of Sami reindeer herders. Even in 2016, 30 years after the accident, high radiation levels were still detected in reindeer. Reindeer herders still undergo regular whole-body monitoring, and the Norwegian government still compensates herders for unmarketable reindeer meat. The road to the Sami women's goal of zero becquerels remains long. Hence, food security is an ongoing problem for the Sami people. The food security of Sami reindeer herders encompasses not only food access, but also preventative measures to mitigate the risks surrounding food. As this chapter has demonstrated, situations may arise in which protecting traditional food customs exposes the Sami to further health risks. We have seen that women in Sami reindeer community raised their voices for food security from care perspective. A care approach complements the human security approach and is an effective way to cultivate a more inclusive approach to food security.

Note

1 The term 'Lapp' is not used by the Sami themselves. See a report by Pikkarainen and Brodin in 2008. Here, the author cites the original text.

References

Åhman, B. 1999. Transfer of radiocaesium via reindeer meat to man – effects of countermeasures applied in Sweden following Chernobyl accident. *Journal of Environmental Radioactivity* 46(1): 113–120.

Åhman, B. 2007. Modelling radiocaesium transfer and long-term changes in reindeer. *Journal of Environmental Radioactivity* 98: 153–165.

Andrews, S. *Women in Saami Society*. Retrieved from https://www.laits.utexas.edu/sami/dieda/hist/women.htm.

Antelet samer i Sapmi. Retrieved from http://www.samer.se/samernaisiffror.samer.se/samernaisiffror.

Axelsson, P. and P. Sköld. 2006. Indigenous populations and vulnerability. Characterising vulnerability in a Sami context. *Anales de démographie historique* 1(111): 115–132.

Bacon, P., C. Hobson, and R. Cameron. (eds.) 2014. *Human Security and Natural Disasters*. London: Routledge.

Beresford, N.A., S. Fesenko, A. Konoplev, L. Skuterud, J.T. Smith, and G. Voigt. 2016. Thirty years after the Chernobyl accident: What lessons have we learnt? *Journal of Environmental Radioactivity* 157: 77–89.

Blackwell, M. 2003. Effects of the Chernobyl Disaster on Sámi life. Retrieved from https://www.laits.utexas.edu/sami/dieda/socio/chernobyl.htm.

Blomqvist, L., R. Mustonen, O. Paakkola, and K. Salminen. 1991. Economic and social aspects of the Chernobyl accident in Finland. STUK-B-VALO 64 (AKTU-280), Helsinki: Finnish Centre for Radiation and Nuclear Safety.

Brustad, M., C.L. Parr, M. Melhus, and E. Lund. 2008. Dietary patterns in the population living in the Sami core areas of Norway – the SAMINOR study. *International Journal of Circumpolar Health* 67(1): 84–98.

Dahlgaard, H. (ed.). 1994. Nordic radioecology. The transfer of radionuclides through Nordic ecosystems to man. *Studies in Environmental Science* 62: 9 (Figure 2 is reprinted with permission from Elsevier).

Definition of a Sámi. In Act on the Sámi Parliament, translated by the Ministry of Justice, Finland. Retrieved from https://www.finlex.fi/fi/laki/kaannokset/1995/en19950974.pdf.

Drozdovic, V., A. Bouville, N. Chobanova, V. Filistovic, T. Ilus, M. Kovacic, I. Malatova, M. Moser, T. Nedveckaite, H. Völkle, and E. Cardis. 2007. Radiation exposure to the population of Europe following the Chernobyl accident. *Radiation Protection Dosimetry* 123: 515–528.

Food and Agricultural Organisation. 1996. *Rome declaration on world food security and world food summit plan of action*, World Food Summit, 13–17 November 1996, Rome.

Food and Agricultural Organization. 2016. *Human security & food security*. Retrieved from http://www.fao.org/3/a-i5522e.pdf.%20Accessed%2020%20March%202019.

Gilligan, C. 1993. *In a Different Voice*. Cambridge, MA: Harvard University Press.

Helander, E. and K. Kailo (eds.). 1998. *No beginning, no end: The Sami speak up*, Alberta/Kautokeino: The Canadian Circumpolar Institute/ Nordic Sami Institute.

Hornby, A., S. and J. Turnbull. 2010. *Oxford Advanced Learner's Dictionary 8th edition*. Oxford: Oxford University Press.

Horstkotte, T. and Käyhkö, J. 2017. *Reindeer husbandry under global change in the tundra region of Northern Fennoscandia*, Department of Geography and Geology, University of Turku. Retrieved from https://www.researchgate.net/publication/314239597_Reindeer_husbandry_under_global_change_in_the_tundra_region_of_Northern_Fennoscandia.

Hove, K., H. Lönsjö, I. Andersson, R. Sormunen-Cristian, H.S. Hansen, K. Indridason, H.P. Joensen, V. Kossila, A. Liken, S.M. Magnússon, and S.P Nielsen. 1994. Radiocaesium transfer to grazing sheep in Nordic environments. *Studies in Environmental Science* 62: 211–277.

Institut de Radioprotection et de Sûreté Nucléaire (IRSN). 2011. *Chernobyl 25 years on*.

Irgens, L.M., R.T. Lie, M. Ulstein, T.S. Jensen, R. Skjaerven, F. Sivertsen, J.B. Reitan, F. Strand, T. Strand, and F.E Skjeldestad. 1991. Pregnancy outcome in Norway after Chernobyl. *Biomed Pharmacother* 45(6): 233–241.

Kurttion, P., E. Pukkala, T. Ilus, T. Rahola, and A. Auvinen. 2010. Radiation doses from global fallout and cancer incident among reindeer herders and Sami in Northern Finland. *Occupational and Environmental Medicine* 67: 737–743.

Larsen, J. N. and G. Fondahl (eds.). 2014. *Arctic Human Development Report: Regional Processes and Global Linkages*, Nordic Council of Ministers.

Larsson, S. 2004. Legislating gender equality. *Cultural Survival Quarterly Magazine*, December 2004. Retrieved from https://www.culturalsurvival.org/publications/cultural-survival-quarterly/legislating-gender-equality-sara-larsson.

Laugier,S., P. Molinier, and P. Paperman. 2009. *Qu'est-ce que le care?* Paris: Payot and Rivages.

Laugier, S. 2015. The ethics of care as a politics of the ordinary. *New Literary History* 46(2): 217–240.

Liland, A., J. Lochard, and L. Skuterud. 2009. How long is long-term? Reflections based on over 20 years of post-Chernobyl engagement in Norway. *Journal of Environmental Radioactivity* 100: 581–584.

Magowan, F. and L. Wrazen (eds.). 2013. *Performing Gender, Place, and Emotion in Music*: Global Perspectives. New York: Rochester, Suffolk: Boydell & Brewer.

Marcus, F.R. 1997. Half a century of Nordic nuclear co-operation. Nordic Committee for Nuclear Safety Research.

Mehli, H., L. Skuterud, A. Mosdøl, A. Tønnesen. 2000. The impact of Chernobyl fallout on the southern Saami reindeer herders of Norway in 1996. *Health Physics* 79(6): 682–690.

Mustonen, T. and G. Jones. 2015. Reindeer herding in Finland: A report for Trashumancia y Naturaleza. Retrieved from www.efncp.org/download/Reindeerherdingin Finland.pdf.

Nillson, L. M., A. Winkvist, M. Brustad, J.H. Jansson, I. Johansson, P. Lenner, B. Lindahl, and B. Van Guelpen. 2012. A traditional Sami diet score as a determinant of mortality in a general northern Swedish population. *International Journal of Circumpolar Health* 71: 18537. Retrieved from https://www.researchgate.net/publication/224957342_A_traditional_Sami_diet_score_as_a_determinant_of_mortality_in_a_general_northern_Swedish_population.

Nordic Council of Ministers. 1992. Food safety after nuclear accidents: A Nordic model for national response. *Nord* 33: 1–46.

Pikkarainen, H. and B. Brodin. 2008. *Discrimination of the Sami – the rights of the Sami from a discrimination perspective*. DO report no.1.

Radio Free Europe/Radio Liberty. 2016. *A Norwegian village and its radioactive reindeer*. Retrieved from https://www.rferl.org/a/a-norwegian-village-and-its-radioactive-reindeer/27695926.html.

Ramnarine, T. K. 2013. "In our foremothers' arms": Goddesses, feminism, and the politics of emotion in Sámi songs. In Magowan, F. and L. J. Wrazen (eds.). *Performing gender, place, and emotion in music: global perspectives*. New York: University of Rochester Press. 162–184.

Reindeer Husbandry. Retrieved from https://www.regjeringen.no/en/topics/food-fisheries-and-agriculture/reindeer-husbandry/reindeer-husbandry/id2339774.

Reindeer Husbandry in Sweden. Retrieved from http://www.reindeer-husbandry.uit.no/online/Final_Report/sweden.pdf.

Rekacewicz, P. 2007. *Radiation from Chernobyl*. Retrieved from https://www.grida.no/resources/7418.

Roto, J. 2015a. *Reindeer herding areas and districts in Norway*. Retrieved from http://www.nordregio.org/maps/reindeer-herding-areas-and-districts-in-norway.

Roto, J. 2015b. *Reindeer herding districts in Finland*. Retrieved from http://www.nordregio.org/maps/reindeer-herding-districts-in-finland.

Sami in Sweden. Retrieved from https://sweden.se/society/sami-in-sweden.

Sami Parliament (National Sami Information Centre). 2005. *The Sami – an Indigenous People in Sweden*. Edita Västra Aros. Retrieved from http://www.samer.se/2137

Sen, A. and S. Ogata. 2003. *Human Security Now*. Commission on Human Security. Retrieved from https://reliefweb.int/sites/reliefweb.int/files/resources/91BAEEDBA50C6907C12556D19006A9353-chs-security-may03.pdf.

Skuterud, L. and H. Thørring. 2012. Averted doses to Norwegian Sámi reindeer herders after the Chernobyl accident. *Health Physics Society* 102(2): 208–216.

Snihs, J.O. 1996. *Contamination and Radiation Exposure. Evaluation and measures in the Nordic countries after the Chernobyl accident*. Stockholm: Swedish Radiation Protection Institute.

Statens strålskyddsinstitut 1986. Chernobyl – its impact on Sweden, National Institute of Radiation Protection report, Vol. 12. Retrieved from https://www.stralsakerhetsmyndigheten.se/contentassets/66f4f029351d4baea989543f24414795/198612-chernobyl---its-impact-on-sweden

Steinlien, O. 1989. The Sami law: A change of Norwegian government policy toward the Sami minority. *Canadian Journal of Native Study* 9(1): 1–14.

Stephens, S. 1987. Chernobyl fallout: A hard rain for the Sami. *Cultural Survival Quarterly Magazin*, June. Retrieved from https://www.culturalsurvival.org/publications/cultural-survival-quarterly/chernobyl-fallout-hard-rain-sami.

Stephens, S. 1995. Physical and cultural reproduction in a post-Chernobyl Norwegian Sami community in Ginsburg. In Ginsburg, Rapp, Reiter, and Rapp (eds.). *Conceiving the New World Order: The Global Politics of Reproduction*. Berkeley: University of California Press, pp. 270–290.

Strand, P. 1994. Radioactive fallout in Norway from the Chernobyl accident. *Norwegian Radiation Protection Authority Report*, Norwegian Radiation Protection Authority, Vol. 2.

Strand, P., L.I. Brynildsen, O. Harbitz, and U. Tveten. 1990. Measures introduced in Norway after the Chernobyl accident. A cost benefit analysis. *International Symposium on Environment Contamination Following a Major Nuclear Accident* (IAEA/SM/306/36).

Strand, P., T.D. Senlaes, E. Boe, O. Harbitz, and A. Andersson-Sorlie. 1992a. Chernobyl fallout: International doses to the Norwegian population and the effect of dietary advice. *Health Physics Society*. 63(4): 385–392.

Strand, P., T.D. Selnaes, and J.B. Reitan. 1992b. Area and time distribution of external and internal doses from Chernobyl fallout: The lack of correlation in Norway. *Health Physics Society*. 62(6): 512–518.

Szpak, A. 2017. Human security of the indigenous peoples in the Arctic. The Sami case. *International Studies Interdisciplinary Political and Cultural Journal* 20(1): 75–96.

Tanoi, K., S. Miura, E. Forssell-Aronsson, and C. Bradshaw, eds. 2016. Sweden–Japan Radioecology Workshop for Students. The workshop was located in Tokyo and hosted by University of Tokyo. Retrieved from https://www.a.u-tokyo.ac.jp/rpjt/20160519.pdf.

The Sámi Act. Retrieved from https://www.regjeringen.no/en/dokumenter/the-sami-act-/id449701.

The Sami in figures. Retrieved from http://www.samer.se/4529.

The Sámi in Finland. Retrieved from https://www.samediggi.fi/sami-info/?lang=en.

The Sami Parliament. Retrieved from https://www.sametinget.se/english.

Tronto, J. 1993. *Moral Boundaries. A Political Argument for an Ethic of Care*. London/New York: Routledge.

United Nations. 1975. *Report of the World Food Conference*, Rome, 5–16 November 1974. New York.

United Nations Declaration on the Rights of Indigenous Peoples (A/RES/61/295), Article 24. Retrieved from https://undocs.org/A/RES/61/295.

United Nations Development Programme. 1994. *Human Development Report*. New York/Oxford: Oxford University Press.

United Nations Educational, Scientific, and Cultural Organisation. 1996. *Ten years after the Chernobyl Catastrophe*, 121. Retrieved from http://unesdoc.unesco.org/images/0012/001237/123781eo.pdf.

Zojer, G. and K. Hossain. 2017. Rethinking multifaceted human security threats in the Barents region: A multilevel approach to societal security. *Juridica Lapponica* 42: 1–87.

4 Traditional nutrition of Indigenous Peoples in the Arctic zone of Western Siberia

Challenges and impact on food security and health

Elena Bogdanova, Andrey Lobanov, Sergei Andronov, Andrei Popov, Ruslan Kochkin and Ildikó Asztalos Morell

Acronyms

FAO: Food and Agriculture Organisation of the United Nations; RAIPON: Russian Association of Indigenous Peoples of the North; YNAO: Yamal-Nenets Autonomous Okrug.

Introduction

Food security is an important component of national economic security. The Food and Agriculture Organisation of the United Nations (FAO) and the Doctrine of Food Security of the Russian Federation (2010) use this concept to describe the state of the national or global food market. Food security is 'the conditions under which people have physical, social, and economic access to adequate, safe, and complete nutrition that satisfies the needs and preferences in their diet and nutrition for actively healthy living' (Doctrine of Food Security of the Russian Federation, 2010). Its constituent elements are food independence, physical, and economic accessibility of food, and the safety of available food (Rome Declaration on World Food Security and the World Food Summit Action Plan, 1996; Declaration of the World Summit on Food Security, 2009; Doctrine of Food Security of the Russian Federation, 2010).

Policy documents prioritise food security challenges in the Russian Federation, and in some regions (especially in the Arctic region) these challenges are urgent. Professor V.A. Ivanov states that, due to outflow of products the population of the Arctic in Russia is deprived from sufficient access to the products of the region. Thus, Russia does not ensure food security (Ivanov, 2015, pp. 152–174). This problem can require emergency measures at the local and federal levels.

In the Russian Federation, food security should be diversified at the regional level. This is especially relevant in the Arctic zone of the Russian Federation, where quality of life differs significantly between settled populations and the nomadic Indigenous Peoples who live in the tundra. This

DOI: 10.4324/9781003057758-4

aligns with the FAO Policy on Indigenous and Tribal Peoples (Rome, 2015) (FAO, 2015), which states that, although Indigenous Peoples make up about 5% of the world's total population (and 15% of the total population of the Russian Arctic), about 15% of them live below the poverty line (Rosstat). The FAO Policy declares:

> The adversities faced by Indigenous Peoples have grown in the last few decades, but so too have the recognition of and appreciation for their potential contributions to sustainable development and natural resources management. Protecting the livelihood systems and specialised knowledge that are held within these communities will reverse the steady erosion of indigenous cultures but may also bring novel solutions to the fight against food insecurity and malnutrition, poverty, and environmental degradation.
>
> (FAO, 2015, p. 2)

In 2018, the FAO prepared a regional overview of the state of food security problems in Europe and Central Asia in 2017 (The State of Food Security and Nutrition in Europe and Central Asia, 2018). However, the FAO's findings for Russia are based primarily on statistical data collected from the regions of the Russian Federation. They therefore fail to reflect the real differences in food security between urban and rural areas and between the permanent populations in municipalities and Indigenous Peoples following a traditional lifestyle. There is insufficient data on the quality of life and food security of Indigenous Peoples in the Russian Arctic (SLiCA, 2015; Anderson et al., 2016, pp. 1–27; Kozlov et al., 2018).

However, our research shows that development of Indigenous communities in this region correlates to global factors affecting food security identified by the FAO and the World Health Organisation. Rapidly growing fuel prices have a dual effect, increasing the cost of production and of sales. The growing demand for food in emerging economies and the impact of climate change on agriculture, fishing, hunting, and gathering also affect food security in the Russian Arctic. This is supported by the document *Food Security in the Arctic Region* developed by the Inuit Circumpolar Council (Canada) (May 2012), which identifies two main problems concerning food insecurity of Indigenous Peoples living in the Arctic region: (1) high prices, often combined with economic vulnerability, and (2) reduced consumption of products made from the Arctic flora and fauna (Inuit Circumpolar Council, 2012). The situation is further complicated by global climate change.

Food sovereignty movements challenge the issue of food security (the right to adequate amounts of quality food) and the social relationships of production and consumption. Food sovereignty critically assesses the ability of people, communities, and nations to control their own food systems (markets, modes of production, food cultures, and environments) and to provide an alternative to the neoliberal model of agriculture (Desmarais, 2002, pp. 91–124; Wittmanm, 2009, pp. 120–130). This neoliberal model is associated with the expansion of the 'third food regime', which refers to the withdrawal of the state and the expansion of market forces via integration with the global food market and the lifting of regulations on the free trade of food, along with the privatisation of food security and the domination of the corporate food regime (Friedmann et al., 1989, pp. 93–117). The third food regime brought about the increasing displacement of local family producers (Van der Ploeg, 2013). Reindeer herder economies integrated into market economies by selling their products to local and external markets; this caused them to become more dependent on complementary sources of food in the form of products produced and distributed to local vendors by external large-scale market ones. Rising fuel costs are one reason for the increasing need for cash incomes.

However, the claims of food sovereignty movements do not completely correspond to Indigenous issues. Lauren Kepkiewicz, Dawn Morrison, Annette Desmarais, and Hannah Wittman point out that food sovereignty in Canada remains disproportionately focused on agricultural production and state solutions while failing to adequately engage with Indigenous food systems based on hunting, fishing, and gathering (Morrison, 2011, pp. 97–114; Desmarais et al., 2014, pp. 1153–1173; Kepkiewicz, 2017, pp. 164–180). Food sovereignty issues for Indigenous People are intertwined with the colonial roots of the regulations around land use by Indigenous People and colonial settlers.

Moreover, food sovereignty for Indigenous Peoples also concerns specifics of their lifestyle. Due to the transformation of the traditional lifestyle of Indigenous Peoples (replacing nomadic lifestyles with settled ones), those who are still involved in traditional occupations (reindeer herding, fishing, hunting, gathering, etc.) have a strong impact on the food sovereignty of Indigenous and non-Indigenous communities since they provide traditional food. According to Lauren Kepkiewicz, not only do indigenous food systems form the basis of Indigenous Peoples' ability to sustain and nourish themselves, but indigenous food systems form the basis of all people's food systems (Kepkiewicz, 2015, pp. 185–198). Dawn Morrison points to Indigenous food sovereignty as a restorative framework for the transformation of food systems to which all people and cultures – Indigenous and non-Indigenous – can relate (B.C. Indigenous Food System Network, 2008, pp. 1–22).

This is relevant to the present case because the Indigenous People who live in national settlements in Western Siberia still depend to a significant degree on traditional sources of nutrition. So, the issues of Indigenous Peoples' food security and food sovereignty are both global and specific, and these issues align with many geographical, cultural, and socio-economic specifics.

Since the concern for food sovereignty emerged from the social movements of producers, it is worthwhile to explore how food sovereignty emerges in the relationships among the state, the market, and producers. P. Clark juxtaposes ideas of food sovereignty that promote a '"*Westphalian" state-centric notion* [with] those implying a more *pluralistic notion of sovereignty* over territories by non-state actors (Clark, 2016, pp. 183–205).

As this chapter will show, in Russia, the state takes an active role by regulating reindeer herd sizes and access to grazing to protect traditional lifestyles. Reindeer herders have organised to 'fight' for more influence over national and international policies. Meanwhile, reindeer herding can be also seen as a continuation of the 'silent resistance' of the 'people's economies' (Visser et al., 2015, pp. 513–528) of the state socialist period. Reindeer herders retained an Indigenous, autonomous, household-based food system parallel to the dominant state-run, socialist, planned economy. Following the postsocialist transition, reindeer herding has continued to have a strong standing; it embraces an Indigenous food system even during its emergent market integration.

This chapter describes current trends related to the food security and food sovereignty of the Indigenous population in the Arctic zone of Western Siberia (Yamal-Nenets Autonomous Okrug – YNAO) (Figure 4.1).

This region is the part of the High North, and more than half of its territory is located above the Arctic Circle. There are three climatic zones: (1) the Arctic, (2) subarctic zones, and (3) the northern part of the Western Siberian lowlands. This chapter considers only the Arctic region. Its climate is determined by the permafrost. It is characterised by long, cold, severe winters (up to eight months) with strong magnetic storms, frosts, frequent snowstorms, and a short summer (50 days). The average annual temperature is about −10°C. In winter, low temperatures can sink to −70°C; in summer (in July), the temperature can rise to +30°C. This severe climate makes industrial agriculture impossible, but the sustained cold temperatures provide the necessary conditions for reindeer herding.

The Indigenous population's food security and sovereignty are based on maintaining traditional nutrition, which is an important part of the nomadic Indigenous population's traditional culture and the Indigenous economies in the Arctic zone of Western Siberia. Traditional nutrition effectively prevents health problems and allowed Indigenous Peoples of this region to adapt to the cold and geomagnetic activity of the High North. 'Traditional nutrition' is a diet that contains more than 50% local, traditional products (i.e., venison, local fish, etc.) (Lobanov et al., Traditional nutrition…, 2018, pp. 32–33). Dramatic transitions away from Indigenous nutrition have significantly affected the prevalence of respiratory diseases, circulation problems, and obesity, and reductions in the consumption of traditional foods lead to a decrease in adaptation (Andronov et al., The impact…, 2018, PA796; Andronov et al., The role…, 2018, pp. 142–144; Bogdanova et al., 2018, pp. 120–126; Lobanov et al., Study…, 2018, pp. 31–32).

Food security issues differ for different Indigenous populations (fishermen, hunters, or nomadic reindeer herders). Food security concerns access to natural resources as well as obtaining hunting licenses and fishing quotas for Indigenous communities. This study investigates whether national policies and the current trends in reindeer herding in the Arctic zone of Western Siberia reflect issues of food security and sovereignty. It considers four main aspects of these topics: policies, environmental issues, economic and health impacts.

Figure 4.1 The territory of the Yamalo-Nenets Autonomous Okrug.

The chapter takes a multidisciplinary approach that draws on methods used in the fields of law, sociology, economics, and medicine. Research was conducted using the actor–network theory, a sociological method. A comprehensive analysis of socioeconomic factors affecting reindeer herding maintenance was also conducted. The primary sources used in this study include data collected from interviews and surveys of 590 Indigenous Peoples (reindeer herders) during expeditions to the Nyda settlement, the Nydinskaya tundra, the Tazovsky settlement, the Tazovskaya tundra, the Nakhodka tundra, the Gyda and Gydansky settlements, the Yavai-Salinskaya tundra, the Seyakha settlement, the Seyakhinskaya and Tambeyskaya tundras (located along the southern coast of the Ob Bay), the northeast coast of the Yamal Peninsula, and the Tazovsky and Gydansky Peninsulas. Data were collected in summer (August) and winter (March and November) in 2014–2017 (Figure 4.2). Fieldwork was conducted by the researchers of the YNAO (Yamal-Nenets Autonomous Okrug) Arctic Scientific Research Centre.

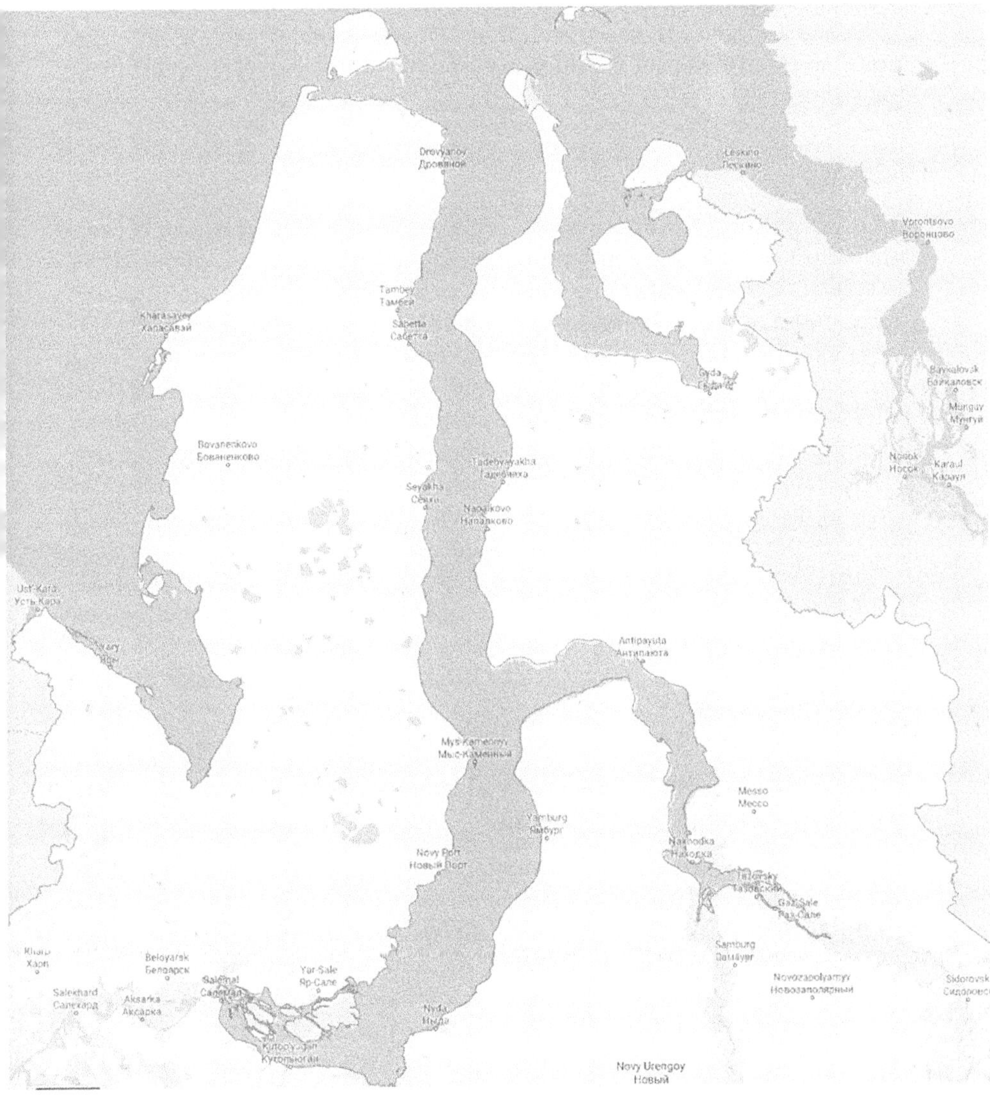

Figure 4.2 The territory of the Arctic climatic zone of the Yamalo-Nenets Autonomous Okrug.

Two hundred and fifty-two semi-structured interviews were conducted based on the interview guide developed and approved by the YNAO Arctic Scientific Research Centre. Two hundred and thirty-eight surveys with fixed questions were received. The results were analysed using STATISTICA 6.

The selection criteria for the respondents were involved in reindeer herding, Nenets origin (at least one parent was of Nenets origin or a participant was a Nenets speaker), nomadic or semi-nomadic lifestyle, and participation in a medical examination conducted by the YNAO Arctic Scientific Research Centre. All participants over 18 years of age provided written informed consent. Participants' personal data were anonymised, numbered, and added to the databases of the Northern Arctic Federal University project team.

Secondary sources used in the study consisted of official information requested from local authorities, public statistical data, and official government reports.

Settings

The Arctic zone of Western Siberia (YNAO) is a unique territory because almost 1/3 of the Indigenous population of the Russian Arctic (about 42,000 people) reside there, including Nenets, Khanty, Selkups, and Komi-Zyryans. Almost half of the residents are still nomadic (about 22,000 people). The culture, health, and social well-being of Indigenous Peoples are strongly linked to a traditional lifestyle and the consumption of traditional food. Therefore, it is essential to encourage traditional occupations (reindeer herding, fishing, hunting, etc.) in this region in order to preserve the Indigenous population.

The geographic location of YNAO, which is mostly above the Arctic Circle, significantly impacts traditional occupations in this region; its conditions are the most appropriate for reindeer herding.

Current trends in the traditional economy of reindeer herders in YNAO are fundamentally different from economic trends in other Arctic regions. Worldwide, the number of people following a nomadic lifestyle is decreasing, but in YNAO, it is increasing, as is the number of young men engaged in reindeer herding.

Policies affecting the food security of Indigenous Peoples in the Arctic zone of Western Siberia

The first legislative initiatives to support reindeer herding were in the 1920s and 1930s. During WWII, the Soviet government adopted new legislation aimed at preserving and developing the reindeer industry. However, only since the 1990s it has national policy focused on maintaining the Indigenous population and their traditional lifestyle. Efforts in this direction have led to the formation of associations of Indigenous Peoples and a set of federal and local (municipal level) acts, including *On the Territorial Public Self-Government in Yamal-Nenets Autonomous Okrug* (Law No. 44, 9 December 1996), *On Reindeer Herding* (Law No. 46, 2 November 1998;

Law No. 34-ZAO, 6 June 2016, ed. 2 October 2018), and *On the General Principle of Organising Communities of Small Indigenous Peoples of the North, Siberia, and the Far East of the Russian Federation* (Federal Law No. 104-FL, 20 July 2000). However, issues related to reindeer husbandry are mostly regulated on the local level because the draft federal laws, such as *On Northern Reindeer Herding* (No. 97700749-2, 1997 and No. 220824-3, 2002), were rejected by the State Duma of the Russian Federation.

The regional YNAO law *On Reindeer Herding* (Law No. 34-ZAO, 6 June 2016, ed. 02 October 2018) establishes two types of reindeer husbandry: individual reindeer herders and reindeer herding households. The first is a collaboration of Indigenous or non-Indigenous Peoples of the North, or other persons who are not members of the Indigenous Peoples of the North but are permanent residents of the Okrug and follow the traditional lifestyle of Indigenous Peoples of the North. They engage in traditional economic activities and traditional occupations in the areas of traditional residence and traditional economic activities of the Indigenous Peoples of the North (On Reindeer Herding, 2016, article 3, par. 5). This definition partly correlates to 'a peasant (farm) household' (Civil Code of the Russian Federation, 1994, article 86.1), which describes 'citizens conducting joint activities in agriculture without establishing a legal entity on the basis of an agreement for running a peasant (farm) household or entitled to create a legal entity – a peasant farm' (Civil Code of the Russian Federation, 1994, 23). In contrast, reindeer herding husbandry refers to individual entrepreneurs and legal entities that herd reindeer as a business (On Reindeer Herding, 2016, article 3, par. 6). However, none of these legal definitions provides quantitative measurements to determine the status of a reindeer herder.

In the late 1980s and 1990s, the main stakeholders protecting the rights of Indigenous Peoples appeared. On 12 December 1989, the organisation '*YASAVEY*' was founded. In 1998, it was renamed *the Association of the Nenets People of Nenets Autonomous Okrug 'YASAVEY'*. This association joined *the Russian Association of Indigenous Peoples of the North 'RAIPON'*, which was founded in March 1990 at the Congress of the Peoples of the North. At first, the Association '*RAIPON*' was called the Association of the Peoples of the North of the USSR and united 26 indigenous minorities from the North. On 24 November 1993, it was registered as a socio-political movement – the Association of Indigenous Minorities of the North, Siberia, and the Far East of the Russian Federation. This association lobbies for the interests of Indigenous Peoples in the Russian Federation. It works in collaboration with the chambers of the Federal Assembly of the Russian Federation, the government, and the President of the Russian Federation, and it participates in work on legislation designed to ensure the rights of Indigenous Peoples and to help them to maintain their lifestyle and develop their economy. It has a special consultative status with the UN Economic and Social Council (ECOSOC), and is an active participant in the sessions of UN working groups on Indigenous issues. It helped to develop the draft Declaration on the Rights of Indigenous Peoples and also participates in the Commission on Human Rights and the UN Permanent Forum on Indigenous Issues. '*RAIPON*' is a permanent member of the Arctic Council, which was established by eight Arctic countries: Norway, Denmark, Sweden, Finland, Iceland, Canada, the US, and Russia. Its experts take part in various Arctic Council working groups and programs. These activities make '*RAIPON*' the strongest stakeholder involved in protecting the civil rights of Indigenous Peoples in Russia. It has had a significant impact on governmental programs and legislative initiatives, including:

- *The Conception of Sustainable Development of Indigenous Minorities of the North, Siberia, and the Far East of the Russian Federation*, approved by governmental decree of the Russian Federation on 4 February 2009 (No. 132-P);
- *The Conception of Sustainable Development of Indigenous Peoples of the North in Yamal-Nenets Autonomous Okrug*, approved by a resolution of the YNAO Legislative Assembly on 9 December 2009 (No. 1996);
- *The Strategy of Socio-Economic Development of YNAO until 2020*, approved by a decision of the YNAO Legislative Assembly on 14 December 2011 (No. 839);
- *The National Program of the Indigenous Peoples of the North in YNAO*, approved by a decree of the governor of YNAO on 28 December 2017 (No. 132-PG);
- The YNAO state program *Development of the Agricultural Complex, Fisheries, and Regulation of Markets for Agricultural Products, Raw Materials, and Food in 2014–2021*, approved by YNAO governmental decree on 26 November 2013 (No. 964-P, ed. 16 August 2018);
- The complex program *Sustainable Development of Indigenous Peoples of the North in YNAO in 2018–2020*, approved by YNAO governmental decree on 12 December 2018 (No. 1271-P).

However, local policies governing reindeer herding and husbandry still have many gaps that affect the efficiency of the Indigenous economy. According to Konstantin Filant, a researcher at the YNAO Arctic Scientific Research Centre, the legislation on reindeer herding in Russia is insufficient. The laws are poor quality because they are formulated using terms from ethnology and history rather than clear legal language (Filant, 2017, pp. 31–47). Furthermore, the legislation is neither based on the existing norms of land and civil law, nor does it aim to protect citizens. In addition, its focus on Indigenous People's rights overlooks the need to consider environmental issues as well. Therefore, the number of reindeer permitted per herd under current legislation is not based on pasture capacity.

The next section of this chapter considers some important issues about the relevance of these regulations. The issue of decreasing pasture capacity in tundra areas is urgent, and the impact of existing models of reindeer herding on the environment must be addressed. There is a conflict between reindeer herders' decreasing standards of living (and the herders' efforts to survive and increase their profits by increasing the size of their herds) and the growing degradation of reindeer pastures in the tundra (which affects reindeer's health). Another issue is herders' growing need to get involved in market activities because they have little or no access to the necessary facilities for reindeer slaughter and the storage of reindeer products.

The interaction between grazing capacity and models of reindeer herding

Reindeer husbandry in YNAO today is an example of the successful integration of traditional occupations into the modern economy. It is the main traditional occupation and business of the nomadic Indigenous Peoples of the Tazovskaya, Messoyakhinskaya, Antipayutinskaya, and Tanamskaya tundras (Figure 4.2), most of whom work in commodity production of reindeer meat and antlers. The reindeer herders of the Tazovsky district differ widely in methods of cooperation and economic interests. They can be divided into three groups by the number of reindeer in a herd: small-scale herders (fewer than 100 reindeer per herd), middle-sized herders (100–999 reindeer per herd), and large-scale herders (over 1,000 reindeer per herd) (see Table 4.1).

Many communities in the tundra today are shifting from a subsistence economy (in which the primary goal is providing a family with food) into commodity production (in which the goal is to earn income from sales of a product). Large-scale reindeer herding is well-suited to this new model of Indigenous business. For a long time, reindeer herding involved temporary collaborations of relatives based on informal agreements. Only in 2019, the first reindeer farm was registered in the Seyahinskaya tundra in YNAO. This official legal status allows herders to apply for subsidy programs and other government support for breeding reindeer and processing meat.

Table 4.1 The structure of reindeer herding households in the Tazovskiy region in 2018

Model of reindeer herding	*Transport reindeer*	*Reindeer for food*	*Other reindeer*
LARGE-SCALE			
Mean number of reindeer per herd	2060	40	4900
Share in a herd, %	29.42	0.57	70.01
MIDDLE-SCALE			
Mean number of reindeer per herd	55	18	232
Share in a herd, %	18.03	5.09	76.88
SMALL-SCALE			
Mean number of reindeer per herd	30	10	110
Share in a herd, %	20	6.6	73.4

Unfortunately, ecologists conclude that further extensive growth could cause an environmental catastrophe due to overgrazing (Kryazhimskiy et al., 2011, pp. 323–333; Yuzhakov, 2017, pp. 131–137). This would, of course, endanger the population's primary food source (Morozova and Magomedova, 2006, pp. 235–249). However, local governments still pay little attention to pastures' capacity to support reindeer. Most reindeer herders in the Tazovskaya, Tanamskaya, Gydanskaya, and Yuribeyskaya tundras are complaining about the reduction in quality of winter reindeer pastures. The residents of the Esseyahinskaya and Yavai-Salinskaya tundras (Figure 4.2) are not yet experiencing these problems.

Expeditions organised by the YNAO Arctic Scientific Research Centre have found the most significant damage to pastures in the Antipayutinskaya tundra near 'waiting' areas (temporary camps) close to the settlements where reindeer herders make purchases (i.e., in Yuribey, Tanama, Antipayuta, and Gyda); in fishing areas on the gulf of the Ob River (north of Antiputa, on the western shore of the Yambuto lake, and on the shore of the Gydan Bay near Yuribey); and near oil and gas deposits where nomadic people sell fish and venison and buy gasoline (mostly at the deposits of the Messoyakhinskaya group). Overgrazing is accompanied by soil degradation around populated areas near infrastructure facilities, and the consequence of overgrazing is a lack of nutrition for reindeer, which leads to deteriorating health (Figure 4.3).

However, more than 80% of the surveyed reindeer herders do not consider the situation to be critical. They think that 'a reindeer finds its own food; the main thing is not to interfere with them', and 'the tundra will regulate itself'. They are convinced that reindeer are adapting to changes in their food supply.

Figure 4.3 The damaged teeth of a three-year-old reindeer as a result of eating low lichen cover in the areas of overgrazing in the Messoyahinskaya tundra, the Lucemer-Yakh camp, 2017.

In recent years, climate change (which has caused hot summers and a long snowless season) has led to an increase in the biomass of herbaceous plants. In addition, the thawing of permafrost has resulted in the formation of small ravines, improving drainage in the territory, and creating better conditions for plant growth. These climatic changes have led to an increase in grassy plants in reindeer's winter diets (Morozova and Magomedova, 2006, pp. 235–249; Kryazhimskiy et al., 2011, pp. 323–333; Yuzhakov, 2017, pp. 131–137). Less than 20% of the surveyed reindeer herders were worried about climate change and said that they had moved further north with their herds because increasing temperatures were uncomfortable and harmful for reindeer.

Reindeer herders with herds of up to 300 reindeer believe that large-scale reindeer herding is the main problem. After a large herd of 2,000–3,000 reindeer grazes, the tundra needs a long time to recover. This aligns with the 'tragedy of the commons' proposed by Garrett Hardin (Hardin, 1968, pp. 1243–1248). Nobody wants to lose profit, and common natural resources are easily exhausted. This results in pasture degradation, which could be tragic for the tundra and reindeer herding.

The owners of large herds, meanwhile, believe that small-scale herders cause more damage since they follow shorter migratory paths and often stay at settlements and trading stations for long periods or even cease to migrate altogether. However, reindeer herders strongly oppose governmental regulations limiting the number of reindeer in a herd, and there are still no legal, economic, or social precedents for herd reduction under discussion by regional authorities.

The solution to overgrazing issues can be found in regulating nomadic reindeer routes based on the old Soviet system. Under this system, all reindeer herders were required to move from winter pastures to summer ones by crossing the rivers that flow east–west near Antipayuta, Tanama, and Gyda settlements. Younger reindeer herders find this model fair, effective, and easy to understand. However, there are some doubts whether it would work in the Tanamskaya and Messoyakhinskaya tundras. Many reindeer herders in these districts do not go far from the oil and gas deposits.

Marketing capacities of reindeer herders

Changes in export trends also impact traditional food consumption. Local venison and fish are necessary to maintain the health of reindeer herding families and also of Indigenous Peoples who live in settlements or towns. Until recently, a substantial portion of venison was sold in the nearest settlements or urban areas, but this has changed dramatically in recent years. In the national villages that do not have their own facilities for the slaughter and storage of reindeer products, venison consumption decreased by 20%–48% over the last five years (2014–2018). Such facilities only work with large exporters and are mostly located in the towns. In Canada, Indigenous communities have faced similar problems: a lack of freezers to store food and insufficient access to markets with cheap, high-quality food have become barriers to food security (Chan et al., 2006, pp. 416–431). However, in the Arctic zone of Western Siberia, developing the infrastructure to store the production of Indigenous businessmen (local venison and fish) in the national settlements could shift the current export trends, increasing the availability of traditional food to residents of this territory.

Strict government limitations on informal trade ('trade from sledges', the typical method for selling to local communities), which are considered illegal business activities, also discourage reindeer herders and fishermen from selling their products in national settlements. Instead, they sell in bulk to export companies. At the same time, some reindeer herders focus on the export of antlers and pants, treating meat as an accompanying product. This is due to the lack of storage facilities and to the higher profits on these items, as well as the lack of delay in payment for them. Meanwhile, payment for venison can be delayed for three to six months. Yet another reason for reindeer herders' focus on antler and pants sales is the difficulty in receiving veterinary approval for the sale of venison because of the lack of veterinary facilities near slaughter sites. Companies that export antlers and pants provide reindeer herders with food, fuel, transport, and spare parts for vehicles and other devices on credit for the following year's 'harvest'. All of this makes the sale of antlers and pants more attractive. The issue is complicated still more by the distance of nomadic routes from settlements and the lack of fuel, petrol stations, and other infrastructure along the routes. These factors combined make selling venison inside the Okrug unprofitable.

Currently, residents in settlements in the Arctic zone of Western Siberia consume venison only five to six months of the year. Indigenous residents' insufficient consumption of venison can severely impact their bodies' ability to handle the harsh Arctic conditions, resulting in an increase in circulatory, respiratory, and metabolic disorders (Lobanov et al., Consumption…, 2018, pp. 190–194; Lobanov et al., Risk factors…, 2018, PA799; Lobanov et al., The impact…, 2018, PA796, Lobanov et al., The role…, 2018, pp. 156–157). Reindeer herders prefer to sell venison to shift workers near oil and gas deposits.

There are several reasons for this restricted access to venison. During summer calving, tundra herders deliver almost no meat to relatives living in the settlements. However, the lack of venison is strongly related to recent changes in reindeer herding. Modern reindeer herding is rapidly ceasing to be a subsistence activity and becoming a form of commodity production. According to our research in the Gydanskaya tundra (2014–2017, n = 149), reindeer herders' families use several devices that consume electricity: TVs (58%), satellite phones (26%), and computers (36%). A generator consumes about three litres of gasoline per day. Many families also have two or three snowmobiles and a motorboat. Therefore, a typical reindeer herder's family in the Tazovsky district uses four to eight barrels of fuel per month. To buy fuel, they need money, which can only be obtained by running a commodity production business. This motivates reindeer herders to stay closer to settlements, factories, roads, and fields where they have access to fuel, imported products, medical care, schools, and the Internet. So, they provide settlements with venison for reasonable prices. They also sell other products (antlers and pants) near settlements because export companies offer higher prices when reindeer herders sell near roads or settlements. As a result, over the last 16 years (1992–2017), in the Gydanskaya, Yavai-Salinskaya, Tanamskaya, and Essiahinskaya tundras of the Gydansky Peninsula, the average length of reindeer herders' routes has reduced by almost 50%: in 1992, the average route was 380 km; in 2017 – 180 km.

However, the proximity of temporary nomadic camps to national settlements has also strengthened the transformation of the traditional lifestyle. The nomadic population has access to imported products which partly replace traditional food. However, these imported products are usually low quality and very expensive. Their consumption also contributes to the increased risk of chronic obstructive bronchitis, obesity, and has decreased Indigenous Peoples' level of adaptation to the harsh conditions of the Arctic (Andronov et al., Prediction…, 2018, pp. 54–67; Kochkin et al., 2018, pp. 130–131).

Implications of ecological and marketing concerns about reindeer herding

Regional and federal strategies concerning reindeer herding must consider the social, economic, and ecological aspects of sustainability.

- The overriding social dilemma is whether to maintain the traditional lifestyle and diet or to find new ways to replace the nutritional value of traditional foods for the subsistence of the Arctic population. It is also important to ensure the participation of local communities in issues concerning their development.
- The overriding economic dilemma is whether to preserve traditional farming that focuses on venison production or to maximise the profits of the reindeer husbandry and integrate reindeer husbandry into the world economy by developing the antlers market.
- The major ecological dilemma is whether to foster the transition to modern technologies for feeding reindeer and the development of nomadic reindeer herding or to limit the number of reindeer in a herd.
- Another dilemma is caused by the conflict between protecting the tundra from overgrazing and increasing the welfare of herders.

At the same time, some tactical issues must be addressed:

- There is a lack of legitimate mechanisms for regulating the number of reindeer and their routes. Since there are no legal, economic, or social precedents for herd reduction, the most likely outcome is the natural regulation of the number of reindeer due to adverse weather conditions or a significant economic change that causes antler reindeer herding to become unprofitable. Therefore, it is necessary not only to develop a strategy for regulating the number of reindeer if there are too many but also to prepare for conditions leading to a sharp decline in the number of reindeer. Decreasing the number of reindeer could cause some reindeer herders to go bankrupt, forcing them to exchange this traditional occupation for fishing. This could then lead to overfishing. However, reducing the consumption of traditional food (fish and venison) would lead to an increase in circulatory and respiratory diseases among the Arctic population.

- The problem of overgrazing is extremely politicised. Geobotanical studies have mostly been conducted in areas that are not grazed by reindeer. Therefore, overgrazed areas have not been studied. However, the results of these studies have been extrapolated to large areas with very different landscapes and climatic conditions. Furthermore, all studies were conducted in summer. Since weather conditions and vegetation vary significantly during the snowless period of the year in YNAO, it is quite difficult to generalise these results or to use them to develop forecast models. Overgrazing should be monitored using satellite images, and reindeer migration routes should be studied by attaching GPS trackers to reindeer.
- The reindeer counting mechanism currently in use is ineffective. Since official data on the number of reindeer and on pasture resources are extremely inaccurate, they cannot be used to develop models. There is no precise map of reindeer routes, nor is there reliable information about the owners of most of the reindeer in the Tazovsky district.
- There are no petroleum stations or gasoline storage facilities in the tundra, which leads to extremely high fuel costs and significantly affects the welfare of the people living there. When petrol prices rise or there is a shortage, frequent trips to the settlements become impossible for most nomadic families. This in turn limits the availability of medicine, industrial goods, and imported food. It also limits communication with children studying in settlements and reduces the supply of traditional food to relatives in the settlements. In addition, high fuel prices exacerbate overgrazing, as tundra residents tend remain close to settlements or oil fields so they can buy fuel at lower prices, reducing the cost of delivering goods to tundra.

Special measures are needed to address these issues and to maintain traditional reindeer herding. Reindeer herding should be included in the YNAO state program *Development of the Agricultural Complex, Fisheries and Regulation of Markets for Agricultural Products, Raw Materials and Food in 2014–2021*, which was approved by YNAO governmental decree on 26 November 2013 (No. 964-P, ed. on 16 August 2018). Subsidy programs (included in the budget of this decree) can support reindeer herders and motivate them to comply with the restrictions and recommendations of the authorities. However, we recommend that the criteria for determining the size of a herd be re-evaluated. The most important factors should be: (1) pasture reindeer capacity; (2) the minimum number of reindeers per herd for reindeer herding (as a criterion for maintaining a nomadic or settled lifestyle); and (3) the subsistence minimum (as a criterion for the minimum size of a one family's reindeer herd). Another criterion should be considered when assigning subsidies – the volume of delivered or sold meat. Therefore, government support should be based not only on the number of reindeer per member of the Indigenous family but on the type of reindeer herding, pasture reindeer capacity, the standard of living in the region (subsistence minimum), the average market price of venison, and the demographic characteristics of a family.

Conclusion

In the Russian Arctic, the issues of food security and food sovereignty pose a challenge that threatens the maintenance of Indigenous Peoples' health and traditional lifestyles. Short- and long-term measures are needed. These measures should support the survival of the traditional lifestyle and strengthen the economic position of reindeer herders, who are experiencing increasing market exposure and a need for more cash.

Urgent short-term measures

- Modify communication between local authorities and reindeer herders. Instead of top-down regulations, authorities should collaborate with local communities and reindeer herders to maintain traditional lifestyle, improve the welfare of Indigenous Peoples, and develop reindeer herding.
- Develop systems to monitor the locations of nomadic reindeer herders using satellite phones (which can also enable them to access help in emergency situations). The Ministry of Emergency Situations, regional authorities, veterinary services, Federal Service for Supervision of Consumer Rights Protection and Human Welfare, and hospitals should be able to access the locations of reindeer herders if needed.
- Organise veterinary facilities and venison storage facilities in the national settlements. Indigenous Peoples are drawn into market relationships, but they lack the facilities and infrastructure to profit from their business activities. This lack of infrastructure encourages overuse of nature (overgrazing); proper infrastructure would make venison processing more cost effective.
- Provide reindeer herders with reasonably priced fuel, drinking water, and heated tents in temporary camps near the national settlements

and trading spots. The lack of petroleum stations and storage gasoline facilities forces herders to buy fuel at high prices, and fuel represents a significant portion of a reindeer herders' family budget. Fuel prices force them to limit other expenses (food, goods, medications, travel to national settlements, etc.), which significantly impacts their welfare.

- Organise mobile slaughter and supplemental reindeer feeding facilities in the tundra along the nomadic routes of reindeer herders based on the data from the monitoring system.

Long-term measures

- Implement risk management to avoid overgrazing; monitoring nomadic reindeer herders' routes is a part of this.
- Develop a year-round cycle of reindeer herding and husbandry (*slaughter – storage – processing/production – logistics/delivery – sale*) in the tundra using new methods for the complete processing of reindeer beef. Subsidise the harvest and delivery of reindeer products and increase their profitability. Train Indigenous Peoples to harvest and process reindeer beef according to bioproduction standards.
- Regulate nomadic reindeer routes in the areas affected by overgrazing by monitoring these routes (with GPS trackers attached to reindeer). Determine nomadic routes in close collaboration with reindeer herder representatives. Require reindeer herders to follow assigned nomadic routes to receive subsidies.
- Subsidise fuel for reindeer herders and develop infrastructure to facilitate the economic empowerment of reindeer herders.
- Support reindeer herding in the forest zone of YNAO. Encourage small-scale reindeer herders to relocate to the southern areas of the district where overgrazing is not a problem and there is more infrastructure in place.
- Monitor the consumption of traditional food, and the health and social welfare of the Indigenous population in the Russian Arctic. Explore ways to optimise and enrich their diet while reducing the consumption of traditional food.

References

Anderson, I., B. Robson, M. Connolly et al. 2016. Indigenous and tribal peoples' health (The Lancet–Lowitja Institute Global Collaboration): A population study. *Lancet* 386: 1–27.

Andronov, S., A. Lobanov, A. Popov et al. 2018. Prediction of arterial hypertension in the inhabitants of the Arctic region of Western Siberia. *Current Problems of Health Care and Medical Statistics* 1: 54–67.

Andronov, S., A. Lobanov, A. Popov et al. 2018. The impact of traditional nutrition on reduction of the chronic nonobstructive bronchitis risk in the Indigenous

Peoples living in tundra of the Arctic zone in Western Siberia, Russia. *European Respiratory Journal* 52(62): PA796.

Andronov, S.V., A.A. Lobanov, R.A. Kochkin et al. 2018. The role of traditional nutrition in maintaining the health of the indigenous population in the Arctic zone of Western Siberia. In *Maintenance of the Indigenous Population in the Arctic zone of the Russian Federation in the context of lifestyle transformation and climate change.* Proceedings of the All-Russian Scientific-Practical Conference with International Participation, Arkhangelsk. 142–144.

Arctic Council. Sustainable Development Working Group. 2015. *SLiCA: Arctic living conditions – Living conditions and quality of life among Inuit, Saami and Indigenous Peoples of Chukotka and the Kola Peninsula.*

B.C. Indigenous Food System Network Working Group on Indigenous Food Sovereignty. 2008. *Final Activity Report*: 1–22.

Bogdanova, E.N., A.A. Lobanov, S.V. Andronov et al. 2018 On the issue of food security of the Indigenous Peoples in the Arctic region. In *Maintenance of the Indigenous Population in the Arctic zone of the Russian Federation in the conditions of transformation of lifestyle and climate change.* Proceedings of the All-Russian Scientific-Practical Conference with International Participation, Arkhangelsk. 120–126.

Chan, H.M., K. Fediuk, S. Hamilton et al. 2006. Food security in Nunavut, Canada: Barriers and recommendations. *International Journal of Circumpolar Health* 65(5): 416–431.

Clark, P. 2016. Can the state foster food sovereignty? Insights from the case of Ecuador. *Agrarian Change* 16(2): 183–205.

Desmarais, A.A. 2002. 'The Via Campesina: Consolidation and international peasant and farm movement. *Journal of Peasant Studies* 29(2): 91–124.

Desmarais, A.A. and H. Wittman. 2014. Farmers, foodies and First Nations: Getting to food sovereignty in Canada. *Journal of Peasant Studies* 41(6): 1153–1173.

Filant, K.G. 2017. Legal support of the mechanisms of state regulation of the reindeer complex. In *To Help the Reindeer Breeder, Hunter, and Fisherman*, 31–47, Legislative Assembly of Yamalo-Nenets Autonomous Okrug, Salekhard.

Food and Agriculture Organisation of United Nations. 1996. *Rome Declaration on World Food Security and the World Food Summit Action Plan*, Rome.

Food and Agriculture Organisation of United Nations. 2009. *Declaration of the World Summit on Food Security*, Rome.

Food and Agriculture Organisation of United Nations. 2015. *FAO Policy on Indigenous and Tribal Peoples*, Rome, Italy.

Food and Agriculture Organisation of United Nations. 2018. *The State of Food Security and Nutrition in Europe and Central Asia*, Rome, Italy.

Friedmann, H. and P. McMichael. 1989. Agriculture and the state system: The rise and fall of national agriculture, 1870 to the present. *Sociologia Ruralis* 29(2): 93–117.

Government of Yamal-Nenets Autonomous Okrug of the Russian Federation. 2016. *On Reindeer Herding*, 34-ZAO, ed. 02 October 2018.

Hardin, G. 1968. The tragedy of the commons. *Science* 162(3859): 1243–1248.

Inuit Circumpolar Council. 2012. *Food Security across the Arctic. Background paper of the Steering Committee of the Circumpolar Inuit Health Strategy*, Canada.

Ivanov, V.A. 2015. Food security: Arctic specificity. Corporate governance and innovative development of the economy of the North. *Bulletin of the Research Centre for Corporate Law, Governance and Venture Capital Investments* 2: 152–174.

Kepkiewicz, L. 2015. Pedagogy lost? Possibilities for adult learning and solidarity in food activism. *Studies in the Education of Adults* 47(2): 185–198.

Kepkiewicz, L. 2017. Understanding food sovereignty in Canada: Settler colonialism and Indigenous–-settler alliances. In A.A. Desmarais, P. Claeys, and A. Trauger (eds.), *Public policies for food sovereignty: Social movements and the state*, 164–180. New York: Routledge.

Kochkin, R., A. Lobanov, S. Andronov et al. 2018. The effect of the consumption of trace elements on the central nervous system under conditions of geomagnetic disturbances. In *Proceedings of the VIII International Congress 'Weak and super-weak fields and radiations in biology and medicine. Low-intensity impacts'* (10–14.09.2018), 130–131. Saint-Petersburg.

Kozlov, A., G. Vershubsky, and M. Kozlova. 2018 Indigenous Peoples of Northern Russia: anthropology and health. *Circumpolar Health Supplements* 1.

Kryazhimskiy, F.V., K.V. Maklakov, L.M. Morozova, and S.N. Ektova. 2011. System analysis of biogeocenoses of the Yamal Peninsula: Simulation modelling of the impact of landfall reindeer herding on vegetation cover. *Ecology* 5: 323–333.

Lobanov, A., S. Andronov, A. Emelyanov et al. 2018. Risk factors for chronic bronchitis in indigenous inhabitants of the Arctic zone in Western Siberia, Russia. *European Respiratory Journal* 52(62): PA799.

Lobanov, A., S. Andronov, R. Kochkin et al. 2018. The role of traditional nutrition of Indigenous People in the Arctic zone of Western Siberia in managing the risks of hypertension. *EFSA Journal Supplement: Science. Food. Society*: 156–157.

Lobanov, A., S. Andronov, A. Popov et al. 2018. The impact of traditional nutrition on reduction of the chronic nonobstructive bronchitis risk in the Indigenous Peoples living in tundra of the Arctic zone in Western Siberia, Russia. *European Respiratory Journal* 52(62): PA796.

Lobanov, A., E. Bogdanova, S. Andronov et al. 2018. Consumption of traditional food by Indigenous Peoples in the conditions of industrial development of the Arctic zone in Western Siberia. In N.G. Dyachenko (ed.), *The future of the Arctic starts here*, 190–194. Apatity: Publishing House of the MAGU.

Lobanov, A.A., E.N. Bogdanova, S.V. Andronov et al. 2018. Study of the traditional nutrition of residents of the Arctic zone of Western Siberia. *Nutrition Issues* 87(S5): 31–32.

Lobanov, A.A., E.N. Bogdanova, S.V. Andronov et al. 2018. Traditional nutrition – criteria of the term. *Nutrition Issues* 87(S5): 32–33.

Ministry of Agriculture of the Russian Federation. 2010. *Doctrine of Food Security of the Russian Federation,* No. 120.

Morozova, L.M. and M.A. Magomedova. 2006. The influence of long-term grazing of deer on the resource potential of vegetation cover. In *Yamal Peninsula: Vegetation cover*, 235–249. Tyumen: City Press.

Morrison, D. 2011. Indigenous food sovereignty: A model for social learning. In H. Wittman, A. Aurélie Desmarais, and N. Wiebe (eds.), *Food sovereignty in Canada: Creating just and sustainable food systems*, 97–114. Winnipeg: Fernwood Publishing.

State Duma of the Russian Federation. 1994. *Civil Code of the Russian Federation (Part One)*, 51-FL, ed. 08.03.2018.

Van der Ploeg, J.D. 2013. *Peasants and the art of farming: A Chayanovian manifesto.* Halifax: NS Fernwood.

Visser, O., N. Mamonova, M. Spoor, and A. Nikulin. 2015. 'Quiet food sovereignty' as food sovereignty without a movement? Insights from post-socialist Russia. *Globalizations* 12(4): 513–528.

Wittmanm, H., 2009. Reframing agrarian citizenship: Land, life and power in Brazil. *Journal of Rural Studies* 25: 120–130.

Yuzhakov, A.A. 2017. Northern reindeer herding in the XXI century: Genetic resource, cultural heritage and business. *Arctic: Ecology and Economy* 2(26): 131–137.

5 Dietary issues in contemporary Greenland

Dietary patterns, food insecurity, and the role of traditional food among Greenlandic Inuit in the twenty-first century

Peter Bjerregaard, Ingelise Olesen, Tine Curtis, and Christina V.L. Larsen

Introduction

The population of Greenland was 55,992 as of 1 January 2019. Of these, an estimated 51,300 (92%) were Indigenous Greenlanders (*Kalaallit* or Inuit) (Young and Bjerregaard 2019). A number of dietary studies have been conducted in Greenland in the last 65 years. Greenlanders' diets consist of a mix of locally harvested food and imported food; the importance of the former has been decreasing for more than a century (Table 5.1).

In the early twentieth century, most Greenlandic Inuit ate a traditional diet that consisted mostly of what could be harvested locally from the sea. This traditional diet, or *kalaalimernit*, comprised marine mammals, fish, and wild fowl supplemented with terrestrial species, particularly caribou (*Rangifer tarandus*). Some local plants, such as seaweed and berries (*Vaccinium uliginosum* and *Empetrum nigrum*) were also consumed. The traditional diet then was much more diverse than today and included the animals' meat, fat, and organs. The foods were always boiled or eaten raw, while today they are also fried and baked. According to the most recent population health

Table 5.1 Energy percent of locally harvested food in Greenland (Bertelsen 1937; Uhl 1955; Jeppesen and Bjerregaard 2012; Larsen et al. 2019).

Year	*Town (%)*	*Village (%)*
1900	82	
1930	38	
1953	21	45
2005	17	32
2018	14	21

DOI: 10.4324/9781003057758-5

survey (2018), the locally harvested food items that provided the most energy were blubber (on seal meat or eaten with dried fish), Greenland halibut (*Reinhardtius hippoglossoides*), caribou meat, seal meat, and dried fish and meat (Bjerregaard unpublished).

Data and methods

The chapter explores the food preferences of Inuit in Greenland, the role of *kalaalimernit*, and the prevalence and social variation of food insecurity in Greenland. Among other instruments, our studies among the Inuit in Greenland used questions from the Household Hunger Scale (Ballard et al. 2011) alongside qualitative methods to explore food insecurity. Two specific research questions were posed.

- First, *kalaalimernit* make up only 20% of the energy intake of Greenlandic Inuit (Jeppesen and Bjerregaard 2012), yet it includes the most popular food items. Why, then, are these foods not consumed more often?
- Second, is food insecurity among the Greenlandic Inuit due to a lack of access to traditional food or a lack of access to any food at all?

Data sources

The chapter describes the published and unpublished results from three quantitative and two qualitative studies among the Greenlandic Inuit. In 1993, a countrywide population health survey of 1,580 Greenlandic Inuit included quantitative information about food preferences (Bjerregaard et al. 1995). Subsequent population health surveys in 2014 and 2018 included 2,120 and 2,436 Greenland Inuit, respectively, and collected dietary information through food frequency questionnaires (Dahl-Petersen et al. 2016; Larsen et al. 2019). The surveys also included questions about the perception of food insecurity (Ballard et al. 2011). A qualitative study from 1997 to 1999 focused on the socio-cultural meanings of traditional Greenlandic food (Curtis et al. unpublished), and another qualitative study from 2016 to 2017 specifically targeted food insecurity (Olesen et al. unpublished).

Preferred food items

Only the 1993 survey collected information about the food preferences of Greenlandic Inuit. Respondents were given a list of 12 locally harvested food items and 12 imported food items and asked to rate each food item on a scale from 1 (dislike) to 5 (like very much). Traditional food items were rated significantly higher (mean score 4.67, SD 0.43) than imported foods (4.10, SD 0.59). Seven locally harvested food items were on the top ten list: mattak (muktuk, whale skin), dried cod, Brünnich's guillemot (*Uria lomvia*), crowberries, qullukkat (freeze-dried Greenland halibut), seal meat, and whale meat. Only three imported food items were among the top ten: potatoes, rye bread, and vegetables. All age groups rated traditional food highly, but the preference for traditional food increased with age (Pars et al. 2001).

The role of traditional Greenlandic food – *kalaalimernit* – in the contemporary diet

From 1997 to 1999, 21 focus group interviews about perceptions of traditional Greenlandic food were conducted in four towns and six villages in Greenland (Curtis et al. unpublished).

Most participants agreed that traditional food is very healthy. One participant said, 'It is said that the food we eat has a large impact on our health. Here in the village, we eat more traditional food than they do in the towns; they eat food with a lot of bad additives'.

The issue of the possible contamination of the food was brought up by participants in most of the focus groups. Heavy metal contamination was mentioned several times, as were stories about presumably contaminated animals. A certain scepticism about research on contaminants was expressed, and it was concluded that contamination is not a factor in deciding whether or not to eat Greenlandic food. Rather, the impact of eating traditional food was very often discussed in terms of satiety, strength, and warmth. As one said, 'When we eat Danish food (which of course we also like as it's something different), we find that it doesn't give much strength. When you've had Danish food, you get hungry immediately afterwards [other participants express agreement]'.

The focus groups had lengthy discussions on problems of availability and accessibility. These problems relate to hunting – which is affected by the lack of animals or the absence of hunters or fishermen in the family – and to the lack of Greenlandic products in the shops and the perceived high cost of the products available. One participant said,

> They close quite early at the local market where we buy traditional food. Sometimes I don't have time to go. And then, unfortunately, it seems easier to just buy a chicken or something you can quickly fry in a pan.

Another agreed,

> Yes. But that's also because it [*kalaalimernit*] is so expensive. We eat a lot of meat when we eat traditional food. When we eat Danish food, we don't eat as much meat. And there are four of us. So, we need one and a half kilograms of traditional food. It's very expensive.

Even though most participants agreed that they preferred traditional food, they also expressed a desire for variety, either because of personal preferences or for health reasons. However, as the following quote illustrates, a lack of *kalaalimernit* may contribute to the feeling of food insecurity: 'When we don't have [locally harvested] meat we usually say that "we don't have anything to boil," – that we don't have any food, even though we do have Danish food'.

The findings from these focus group interviews correspond to responses to open-ended questions in the 1993 health interview survey. In that study, participants who reported that they did not consume *kalaalimernit* every day were asked why they didn't eat *kalaalimernit* more often. The three most frequent answers were that respondents wished to vary their diets (48%), that *kalaalimernit* were difficult to obtain (45%), and that *kalaalimernit* were expensive (39%). Only a few (1%) stated that a fear of contaminants influenced the frequency of their consumption of *kalaalimernit*, although, at that time, the issue of contaminants in marine mammals had been a topic of public discussion for several years (Pars et al. 2001).

Surveys of food insecurity in Greenland

This section presents unpublished analyses of data from a population health survey in 2018. In two population health surveys in Greenland (in 2014 and 2018), food (in)security was assessed with the question: 'In the past twelve months, was there ever no food to eat of any kind in your house because of a lack of resources to obtain food?' These surveys were the first to examine food security in a nationwide survey in Greenland. In the 2018 survey, two additional questions about food security were included: 'In the past twelve months, did you or any household member go to sleep at night hungry because there was not enough food?' and 'In the past twelve months, did you or any household member go a whole day and night without eating anything because there was not enough food?' For all three questions, participants could answer yes or no. These questions were adapted from the Household Hunger Scale, which asks the same questions about a period of four weeks rather than twelve months (Ballard et al. 2011). The two questions added to the 2018 survey did not significantly change the prevalence of food insecurity. In the 2018 survey, 7.9% of respondents experienced food insecurity based on one question and 8.8% based on all three questions. For comparisons the result of one question was used.

As mentioned above, a lack of *kalaalimernit* may contribute to the perception of food insecurity. However, the survey questions did not specifically address this, resulting in a narrow economic definition of food security.

The prevalence of food insecurity decreased from 2014 to 2018 (relative risk 0.69; 95% confidence interval 0.57–0.85). In 2018, reported food insecurity was slightly higher among men than among women (9.5% and 7.6%; $p = 0.08$), and food insecurity decreased with age from 15.8% among 15- to 24-year-olds to 3.5% among participants 60 years old and up ($p < 0.0001$).

Food security varied significantly by wealth. Among the poorest one-fifth of the participants, food insecurity was reported at 20%. Among the wealthiest one-fifth, only 0.5% reported food insecurity (Larsen et al. 2019). Food insecurity also varied significantly among regions. In East and North Greenland, 21% reported food insecurity, compared with 7% in West Greenland ($p < 0.0001$). In West Greenland, food insecurity was higher in villages on the southwest coast than in other towns and villages. Some of this geographic variation disappeared after adjustment for individual wealth or employment.

Mattak was the most popular food item and also the most expensive. Mattak consumption showed a clear socioeconomic gradient; the least affluent ate less mattak than wealthy participants. Assuming the preference for mattak is similar across socioeconomic groups, it is likely that the inability to eat as much as one would like of a favourite food item is one aspect of food insecurity that was not captured by the survey.

Qualitative study of food insecurity

In order to increase our understanding of food insecurity among Greenlandic Inuit, 19 qualitative interviews were conducted in 2016 and 2017. Interviewees were selected from the participants in the 2014 population health survey who answered yes to the question 'In the past twelve months, was there ever no food to eat of any kind in your house because of lack of resources to get food?' All interviewees lived in Tasiilaq (a small town in East Greenland) or Nuuk (the capital) in West Greenland). During the interviews, the interviewer attempted to discuss the role of *kalaalimernit* in food insecurity with the participants. However, due to the magnitude of the economic aspects of food insecurity, most interviews eventually focused on how to buy enough nutritious food on a limited budget.

For some participating families in Tasiilaq and Nuuk, obtaining decent food, especially for their children, was a daily fight. The main meal of the day often consisted of cheap solutions like breakfast cereal, pancakes, or plain pasta for the children while the parents ate very little or went hungry. For these participants, the national food recommendations (eat a variety of meat, fish, vegetables, and fruit) were far removed from daily life.

Most participants from Tasiilaq were unemployed, and many saw a lack of money as the main cause of food insecurity. One said,

> … a welfare payment doesn't last fourteen days, and they have reduced the welfare payments for the whole country. So, when groceries are so very expensive here [in Tasiilaq], the welfare payment doesn't go far at all; it's very hard.

A study of prices in Greenland found that (weighted for the average composition of meals) groceries cost 16% more in Tasiilaq than in Nuuk (Bjerregaard 2019 unpublished). However, participants from Nuuk voiced similar complaints. Not only unemployed participants, but also families subsisting on minimum wages said that they were unable to make ends meet.

For families with low salaries and many children and family members, the high cost of food was a serious problem. One mother said,

> Well, I don't eat very much so there is enough for my children. Sometimes I only eat twice a day. I eat in the morning and then I have a cooked dinner, and then I go to bed without anything else so my children won't starve.

Most participants agreed that welfare payments were too low. Many also complained about the attitudes of the staff in the municipal offices responsible for welfare payments. They felt humiliated and lost confidence in the authorities because staff were slow to act and often turned down requests for additional money.

Several participants admitted that one factor contributing to food insecurity was the purchase of alcohol and tobacco, which are both expensive in Greenland, instead of food. Marihuana and gambling addictions were mentioned as well. One participant said, 'I am hungry, but then I tell myself that on Monday I'll get my money (...) then I tell myself that on Monday I'll buy some food, but when Monday comes … then it is only alcohol and cigarettes'.

Igaassat, which is cooked or prepared food, similar to the British concept of 'a proper meal' (Murcott 1982), was mentioned in all interviews in Tasiilaq. Not all participants in Nuuk talked about it. *Igaassat* is viewed as healthy and tasty but expensive. One participant commented, 'Food that can be cooked is very expensive, so I don't buy a lot of that – meat and fish and things like that. I mostly buy rye bread and cereal'.

In this context, the lack of *kalaalimernit* was also identified as contributing to food insecurity. One said, 'It may sometimes be a problem not being able to eat *kalaalimernit*, guillemot, or seal … now I buy the cheapest food, bulk packages for 14 days or 28 days, something for the freezer'.

Another woman expressed similar shopping preferences, saying, '… I buy the food that makes you feel full, like rye bread, frozen cold cuts, and the cheapest food that you can cook'.

Some participants relied on help in the form of food or money from friends or family members, but many didn't have this option because their family relationships were strained or because they didn't want to beg. They felt ashamed to have to rely on gifts of food or to beg for food. Participants in Tasiilaq and Nuuk often indicated that they had given up and tried to make their children accept this aspect of their lives as well. One said, 'Now and then the children get a little bit angry or sad, especially the youngest one, but then I usually tell them that we can't do anything about it, that this is the way things have turned out for us'.

Conclusions

Participants in this study prefer locally harvested food, *kalaalimernit*, to imported food. *Kalaalimernit* are important to Greenlandic Inuit for a variety of reasons, but imported food is also popular because it is cheaper and easier to prepare and due to a wish for variety, among other reasons. Participants in focus groups mentioned contamination of Greenlandic food, but this issue didn't play a role in their food choices. In a qualitative study among the general population, a lack of access to Greenlandic food was sometimes perceived as food insecurity. In another qualitative study of participants who had previously reported experiencing food insecurity, the main concern was a shortage of food altogether due to a lack of money. The food insecurity captured by the Household Hunger Scale questions in the population surveys was an economic insecurity: the inability to purchase enough food. This type of food insecurity was associated with socio-economic conditions and was reported much more often in East Greenland than in West Greenland, even after adjustment for social and economic variables.

The first research question asked why Greenlandic Inuit do not eat *kalaalimernit* more often. In general, the seemingly low consumption of *kalaalimernit* is by choice and not due to a lack of access or money. Most participants expressed a desire for variety in their diets and praised the ease of preparing imported food.

The second question asked whether food insecurity among Greenlandic Inuit is a lack of *kalaalimernit* or a lack of access to food. While the qualitative studies considered here hint that a lack of *kalaalimernit* might contribute to food insecurity, the overwhelming narrative addressed the lack of money to buy sufficient nutritious food for a family.

Additional issues to explore in future studies are the roles of accessibility and purchasing power in the geographical variation in food security. Variations in dietary patterns based on degree of food security should also be investigated further by including food prices in the analyses of the data from the current studies. Finally, the role of choosing tobacco and alcohol over food (mentioned by one interviewee here) in food insecurity deserves further investigation.

References

Ballard, T., J. Coates, A. Swindale, and M. Deitchler. 2011. *Household hunger scale: Indicator definition and measurement guide.* Washington, DC: Food and Nutrition Technical Assistance III Project (FANTA).

Bertelsen, A. 1937. Grønlandsk medicinsk Statistik og Nosografi. II. Sundhedsvilkaarene i Grønland [Greenland medical history and nosography. II. Health conditions in Greenland. In Danish]. *Meddelelser om Grønland* 117(2): 1–248.

Bjerregaard, P., T. Curtis, F. Senderovitz, U. Christensen, and T. Pars. 1995. *Levevilkår, livsstil og helbred i Grønland* [*Living conditions, lifestyle, and health in Greenland.* In Danish]. Copenhagen: DIKE.

Dahl-Petersen, I.K., C.V.L. Larsen, N.O. Nielsen, M.E. Jørgensen, and P. Bjerregaard. 2016. *Befolkningsundersøgelsen i Grønland 2014. Levevilkår, livsstil og helbred* [*Population health survey in Greenland 2014. Living conditions, lifestyle, and health.* In Danish]. Copenhagen and Nuuk: Statens Institut for Folkesundhed og Departementet for Sundhed.

Jeppesen, C. and P. Bjerregaard. 2012. Consumption of traditional food and adherence to nutrition recommendations in Greenland. *Scandinavian Journal of Public Health* 40: 475–481.

Larsen, C.V.L., C.B. Hansen, C. Ingemann, M.E. Jørgensen, I. Olesen, I.K. Sørensen, A. Koch, V. Backer, and P. Bjerregaard. 2019. *Befolkningsundersøgelsen i Grønland 2018. Levevilkår, livsstil og helbred* [*Population Health Survey in Greenland 2018. Living conditions, lifestyle, and health.* In Danish]. Copenhagen: National Institute of Public Health. SIF's Grønlandsskrifter nr. 30. Retrieved from www.sdu.dk/sif.

Murcott, A. 1982. The cultural significance of food and eating. *Proceedings of the Nutrition Society* 41: 203–210.

Pars, T., M. Osler, and P. Bjerregaard. 2001. Contemporary use of traditional and imported food among Greenlandic Inuit. *Arctic* 54: 22–31.

Uhl, E. 1955. Nogle Undersøgelser af Grønlandske Levnedsmidler og Kostforhold [Some studies of dietary issues in Greenland. In Danish]. *Beretning vedrørende Grønland* 3. Copenhagen: Ministry for Greenland.

Young, K. and P. Bjerregaard. (2019). Towards estimating the Indigenous population in Circumpolar regions. *International Journal of Circumpolar Health* 78: 1, 1653749.

PART II

Multi-disciplinary perspectives on food (in)security

6 Human rights begin with breakfast

Maintenance of and access to stable traditional food systems with a focus on the European High Arctic

Kamrul Hossain and Noor Jahan Punam

Introduction

Food systems include a range of activities, from the production to the consumption of food products. Traditional food systems (TFSs) consist of activities in which food is collected or produced locally from available natural resources, shared and consumed locally as part of customary food practices, and processed for wider distribution within and beyond the local region (Kuhnlein, 2009). Across the Arctic, TFSs comprise traditional activities such as berry picking, hunting, reindeer herding, and fishing along with processing, distributing, and consuming these food products. Due to climate change and increasing human activities in the region (such as extractive industrial developments, tourism, and maritime activities), the maintenance or preservation of stable TFSs is facing serious challenges. The people residing in this region, including the Indigenous population, are key players in the region's TFS. For them, the maintenance of stable TFSs is vital. It is also connected to their sociocultural, emotional, and psychological values and to their local and traditional knowledge (Kuhnlein et al. 2013). In addition to establishing stable food systems, maintaining TFSs is sustainable, conserves ecological systems, and enables ecosystems to function. Stable food systems also ensure food security (International Union for Conservation of Nature [IUCN] 2013), which enhances human security (Food and Agricultural Organisation of the United Nations [FAO] 2016). Against this background, this chapter examines TFSs and their importance to local Indigenous and non-Indigenous peoples in the European High Arctic (EHA) region. It seeks to pinpoint the position of TFSs within a human rights framework. In doing so, it places the right to traditional food not only under the general right to food provision, but under a broader human rights framework. This broad framework designates food as an individual right like the right to health and to practice one's culture and creates a collective dimension where food becomes a community property (and/or a collective concern). In this approach, stable food systems require integrating local and

DOI: 10.4324/9781003057758-6

traditional knowledge and ensuring meaningful community participation in protecting and promoting the systems. Therefore, this chapter uses a human rights framework to argue that protecting and promoting TFSs must include the right to self-determination. This right guarantees food sovereignty in a TFS. In sum, this chapter examines the relationship between an overarching human rights framework and TFSs in the context of the EHA. The chapter is divided into six sections including the introduction and the conclusion. The next section identifies the connections among food, population, and the EHA region.

European High Arctic (EHA): the region, population, and food

The European High Arctic (EHA) refers to the European North and the European portions of the Arctic (Tamnes and Offerdal 2014, 5). This area includes the northernmost parts of Norway, Sweden, and Finland and the north-western part of Russia (Stepien and Koivurova 2017). The population in this region is comprised of Indigenous Peoples and local communities. The Indigenous population includes Sami, Nenets, and Veps, among others (Barents Info a). Non-Indigenous local people include Norwegians, Swedes, Finns, and Russians, along with a small number of immigrants from other nationalities. There are at least 90,000 Sami in the region (Barents Info b). Most of them live in Norway (50,000–65,000), many in Sweden (20,000), and some in Finland (8,000) and Russia (2,000). Many other Indigenous Peoples live in the Russian part of the European Arctic; approximately 283,975 Nenets, Pomors, Veps, and Komi (the Komi are not considered Indigenous Peoples under Russian law) live in Russia (Barents Info a). Indigenous inhabitants of the EHA region, particularly those in the three Scandinavian countries, often cross national and international borders to engage in traditional livelihood activities, which include herding, hunting, fishing, and foraging (Hossain, Raheem, and Cormier 2018). These activities comprise family traditions and build relationships in communities (Hossain, Raheem, and Cormier 2018).

In the Nordic countries in general, the food culture creates a 'strong symbolic bond' between nature and food (Bergflodt, Amilien, and Skuland 2012). This bond is linked with the traditional activities mentioned above; even eating outdoors is said to represent Nordic cultural values and practices (Bergflodt, Amilien, and Skuland 2012). This is also true for the rests of the EHA. The food culture in this region is diverse, yet many traditional virtues and practices relating to food are similar throughout the EHA. In the Arctic regions, abundant food resources can be gathered or accessed from the natural environment. However, available food resources differ in different Arctic countries. EHA tourist industries promote specific local foods in each country. For instance, in Northern Norway, cod, coley, Atlantic halibut, herring, haddock, and red bream are available from the sea (Nord Norge 2018a). Fresh fish is a source of food, but dried fish is also enjoyed by the locals and is Norway's oldest export (Nord Norge 2018b). The forests, mountains, and plateaus in the north are home to reindeers and elks (Nord Norge 2018b). Cloudberries, blueberries, and lingonberries are also abundant in the forests and are often served as a side dish or dessert (Nord Norge 2018b). In the Swedish part of the EHA, salmon, grayling, whitefish, and roe are found in the clear waters, and traditional dishes include reindeer, moose, grouse, and capercaillie. These dishes are also served with berries, often freshly picked. The tourist industry often promotes these local foods (Heart of Lapland 2018). Like in Northern Norway, in the Swedish part of the EHA, cranberries, blueberries, cloudberries, and Arctic raspberries are often served as side dishes or desserts, in drinks, or even eaten fresh (Heart of Lapland 2018). In Northernmost Sweden, the local cuisine may vary from place to place. The food culture of this region is directly connected to nature, and the same is true for Northern Finland (Bergflodt, Amilien, and Skuland 2012). In the Finnish part of the EHA, for instance, reindeer is an essential part of the local cuisine (Reindeer Herders' Association 2014); in fact, no part of the reindeer remains unused. The meat is sautéed or served in fillets, cold cuts, jerky, mince, and sausages, among other forms (Reindeer Herders' Association 2014). Game birds such as capercaillie, willow grouse, black grouse, and waterfowl also form part of the cuisine of Northern Finland. Elk meat is also popular in this area, and bear meat is considerably less popular but still eaten (Visit Rovaniemi 2018). Tourist information about Finnish Lapland highlights the access to fish such as salmon, trout, pike perch, whitefish, arctic char, grayling, and vendace. Berries are also common in Northern Finnish forests, including bilberry, cloudberry, lingonberry, and cranberry. These foods are prepared and eaten in similar ways in Northern Norway, Sweden, and Finland (Visit Rovaniemi 2018). In the Russian part of the EHA, wild mushrooms and berries are often gathered (Hossain, Raheem, and Cormier 2018). Reindeer meat also forms part of the food system in the Russian Arctic. Reindeer hides are often salted to preserve them (Karpukhin 2016). A variety of fish are found in the Russian Arctic, including haddock, cod, pollock, capelin, poutassou, herring, grouper, and mackerel (Stupachenko 2018). Atlantic salmon can also be found in the Kola Peninsula (Aas, Klemetsen, and Einum 2011). Fish can be smoked or cooked in soups.

Traditional food systems

Conceptualising TFS

The TFS of the EHA is similar in many ways to the food system promoted by the tourism industry, but there are also some differences (see Chapter 7 of this book). To understand TFSs, it is necessary to first define a food system. Global Environmental Change and Food Systems (GECAFS, an interdisciplinary research program), categorises food systems based on whether they encompass all activities related to the production, processing, distribution, preparation, and consumption of food. These activities contribute to an essential interrelationship between humans and food. This interrelationship broadly defines the link between food security and food systems. Food security refers to availability, access, utilisation, and stability of food (FAO 2008a). Availability is ensured by production, distribution, and exchange. Access means affordability, allocation, and preference. Utilisation refers to the additional values embedded in food, including nutrition, social values, and food safety. Food security contributes to broader human security, including environmental, health, community, and economic security. In food systems, interactions among biological, geographic, and physical structures and between these structures and human environments influence food-related activities and their outcomes (Ericksen 2008). As a result, food systems encompass all food processes from production to consumption as well as the influence of those processes on society, politics, the economy, the culture, and the environment (Capone 2014). Therefore, a food system cannot be understood only by considering the processes it involves; the actors in a food system, along with their roles and interactions, must be examined as well (Ericksen et al. 2009).

The actors in food systems form the largest group of natural resource managers in the world (European Environment Agency 2017). Their contributions to the food systems are integral; they identify problems and implement solutions (European Environment Agency 2017). Food system actors include farmers, fishermen, traders, processors, food industry workers, retailers, and consumers (Westhoek, Berkum, and Hajer 2016). In large-scale industrial production and processing, actors' roles vary – some are more powerful than others. While these actors' actions are integrated in food system in general, a TFS approach offers a different perspective, suggesting that the consumption preferences of grassroots-level actors deserve to be heard. Most actors in the EHA engage in small-scale food production using traditional resources, means, and methods, primarily to meet local needs. Maintenance of the food system plays a role in their subsistence activities (see, e.g., Chapter 10 of this book). However, as the traditional subsistence activities of the local Indigenous and non-Indigenous communities in this region have been replaced by a mixed economy, more and more local foods and their by-products are being processed to attract external markets.

Consequently, a TFS includes local, natural food products that are socially and culturally acceptable (Kuhnlein, 2009). Although this definition has often been used in connection with Indigenous Peoples, in this chapter, it will be used to include Indigenous and non-Indigenous residents. Here, the EHA's TFS refers to processes that use local food sources, such as berry picking, hunting, reindeer herding, and fishing, as well as processing these food items for consumption. In short, collecting food directly from nature and then preparing and consuming that food is the TFS. The cultural, social, political, economic, and environmental outcomes of these processes are also part of the TFS, and local actors and their roles in the maintenance of food systems are the most important elements of a TFS. The EHA is characterised by robust interactions amongst traditional cultures and the lands used and/or otherwise occupied by Indigenous and non-Indigenous local peoples (Paci et al. 2004). As a result, its TFS does not exclude cultural practices or the spiritual aspects of food practices; rather, the relationships between these practices and food offer a more complete understanding of TFSs in the high Arctic.

Relevance of TFS in the EHA

Maintaining TFSs in the EHA is key to the stability of food systems in this region. Stability is one of the four dimensions of food security – the other three, as mentioned above, are availability, access, and utilisation (Hossain, Raheem, and Cormier 2018). The 1996 FAO definition of food security also includes access to food which is socially and culturally acceptable (Jones et al. 2013). A food-secure community has more human security because TFSs are interconnected with other aspects of society. For example, the nutrition of traditional food provides health security; maintaining traditional and natural food systems complements environmental conservation and preservation, providing environmental security; traditional food is less expensive than imported food, which is often expensive in the high Arctic, contributing to economic security; and finally, the connectedness of traditional food to cultural values creates community security.

The practice of traditional food collection in across the region of the EHA, particularly from natural sources, is consistent over time. The connection between land and food cannot be disregarded – for instance, the growth of berries in wild forests and organised farming are both examples of food cultivated from the land (Elde et al. 2018). In Finland, under the national legislation supporting 'every man's right', anyone may gather wild berries, mushrooms, and apples from the forests for free; this is subject to restriction in conservation areas and on private properties (Ministry of the Environment 2016). Fishing is somewhat restricted; angling and ice fishing are banned in fast-flowing sections of watercourses (Ministry of the Environment 2016). Anyone who pays the state fisheries management fee may lure fish with a single rod anywhere in Finland except in fast-flowing sections, and people under 18 and over 65 can fish for free (Ministry of the Environment 2016). Fishermen who use nets or fish traps must obtain permission and pay the state fisheries management fee (Ministry of the Environment 2016). In Sweden, similar regulations apply to picking berries and mushrooms (Kuruzovic 2019), and Swedish rules and restrictions on fishing also mirror those of Finland (Aas 2008). In Norway, there are no restrictions on gathering berries and mushrooms, but special rules apply to cloudberries in Northern Norway (Kagda, Cooke and Nevins 2017). No license is needed to fish for saltwater species for personal use, but a license is required for all other kinds of fishing (Norwegian Environment Agency 2013).

Reindeer herding and hunting are regulated slightly differently across the three Nordic states. In Finland, hunters must take a hunting examination and pay a state game management fee, after which they can obtain a hunting card (Tuunanen, Tarasti, and Rautiainen 2015). and Residents of Lapland may hunt on state-owned land (Section 8, Hunting Act). The right to practice reindeer herding is not reserved to Sami people, but it can only be done in designated reindeer herding areas (Section 3, Reindeer Husbandry Act). Reindeer herding is not reserved to Indigenous Peoples in Russia, either (Forrest 1997). In Sweden and Norway (Hossain, Raheem, and Cormier 2018), anyone who pays a permit fee for a license may hunt reindeer (Swedish Environmental Protection Agency 2019), but herding rights are reserved to the Sami people (Torp 2013). In Russia, a license is needed for hunting and fishing. Amateurs may hunt animals for fur with a voucher, but for licenced species, such as river beaver, a licence is required by the 1995 Federal Law of the Russian Federation on Wildlife.

Food from traditional or local sources also supports business development in the EHA. Local foods are often used to promote local businesses and services, from tourism businesses and restaurants to other public institutions such as schools, hospitals, offices, and other institutions. Tourist agencies in Finnish, Norwegian, and Swedish Lapland and in the Kola Peninsula, such as Kontiki Finland (Twitter: @kontikifinland), often share traditional recipes and pictures of traditional food on social media. Some of the region's many tourist agencies also organise food tours for their customers. Commercial fisheries are also quite active in this region, and they play a significant role in production of healthy local food. However, these kinds of businesses face challenges due to protective regulatory measures and private consumption (Setälä et al. 1999), especially since private consumption is linked with emotional and traditional values (as discussed elsewhere in this chapter). Restaurants in the EHA also serve wide ranges of traditional foods prepared with local ingredients. For example, Nili Restaurant, a popular restaurant in Rovaniemi, Finland, serves traditional Sami and Finnish food (Nili Restaurant 2019). Even the restaurant at the Arctic Centre in Rovaniemi (which is open to the public) features food prepared from the 'finest raw ingredients sourced directly from Lapland' (Arktikum 2019). Similar approaches to food service are found in many restaurants at universities, schools, and hospitals across the region.

Threats to the TFS in the EHA

The stability of the TFS in the EHA is in jeopardy because of changes in the natural environment due to climate change. Increased human activities such as tourism, mining, and extractive industrial developments and infrastructural changes also disturb the ecosystem in the long run, interrupting the natural processes of food systems (Hassen 2016). Climate change is affecting the stability of Arctic's food systems (FAO 2008b), and climate impacts are visible in all components of local and traditional food systems. For example, extreme increases in temperatures and changes in rainfall patterns have affected the growth of crops, pests, pathogens, and weeds (Myers et al. 2017). The Ministry of the Environment and Statistics Finland predict that moderate increases in temperature will not only enhance crop growth but will also increase the risk of pest outbreaks and diseases in animals (Ministry of the Environment and Statistics Finland 2017). This is true for other areas in this region as well, and these changes alter natural foods systems. The major impact of climate change on food systems is uncertainty; its impacts depend on the connections amongst many physical factors such as temperature and carbon dioxide levels (FCRN 2015). Reindeer populations are expected to face unfavourable impacts from climate change (Ministry of the Environment and Statistics Finland 2017). Milder winters and increases in precipitation result in thicker snow, sometimes with icy layers underneath it (Ministry of the Environment and Statistics Finland 2017). This can make it difficult for reindeer to dig for lichen, which means they require more supplementary food (Ministry of the Environment and Statistics Finland 2017). The short-term impacts of climate change have been felt across the EHA. In Northern Sweden, for example, Sami reindeer herders have been affected by droughts and wildfires. In a 2018 study, these herders expressed concern that some of their herds might not survive the year and that young calves, weakened by the long drought, might not be able to 'follow their mothers to new feeding grounds' (Gerasimova 2018). Climate change affects reindeer directly via thermal stress and also indirectly through increased difficulty in accessing food and water due to rising temperatures (Nuttall et al. 2005). During the summer of 2018, increased temperatures affected berries as well (Pohjanpalo 2018); the drought in Sweden affected production of Scandinavian bilberries, cloudberries, and red lingonberries (Rubin 2018).

In addition, traditional food systems are increasingly threatened by mining industries in the region. For instance, the largest underground ore mine in the world is in the city of Kiruna in the Arctic Circle; it stretches across reindeer migration paths (Gaia Foundation 2014). In 2017, the new mining plan in Kiruna (Sweden) involved relocating all 18,000 inhabitants of the city (Rathi 2017). Another Swedish company, Nickel Mountain, is planning to develop nickel mining in Rönnbäcken; there are concerns that this will damage reindeer migration routes and pastures (Gaia Foundation 2014). This is the topic of a major conflict between the Sami people and the government, but the government continues to support economic interest in the mine. Local Sami organisations claim that the mine will destroy the pasture rotation for about 8,000 reindeers (Environmental Justice Atlas 2010). In the Finnish county of Lapland, Northland Mines, Inc. has decided to start mining in Kolari (Similä and Jokinen 2018). This concerned locals and members of the tourism industry in 2014 (Mainio 2014). Iron mines use chemicals, such as xanthates, which can be very harmful to fish and to microorganisms in water (Similä and Jokinen 2018). The plans for the mine show that its processed waters will be pumped to the Tornio-Muonio River (Similä and Jokinen 2018), a key river for salmon fishing; it goes without saying that this would affect the quality of salmon (Similä and Jokinen 2018). In Murmansk Oblast in Russia, the mining industry has been disruptive; the mine in the town of Nikel remains one of worst polluters in the region (Hossain, Raheem and Cormier 2018). The Kola Company is also located in Nikel, and pollution in these regions has harmful impacts on the local wildlife and environment and on foods such as berries and mushrooms (Hossain, Raheem and Cormier 2018).

Another current cause of concern is the planned Arctic Railway, which the Finnish and Norwegian governments are collaborating to build (Saami Council 2018). Construction of tracks alters the natural environment (Finnish Transport Agency 2018). The natural environment of Northernmost Finland is exceptionally vulnerable, and it is projected that changes would have long-lasting impacts on the area (Finnish Transport Agency 2018). The Arctic Railway Plan poses a major threat to Sami reindeer pastures and will fragment the Sápmi homeland (Saami Council 2018). The primary negative impacts of the railway are predicted to be those affecting reindeer husbandry. The impact will be seen in reindeer herding communities across the region, from Rovaniemi in Finland to the Arctic Ocean via Inari and Kirkenes in Norway (Saami Council 2018). The primary impact on reindeer management will derive from its effect on reindeer pastures; the railroad would make pastures unavailable to reindeer, altering herd ranges (Finnish Transport Agency 2018). The railway would also add toxic residues to reindeer fodder and disturb reindeer grazing (Finnish Transport Agency 2018). Accidents involving reindeers in or around the project area are also expected (Finnish Transport Agency 2018). New reports on the Arctic Railway Plan were published at the end of January 2019; these reports discussed changes made to the previous plan to avoid these adverse impacts (Ministry of Transport and Communications 2019a). A further report published by the Ministry of Transport and Communications in February 2019 explicitly states that the Ministry is aiming to minimise the adverse impacts of the Arctic Railway on reindeer herding and concludes that the Sami people should take part in impact assessments (Ministry of Transport and Communications 2019b).

A further threat to the TFSs in this region is posed by the tourism industry, which is expanding due to worldwide touristic interest in the area (Hossain, Raheem and Cormier 2018). Increased use of helicopters and snowmobiles can disturb the animals, including reindeer. They also cause sound pollution, which may force animals to relocate or cause panic flights by birds (Hossain, Raheem and Cormier 2018). Snowmobiles and helicopters are often used for reindeer herding as well, but tourism is increasing their use, disrupting the natural environment and increasing the carbon footprint of such vehicles in the region (Hossain, Raheem, and Cormier 2018). Berry picking, fishing, and hunting are also popular tourist activities, which puts additional pressure on available resources (Snyder 2007). For example, tourists often visit the Deatnu River Valley, which runs between Norway and Finland, for the area's salmon fishing (Hossain and Petretei 2016). Tourist companies across Finnish Lapland, such as Naamisuvanto Salmon Fishing in Rovaniemi (Visit Rovaniemi 2019), organise guided fishing tours. Furthermore, around 2,000 foreign hunters visit Finland per year. Although they must obtain hunting licences, these hunts are organised by tourist companies in Swedish (Limpopo and Diana Hunting Tours 2019) and Finnish Lapland ('Hunting: Gamefowl permit sales begin' 2019). This alters the natural processes of the food systems, eventually threatening their overall structure.

The natural environment of the EHA has been altered by changes ranging from climate change to increased tourist activities. This alters the stability of EHA food systems, making it more difficult for local Indigenous and non-Indigenous peoples to have adequate access to traditional foods and maintain their traditional livelihoods. As a result, these people must compromise their rights to traditional food.

Human rights frameworks and TFSs: an assessment of the EHA

Human rights framework to analyse TFS

No direct reference to TFSs is found in the framework of human rights. The human right to food encompasses not only the right to physical consumption (which complements the rights to life and health), but also the rights of individuals and communities to cultural, psychological, and spiritual wellbeing. The TFS provides stability and access to locally produced food. The system includes the production, processing, distribution, and consumption of food, as well as all actors involved in these processes, from farmers and producers to consumers. Analysing TFS from the viewpoint of human rights embraces elements that encourage the full enjoyment of many other individual and collective human rights. The international human rights framework consists of a number of instruments. The most fundamental are the Universal Declaration of Human Rights (UDHR), the International Covenant on Civil and Political Rights (ICCPR), and the International Covenant on Economic Social and Cultural Rights (ICESCR). Other instruments that address the right to food, in particular in the EHA, are the International Convention on the Elimination of All Forms of Racial Discrimination (ICERD), the Indigenous and Tribal Peoples Convention (ILO Convention No. 169), the United Nations Declaration on the Rights of Indigenous Peoples (UNDRIP), and the Convention on Biological Diversity (CBD).

The first three instruments – the UDHR, the ICCPR, and the ICESCR, together referred to as the International Bill of Human Rights – endorse the right to food in connection to a standard of living adequate for the health and wellbeing of individuals and families. This endorsement is articulated in Article 25 of the UDHR. Article 11 of the ICESCR highlights the connection between the right to an adequate standard of living and the right to adequate food (para. 1) and also recognises the universal right to be 'free from hunger'. States are therefore obligated to improve the 'method of production, conservation, and distribution of food … by disseminating knowledge … to achieve the most efficient development and utilisation of natural resources' (para. 2[a], ICESCR). The Article also highlights the equitable distribution of global food supplies (para. 2[b], ICESCR). The ICCPR does not directly refer to a right to food. However, the right to life described in Article 6 mentions food as one of the most important elements in the protection of human lives, the promotion of their overall wellbeing, and maintaining their dignity in life. An adequate supply of food that properly fulfils dietary needs only guarantees the most fundamental of all human rights – the right to life. Furthermore, according to General Comment No. 36 on Article 6, the right to life described in Article 6 of the ICCPR is not meant to be 'interpreted narrowly' (ICCPR 2018). Hence, a lack of essentials, such as food, does not create the conditions necessary for humans to not only survive (ICCPR 2018) but to live with dignity as human beings. The General Comment also mentions the resources of Indigenous Peoples, stipulating that any threat to their resources can affect the right to life by jeopardising the dignity of Indigenous Peoples.

Article 27 of the UDHR and Articles 27 and 15(1) of the ICCPR and ICESCR, respectively, address food as a part of culture. Articles 27 of the UDHR and 15(1) of the ICESCR assert the right of individuals to freely take part in the cultural life of a community. Article 27 of the ICCPR addresses the protection of the culture manifested in the practices of minorities which they share and enjoy 'in community with other members of their group'. General Comment No. 23 of Article 27, which is endorsed by the Human Rights Committee (HRC) under the ICCPR, conceptualises the meaning of culture. According to the HRC, culture manifests in many forms, including a particular way of life associated with the use of land resources. This is especially true for Indigenous Peoples, whose cultures may include traditional activities such as fishing and hunting (UNHRC 1994). The Committee on Economic, Social, and Cultural Rights takes a similar approach in General Comment No. 21 (2009) on Article 15(1)(a). According to the Committee, culture encompasses

> *inter alia*, ways of life, language, … belief systems, … natural and man-made environments, food, … [and] traditions through which individuals, groups of individuals, and communities express their humanity and the meaning they give to their existence, and build their world view representing their encounter with the external forces affecting their lives.
> (CESCR 2009)

Since culture is built around food and the food-related habits of a community (among other things), access to food supports individuals' rights to membership in a community. Therefore, the provisions listed here are interpreted to include traditional practices as a part of culture. Food plays a major role in these practices, especially in traditional communities. The UNDRIP lists all the rights that are applicable to Indigenous Peoples, from the practice of culture to participation in decisions relating to their lands, territories, and resources. While the UNDRP does not directly refer to 'food', it acknowledges the rights established under applicable international and domestic law (Article 17), including the improvement of economic and social conditions to promote health and a healthy community (see Articles 21, 23, and 24). The protection of the environment and of the productive capacity of their lands, territories, and resources are also key to maintaining the health of Indigenous Peoples (Article 29). Food practices are an integral part of this.

Articles 47 of the ICCPR and 25 of ICESCR are relevant as they provide for the right of all peoples to enjoy and to fully and freely utilise their natural wealth and resources. The common Article 1 of the ICCPR and the ICESCR is relevant in this context; it describes the rights of all peoples to freely 'pursue their economic, social, and cultural development' (para. 1), and 'to freely dispose of their natural wealth and resources'. It further stipulates the right to not to be deprived of one's 'own means of subsistence' (para. 2). Indigenous Peoples often invoke this article, claiming that they are distinct from the rest of the population in the countries they inhabit and that the right to self-determination therefore applies to them given that they form a distinct group of people. For Indigenous Peoples, self-determination means more autonomy in decisions pertaining to their economic, social, and cultural development. The UNDRIP sets normative standards regarding the rights of Indigenous Peoples; these standards have been endorsed by almost all the states of the world. The exercise of the right to self-determination includes the promotion of traditional food systems that use the lands and resources that Indigenous Peoples own or otherwise occupy. Today, in relation to environment and resource management, other local people also benefit from standard procedural measures which are understood as a form of the right to self-determination. These measures include environmental and social impact assessments and human rights impact assessments. Such measures are becoming more acknowledged in national legal and policy processes and offer locals social and political empowerment in decision-making processes. Therefore, these processes promote human rights standards (Hossain and Petretei 2017).

Other international instruments, such as the CERD and the ILO Convention 169 (1989), provide references for analysing the right to TFSs from various perspectives, such as the viewpoint of non-discrimination and participation in decisions. Article 1 of the ICERD defines racial discrimination, stipulating that no distinctions, exclusions, or preferences are to be made based on race, colour, descent, or national or ethnic origin with a view to impairing equality. However, under some circumstances, special measures should be put in place to ensure fair and adequate development for certain racial groups and for individuals belonging to these groups in order to guarantee full and equal enjoyment of rights (Article 2[2]). Such special measures are especially important for marginalised populations in peripheral regions. The ILO Convention No. 169 provides safeguards for the environment, culture, and property of Indigenous Peoples in independent countries (Article 4). The Convention also recognises the social, cultural, and spiritual values and practices of Indigenous Peoples (Article 5). Article 7 guarantees their right to participate in development processes, and Article 13 ensures respect for the cultural and spiritual values connected to their lands and territories. Articles 14 and 15, respectively, recognise the rights of ownership and possession of lands traditionally occupied by Indigenous Peoples and the right to natural resources pertaining to those lands, which includes the right to participate in the use, management, and conservation of these resources. These provisions necessarily connect TFSs to lands and natural resources. However, the Convention does not apply to the entire EHA region as only Norway has ratified the treaty. Nevertheless, the standards set by the Convention define the norm for practices affecting resources.

The CBD is not a human rights instrument *per se*, but it includes provisions pertaining to the protection of 'bio-cultural rights' which are applicable to the EHA. In its Preamble, the CBD articulates the critical importance of meeting needs for food, health, and other essentials and the importance of sharing genetic resources and technologies. Knowledge and innovation are identified as the criteria for meeting these needs. Article 8(j) addresses the traditional and local knowledge and practices of local Indigenous and non-Indigenous communities that are useful for conservation and the sustainable use of biological diversity. The importance of protecting the traditional knowledge used in traditional food production has been mentioned repeatedly, but there is some legal uncertainty around this issue from the perspective of intellectual property. From a human rights standpoint, the protection of traditional knowledge is key to the human rights of Indigenous Peoples particularly within the scope of the right to enjoy community culture. The CBD therefore interacts with the provisions of the human rights instruments discussed above. Case law jurisprudence also recognises the right to traditional food resources. For example, in the Ogiek case (African Commission on Human and Peoples' Rights v Republic of Kenya 2012), the African Court on Human and Peoples' Rights held that Article 21 of the Charter, which establishes the right of peoples to freely dispose of their wealth and natural resources, was violated because the Ogiek community was 'deprived of their traditional food resources' and also of their 'right to enjoy and freely dispose of the abundance of food produced by their ancestral lands' (201).

The interplay between TFSs and human rights in the EHA

The locally produced foods on which residents of the EHA depend, such as berries, reindeer meat, fish, and cereals, are part of that region's TFS. Local Indigenous and non-Indigenous peoples play an important role in and around this TFS. For residents of the EHA, local, traditional foods are not only key to a healthy diet; they are also part of local emotion and culture, tradition, and spirituality (Hossain, Raheem, and Cormier 2018). For example, berry picking is a popular activity for children and parents – it is a family tradition in the European Arctic (Korpela 2007). Indigenous communities still rely on hunting, herding, fishing, and berry picking for a number of important reasons, including the nutritious value of the food and cultural and social needs (Korpela 2007). To a certain extent, these reasons also apply to non-Indigenous communities in the EHA, particularly the economic and dietary importance of access to traditional foods. Most local, traditional foods, including fish, reindeer, and berries, are nutritionally superior and less expensive than store-bought or imported food products (Korpela 2007). Hence, their consumption supports the right to healthy living conditions, which is implied in the rights to life, health, and a life with dignity.

Traditional food is also an important tool for establishing social relationships according to cultural norms (Nutall et al. 2005). Hunting, herding, fishing, and gathering berries, as well as processing and consuming these foods, are important culturally and socially, especially for Indigenous communities who celebrate these resources (Freeman 2000). In Indigenous communities, such activities define, reinforce, and uphold social relationships and cultural identity (Nuttall et al. 2005; Chapter 7 of this book). They also celebrate the relationships amongst Indigenous Peoples and the animals and the environment they rely on (Callaway 1995). Traditionally, Sami people have a spiritual relationship with reindeer. Whilst acknowledging that humans and animals each have their own place, the Sami people believe that their shared environment creates a social bond between humans and animals (Castro, Hossain, and Tytelman 2016). For certain people, such as reindeer herders, access to traditional reindeer meat is a matter of survival. TFSs enable people to connect with their histories and current culture and provide a way forward for sustainable livelihoods. In the TFS of the EHA, the use of animals stems from the need for survival and is also important for social identity and other cultural reasons (Nuttall et al. 2005). Food from animals is a fundamental requirement for personal, cultural, and mental wellbeing for most Arctic residents. A loss of vitality and decreased health and personal wellbeing have been reported by Indigenous Peoples who cannot consume traditional foods (Wein and Freeman 1992). For Indigenous Peoples in this region, traditional food defines identity by promoting cultural and traditional values which have been passed down for generations (Hossain, Raheem, and Cormier 2018). These cultural and traditional values are further integrated in the management of the natural environment, a task in which traditional knowledge plays a significant role. Traditional knowledge about the relationships among humans, lands, resources, and animals is deeply intertwined with food practices. For both Indigenous and non-Indigenous communities, these practices embody the right to take part in cultural life individually and in community with other members of society.

When analysing the EHA, it is also important to remember that this region is located in the periphery, far from capitals and central administrative hubs. The geophysical and environmental characteristics of this isolated region differ significantly to those of other regions. This reflects the need for the inclusion of local Indigenous and non-Indigenous communities in the governance framework of the EHA. Including local communities as 'subjects' and key actors rather than passive 'objects' in decision-making processes would improve the management of natural resources in this region. The right to decide one's own priorities in the process of development is fundamental to the right to self-determination. Any processes that affect people's lives, beliefs, institutions, or spiritual wellbeing or the lands they occupy or otherwise use affect local communities' socio-cultural development. Therefore, the provisions of human rights instruments discussed above are intrinsically relevant to the management and maintenance of TFSs. Food sovereignty can only be ensured in the EHA when the right to self-determination is promoted in this way.

Conclusion

This chapter began with the phrase 'Human rights begin with breakfast'. Of course, this does not refer to literally placing food on a table. Rather, this chapter examines the importance of every aspect of food to the full enjoyment of human rights. Human rights frameworks do not contain specific reference for food systems or TFSs, but they do articulate the need to ensure adequate food for healthy living conditions. In our view, this goal cannot be realised without a system. As discussed previously, in TFSs, food is gathered or obtained from the natural surroundings and then produced, processed, and distributed. A TFS provides nutritious food that meets the dietary needs of people in a given context, so the availability of and access to local foods promotes healthy living conditions. The three pillars of food security – access, availability, and utilisation – depend on the existence of a TFS. This chapter examines the special characteristics of the TFS of the EHA region, highlighting the importance of this TFS to food security in the region. Access to and maintenance of a stable TSF thus ensures the full enjoyment of human rights for all residents of the EHA.

References

Aas, O. (ed.). 2008. *Global challenges in recreational fisheries*. Oxford: Blackwell Publishing Limited.

Aas, O., A. Klemetsen, and S. Einum (eds.). 2011. *Atlantic salmon ecology*. Chichester: Wiley-Blackwell.

African Commission on Human and Peoples' Rights v Republic of Kenya, No. 006/2012, Judgment, African Court on Human and Peoples' Rights, 6. 26 May 2017. Retrieved 4 December 2019 from http://www.african-court.org/en/index.php/56-pending-cases-details/864-app-no-006-2012-african-commission-on-human-and-peoples-rights-v-republic-of-kenya-details.

Arktikum. 2019. Northern Delights. Retrieved 31 January 2019 from https://www.arktikum.fi/en/cafe.html.

Barents Info (a). Indigenous Peoples in the Barents Region. Retrieved 20 March 2019 from https://www.barentsinfo.org/Contents/Indigenous-people.

Barents Info (b). The Saami. Retrieved 5 November 2018 from https://www.barentsinfo.org/Contents/Indigenous-people/Saami.

Bergflodt, S., B. Amilien, and S. Skuland, S. 2012. Nordic food culture(s) – thoughts and perspectives by way of introduction. *Anthropology of food* [Online]. *Open Edition Journal*, Special Issue 7: 1–17.

Callaway, D. 1995. Resource use in rural Alaskan communities. In D.L. Peterson and D. Johnson (eds.), *Human ecology and climate change: People and resources in the Far North*. Washington, DC: London: Taylor and Francis: 155–168.

Capone, R. et al. 2014. Food system sustainability and food security: Connecting the dots. *Journal of Food Security* 2(1): 13–22.

Castro, D., K. Hossain, and C. Tytelman. 2016. Arctic ontologies: Reframing the relationship between humans and rangifer. *Polar Geography* 38(2): 98–112.

CESCR. 2009. General comment No. 21: Right of everyone to take part in cultural life, para 13.

Elde, S. et al. 2018. *The Arctic as a food-producing region*. Report 10/2018, Nofima. Retrieved 4 December 2019 from https://nofimaas.sharepoint.com/sites/public/Cristin/Rapport%2010-2018.pdf.

Ericksen, P. 2008. What is the vulnerability of a food system to global environmental change? *Ecology and Society* 13(2): 14. Retrieved 4 December 2019 from https://www.ecologyandsociety.org/vol13/iss2/art14/.

Ericksen, P. et al. 2009. Food security and global environmental change: Emerging challenges. *Environmental Science and Policy* 12(4): 373–377.

European Environment Agency. 2017. *Actors in the food chain*. Retrieved 2 November 2018 from https://www.eea.europa.eu/media/infographics/actors-in-the-food-chain-2/view.

FAO. 2008a. *An introduction to the basic concepts of food security*. Rome: EC-FAO Food Security Programme.

FAO. 2008b. *Climate change and food security: A framework document*. FAO, UN. Retrieved 5 October 2018 from http://www.fao.org/forestry/15538-079b31d45081fe9c3dbc6ff34de4807e4.pdf.

FAO. 2016. *Human security and food security*. Retrieved 20 March 2019 from www.fao.org/3/a-i5522e.pdf.

FCRN. 2015. *Food source*. Retrieved 26 January 2019 from https://foodsource.org.uk/book/export/html/6.

Finnish Transport Agency. 2018. *Arctic Ocean railway report*. Finnish Transport Infrastructure Agency. Retrieved 25 January 2019 from https://julkaisut.liikennevirasto.fi/pdf8/lr_2018_arctic_ocean_railway_report_web.pdf.

Forrest, S. 1997. Territoriality and state-Saami relations. University of British Columbia. Retrieved 25 January 2019 from http://arcticcircle.uconn.edu/HistoryCulture/Sami/samisf.html.

Freeman M. (ed). 2000. *Endangered peoples of the Arctic.* Westport, CT: Greenwood Press.

Gaia Foundation. 2014. *Under-mining agriculture: How the extractive industries threaten our food systems.* London: The Gaia Foundation.

Gerasimova, A. 2018. Sweden's reindeer at risk of starvation after summer drought. *Reindeer Herding.* [Web log post]. Retrieved 5 October 2018 from http://reindeerherding.org/blog/swedens-reindeer-at-risk-of-starvation-after-summer-drought/?cn-reloaded=1.

Hassen, Y. 2016. *The impacts of mining on Arctic environment and society from corporate social responsibility and sustainable development perspectives.* Stockholm: Stockholm University.

Heart of Lapland. 2018. *Delightful local cruising – enjoy the taste.* Retrieved 19 November 2018 from http://heartoflapland.com/explore-heart-of-lapland/food-and-drink/.

Hossain, K. and A. Petretei (eds.). 2016. *Understanding the many faces of human security: Perspectives of northern Indigenous Peoples.* Leiden: Brill.

Hossain, K. and A. Petretei. 2017. Resource development and Sámi rights in the Sámi region: Integrating human rights impact assessment in licensing processes. *Nordic Journal of International Law* 86(3): 302–340.

Hossain, K., D. Raheem, and S. Cormier. 2018. *Food security governance in the Arctic-Barents region.* Cham: Springer International Publishing.

Hunting: Gamefowl permit sales begin. *Eräluvat*, 7 June 2019. Retrieved 31 July 2019 from https://www.eraluvat.fi/en/hunting/gamefowl-permit-sales-begin-e2-88-92-check-your-regional-quotas-and-the-instructions-on-how-to-purchase-a-permit.html.

ICCPR. 2018. *General comment No. 36 on Article 6 of the ICCPR*, para. 26.

IUCN. 2013. Food security policies: Making the ecosystem connections. Retrieved 21 March 2019 from http://www.fao.org/forestry/37145-07a8322d5c9e07487af4c8a94caba0291.pdf.

Jones, A. et al. 2013. What are we assessing when we measure food security? A compendium and review of current metrics. *Advanced Nutrition* 4(5): 481–505.

Kagda, S., B. Cooke, and D. Nevins. 2017. *Cultures of the world, 3rd ed.* New York: Cavendish Square Publishing LLC.

Karpukhin, S. 2016. Reindeer herding in Russia's remote Arctic. *Reuters*, 14 December 2016. Retrieved 25 January 2019 from https://www.reuters.com/article/us-russia-reindeer-idUSKBN14317S.

Korpela, S. 2007. Treasures of the boreal forests. This is Finland. Retrieved 5 October 2018 from https://finland.fi/life-society/treasures-of-the-boreal-forests/.

Kuhnlein, H. 2009. Why are Indigenous Peoples' food systems important and why do they need documentation?. In Kuhnlein, H., B. Erasmus and D. Spigelski (eds), Indigenous Peoples' food systems: the many dimensions of culture, diversity and environment for nutrition and health. Rome: FAO.

Kuhnlein, H. et al. (eds.). 2013. *Indigenous Peoples' food systems and well-being.* Rome: FAO.

Kuruzovic, S. 2020. Picking flowers, berries, mushrooms, etc. Swedish Environmental Protection Agency. Retrieved 15 May 2020 from http://www.swedishepa.se/Enjoying- nature/The-Right-of-Public-Access/This-is-permitted/Picking-flowers-berries-mushrooms-etc/.

Limpopo and Diana Hunting Tours. 2019. Moose hunt in Lapland. Retrieved 31 July 2019 from https://www.diana- hunting.com/pdf/597-3.pdf.

Mainio, T. 2014. Mine plan stirs up anxiety in Lapland. *Helsinki Times*, 23 March 2014. Retrieved 27 January 2019 from http://www.helsinkitimes.fi/finland/finland-news/domestic/9926-mine-plan-stirs-anxiety-in-lapland.html.

Ministry of the Environment. 2016. *Everyman's right: Legislation and practice*. Retrieved 15 May 2020 from https://www.ym.fi/download/noname/%7B595923BE-007D-4405-B69C-1748A02055EF%7D/57650.

Ministry of the Environment and Statistics Finland. 2017. *Finland's seventh national communication under the United Nations Framework Convention on Climate Change*. Retrieved 25 January 2019 from https://www.stat.fi/static/media/uploads/tup/khkinv/VII_Climate_Change_16102017.pdf.

Ministry of Transport and Communications. 2019a. *Ministry of Transport and Communications and Sámi Parliament discussed again the Arctic railway report*. Ministry of Transport and Communications 2019.

Ministry of Transport and Communications. 2019b. *Final report of the joint working group between Finland and Norway on the Arctic railway*. Ministry of Transport and Communications 2019.

Myers, S. et al. 2017. Climate change and global food systems: Potential impacts on food security and undernutrition. *Annual Review of Public Health* 38: 259–277.

Nili Restaurant. 2019. Menu. Retrieved 31 January 2019 from http://www.nili.fi/en/food-drinks/rovaniemi-menu.html.

Nord Norge (a). North Norwegian cuisine. Retrieved 15 November 2018 from https://nordnorge.com/en/north-norwegian-cuisine.

Nord Norge (b). The food is half the journey. Retrieved 15 November 2018 from https://nordnorge.com/en/north-norwegian-cuisine/?News=80.

Norwegian Environment Agency. 2013. Right to Roam. Retrieved 31 July 2019 from http://tema.miljodirektoratet.no/Documents/publikasjoner/M86/M86.pdf.

Nuttall, M. et al. 2005. Hunting, herding, fishing and gathering: Indigenous Peoples and renewable resource use in the Arctic. *Arctic climate impact assessment*. Cambridge: Cambridge University Press.

Paci, C.D. et al. 2004. Food security of Northern Indigenous Peoples in a time of uncertainty. Retrieved 5 October 2018 from https://www.rha.is/static/files/NRF/OpenAssemblies/Yellowknife2004/3rd-nrf_plenary-4_pp_paci-et-al.pdf.

Pohjanpalo, K. 2018. Heatwave threatens Nordic crops as Lapland gets new heat record. *Bloomberg*, 19 July 2018. Retrieved 5 October 2018 from www.bloomberg.com/news/articles/2018-07-19/finland-braces-for-power-shortage-as-reactors-fail-amid-heatwave.

Rathi, A. 2017. A Swedish mining company is moving an entire town of 18,000 people – including its buildings – to a new location. *Quartz*, 25 May 2017. Retrieved 6 November 2018 from https://qz.com/991623/a-swedish-mining-company-is-moving-the-entire-town-of-kiruna-with-18000-people-and-several-buildings-to-a-new-location/.

Reindeer Herders' Association. 2014. Reindeer: Poro. Retrieved 31 July 2019 from https://paliskunnat.fi/reindeer-herders-association/wp-content/uploads/2014/12/reindeer_english_web.pdf.

Rönnbäcken nickel mine, Västerbotten, Sweden. 2010. *Environmental Justice Atlas*. Retrieved 6 November 2018 from https://ejatlas.org/conflict/ronnbacken-nickel-mine-vasterbotten-sweden.

Rubin, A. 2018. Scorching summer in Europe signals long-term climate changes. *The New York Times*, 4 August 2018. Retrieved 5 October 2018 from https://www.nytimes.com/2018/08/04/world/europe/europe-heat-wave.html.

Saami Council. 2018. *Statement on the Arctic railway*. Retrieved 7 November 2018 from http://www.saamicouncil.net/en/article/statement-on-the-arctic-railway/.

Setälä, J. et al. 1999. Review of the fish market in Finland. *Boreal Environment Research* 3: 361–370.

Similä, J. and M. Jokinen. 2018. Governing conflicts between mining and tourism in the Arctic. *Arctic Review on Law and Politics* 9: 148–173.

Snyder, J. 2007. *Tourism in the polar regions: The sustainability challenge*. Washington: UNEP.

Stepien, A. and T. Koivurova. 2017. *Arctic Europe: Bringing together the EU Arctic policy and Nordic cooperation*. Retrieved 4 December 2019 from http://julkaisut.valtioneuvosto.fi/bitstream/handle/10024/160217/Artic_Europe_Bringing_together.pdf.

Stupachenko, I. 2018. Can Russia's Arctic deliver on big fishing promises? Seafood Source. Retrieved 25 January 2019 from https://www.seafoodsource.com/features/can-russias-arctic-deliver-on-big-fishing-promises.

Swedish Environmental Protection Agency. 2019. Hunting permit. Retrieved 31 July 2019 from http://www.swedishepa.se/Enjoying-nature/Hunting/Hunting-permit1/.

Tamnes, R. and K. Offerdal (eds.). 2014. Geopolitics and security in the Arctic: Regional dynamics in a global world. In *Routledge Global Security Studies*. London: Routledge: 5–6.

Torp, E. 2013. The legal basis of Sámi reindeer herding rights in Sweden. *Arctic Review on Law and Politics* 4(1): 43–61.

Tuunanen, P., M. Tarasti, and A. Rautiainen (eds.). 2015. *Everyman's rights and the code of conduct on private land: Existing legislation and suggestions for best practices*. Ministry of the Environment.

UNHRC. 1994. *General comment 23: The rights of minorities* (Art. 27), para 3.1.

Visit Rovaniemi. 2018. The pure tastes of Rovaniemi. Retrieved 19 November 2018 from https://www.visitrovaniemi.fi/love/local-food/.

Visit Rovaniemi. 2019. Naamisuvanto salmon fishing. Retrieved 27 January 2019 from https://www.visitrovaniemi.fi/activity-company/naamisuvanto-salmon-fishing/.

Wein, E. and M. Freeman. 1992. Inuvialuit food use and food preferences in Aklavik, Northwest Territories, Canada. *Arctic Medical Sciences* 51: 159–172.

Westhoek, H., S. Berkum, and M. Hajer. 2016. *Food systems and natural resources*. The International Resource Panel, UNEP.

7 Sami identity and traditional livelihood practices

From non-Indigenous to Indigenous food frameworks

Corinna Casi

Introduction

This chapter explores the importance of the traditional food system in Indigenous cultures, particularly in Sami Indigenous communities living in the European High North. Instead of focusing on food culture, this study will investigate the relationship between traditional Sami livelihood practices, their identity, and self-determination as Indigenous People. Food ethics (Zwart 2000; Scrinis 2008; Siipi 2015; Thompson 2015), food sovereignty (Menser 2008; Schanbacher 2010; Werkheiser 2016; Glennie and Alkon 2018), and food security (ICESCR 1966; Legat et al. 1995; FAO 2003; Pimbert 2008; Pinstrup-Andersen 2009; Conway 2012; Murdock and Noll 2015) are the initial frameworks used in this chapter. The study will be further guided by the following research questions:

1 Why are the traditional livelihood practices of reindeer herding and salmon fishing of primary importance in Sami communities?
2 What roles do the frameworks of food security, food ethics, and food sovereignty play in highlighting the relationship between food systems, identity, and self-determination in Sami communities?
3 Why it is important to move from non-Indigenous to Indigenous frameworks?

In response to these questions, this study will briefly explore the general features of the Sami way of living, the significance of reindeer herding and salmon fishing as traditional livelihood practices, and the ties between these practices and traditional food systems. Then, the non-Indigenous theoretical frameworks used in this study will be employed to develop the philosophical reflection about the importance of traditional food systems in Sami culture and the link between food systems and Indigenous identity.

This chapter will also argue that traditional Sami food choices are part of the Sami way of life and therefore interlinked with their integrity, dignity, wellbeing, and self-determination as Indigenous People. The role of food in

DOI: 10.4324/9781003057758-7

traditional Sami culture is not different to its role in many other Indigenous cultures; many Indigenous People perceive the same connection between food and identity. Therefore, this chapter will also compare traditional Sami way of living to those of Native American tribes who engage in similar traditional livelihood practices such as salmon fishing. This comparison will help broaden our understanding of the Sami way of life. Because of their pivotal role in Sami culture, traditional dietary habits fall under the category of 'overriding values', which, according to a deontological perspective, are values so meaningful that they cannot be disregarded nor overlooked. This chapter will conclude with a discussion of the importance of moving from non-Indigenous to Indigenous frameworks to pragmatically empower the Sami people.

Sami traditional livelihoods and Sami Indigenous awareness

The Sami are the Indigenous People situated in the Northernmost Europe, also called the European High North. Sami society is generally organised into small communities spread all over Sapmi, the Sami land, commonly known as Lapland. This region extends across Northern Norway, Sweden, and Finland and the north-western part of thc Russian Federation (UNRIC 2019).

During the eighteenth and nineteenth centuries, the Sami people were vulnerable to the oppressive racial policies instigated by the national governments of the countries enveloping their territories. This situation remained largely unchanged until around the late nineteenth and early twentieth centuries, when the Sami became involved in activism (Nyyssönen 2016, 284). In the period following World War II, the Sami people gradually became more active in defending their own rights and their lands. They also started to demand greater autonomy. Prominent achievements included the establishment of Sami Parliaments in Norway, Finland, and Sweden in the 1980s and 1990s (Greaves 2016, 469; Pedersen 2016, 282). Essential to these struggles are the aims of self-determination and the responsibility to protect Sami lands from exploitation, pollution, and uncontrolled development.

Reindeer herding, also known as reindeer husbandry, is the most common traditional livelihood practice in Sami communities (Åhrén 2016, 281). This activity is combined with and complemented by other livelihood activities: hunting game, such as fox and elk; and fishing, mainly for salmon, whitefish, char, and salmon trout. Other traditional activities include gathering mushrooms, berries such as cloudberries, lingonberries, and bilberries, and a variety of wild herbs, including sorrel, *Angelica archangelica*, and *Mulgedium alpinum* (Nilsson et al. 2011, 307–309; Samer 2019). Traditional Sami knowledge of food systems ensures a sustainable supply of nutritious food through harsh winters. These activities are important for private consumption and subsistence but also for trade. This chapter examines the Indigenous community's food system as a whole. A food system is a multidimensional chain of food production; in the case of Indigenous communities, such as the

Sami, it includes all aspects of food hunting, harvesting and gathering, distribution, preparation, storing, consumption, and recycling or disposing of food waste (Whyte 2016, 6, 16).

For Sami communities living in coastal areas or close to waterways, salmon fishing is a vital livelihood practice (Pedersen 2016, 281), as significant as reindeer herding. However, in recent years, the introduction of trading fishing quotas and the creation of the modern salmon farming industry have greatly reduced the incomes of Sami traditional salmon anglers (Pedersen 2016, 282). Understandably, food obtained through traditional Sami livelihood activities is contingent on various conditions and circumstances in the natural environment. Natural factors such as seasons, the weather, and the climate affect the natural environment and thereby directly impact the traditional livelihood practices of hunting, fishing, and gathering.

In many societies, including that of the Sami, 'longstanding social relations are brought together through the production, preparation, and consumption of food' (Thompson 2015, 75). This means that the Sami peoples' livelihood practices are vital elements of their culture. As such, those practices are not only a means of subsistence; they are parts of complex food systems that form their identity as a community.

Ethics of food and the value of food

In this section, a food ethics framework will be used to outline the value of food in a society and specifically in Sami communities.

Food ethics is a philosophical academic discipline established at the end of the twentieth century. As one branch of applied ethics, it offers a moral analysis and guidance for human behaviour related to food (Mepham 2000, 610). Although this field of study is relatively young, ethical discussions about food and dietary conduct have taken place throughout history. In Ancient Greece, for instance, ethical considerations around food focused on fair measures and temperatures. Contemporary reflections on food ethics examine its production, distribution, preparation, and consumption (Zwart 2000, 113). Several fields of study acknowledge the meaning of food; considerable scholarly works on food ethics have come out of various fields, including anthropology, economics, environmental sciences, and law (Barnhill et al. 2018, 1). The present study examines the value of food and its connections to other ethical questions such as human dignity, identity, and wellbeing. It also points out that traditional livelihood practices in Sami culture are an overriding value.

The value of food can be considered from multiple perspectives: from the standpoint of knowledge values, cultural values, spiritual values, or health values, among others. The value of food includes its social, cultural, and aesthetic aspects and the symbolic meanings of food, all of which significantly impact human wellbeing. Examples include sharing a meal with family and friends to celebrate a happy event (Siipi 2015, 194). Food has the power to create identity and perpetuate culture in a community, as discussed by the renowned food ethics scholar, Paul B. Thompson (2015). The entire process of food production, preparation, and consumption builds connections in local communities (Thompson 2015, 75–76). This view follows the idea that direct contact with and individual commitments to food production create a special bond between people and food. It also promotes a particular value of care for the food in question, as suggested by the food scholar Ian Werkheiser when claiming that 'the experiences of producing food – working with the land, cooperating with one another by necessity, watching something grow and caring for it [...] – encourage particular type of values and culture practices for many who works as producers' (2016, 381).

Therefore, the methods used in food production as well as food 'distribution and consumption affect us indirectly through their social and environmental impacts' (Thompson and MacDonald 2013; Siipi 2015, 194). Evidently, food, and particularly traditional Sami food, is profoundly linked to important spheres of life such as identity, culture, and wellbeing, since the value of food for humans goes beyond its nutritional content (Scrinis 2008; Whyte 2016).

Reindeer herding and Sami identity

The Sami Indigenous People and other local non-Sami communities living in the European High North have been practicing reindeer husbandry for centuries. However, in recent decades, this traditional livelihood activity has come under increased threat due to the deterioration of grazing land and lichen pastures, the main food for reindeers. This gradual diminishing of resources has been triggered by modern forms of land use. Forestry, especially the clear cutting of old-growth forests rich in arboreal lichens, is a major culprit, along with tourism, hydropower generation, nature conservation policies (Kyllönen et. al. 2006, 695–697), mining, and changes due to the climate crisis. For example, recent plans to construct an Arctic railway across northern Norway and Finland could worsen the situation for reindeer herders as the railway could disrupt traditional reindeer migratory paths (Gertz 2017; Staalesen 2018).

The deep connection of the Sami people to their food traditions has been explored by Sami scholar Jon Petter A. Stoor. Stoor et al. (2015) highlight and strengthen the link between 'food production' and Sami identity, focusing on the meaning of reindeer herding. This study investigated the cause of high suicide rates in the Sami community by interviewing Sami people living in Northern Sweden. In Sami culture, life as a herder is 'understood as the best way to preserve Sami identity and to be able to pass the legacy on to the new generation' (Stoor et al. 2015, 3). For the Sami, salmon fishing and the reindeer herding are traditional practices that are as meaningful as their own lives and identities. Stoor's article shows how threats to reindeer husbandry have disrupted the balance between the traditional Sami food system and Sami identity. Some negative impacts of this disruption are disconnection from one's own past and traditions, a loss of the sense of community belonging and, in certain cases, high suicide rates in the Sami community (Stoor et al. 2015). Some Sami who lose their reindeer herds or the right to fish can see no reason to continue living. In this light, the suicide might represent a way to elude the existential void (Stoor et al. 2015, 1, 4). The article clearly demonstrates the importance of reindeer herding to the Sami community, both as a primary source of food and as a source of cultural identity. Moreover, this livelihood activity helps maintain intergenerational links, connecting Sami culture to past, present, and future generations.

Dignity, identity, overriding values, and negative duties

Deontology is a traditional philosophical theory concerned with moral actions and its ethical motivations. According to deontology, human beings have rights, duties, and responsibilities toward their fellow humans. The word 'duty' comes from the Greek word *deon*, from which this theory took its name. Human communities create human rights to 'ensure a certain baseline level of decent treatment for all persons' (Orend 2002, 142). One of the first advocates of deontology was the German philosopher Immanuel Kant, who argued that people are not solely means or tools to our personal goals but rather 'ends in themselves' (Kant 1983, 1996, 210). The theory presents individuals as rational, unique, and therefore deserving of rights and due respect as autonomous individuals (Kant 1996, 210–211) who are free to choose their own identity and way of life. Human life, dignity, and integrity are key values that cannot be ignored or overridden; as such they should be protected. These values cannot be disregarded or usurped as they play a crucial role in ensuring a decent human life.

Following this argument, the professor of philosophy Gary Comstock (2013) defines 'overriding values' as interests, beliefs, and aspects of life that are vital to an individual's identity and worldview. Based on our understanding of the Sami identity and way of life, it is reasonable to opine that their food traditions belong to the sphere of 'overriding values'. Traditional Sami methods of food production are of utmost importance to decisions regarding the kind of life a Sami wants and chooses to live. Therefore, these activities are essential to the Sami's self-determination as Indigenous communities, as Stoor et al.'s article also demonstrates (2015).

If Sami food traditions are overriding values, they can generate negative duties under a decolonial approach. A negative duty is 'a duty not to inflict grievous and unjust harm on another' (Orend 2002, 143). In this context, Nordic states have a negative duty not to pollute the natural environment and not to overuse natural resources in Sami lands.

In summation, Sami food traditions represent a set of overriding values. As such, those values should generate negative duties for the governments of Nordic states to prevent pollution of the Sapmi natural environment in order to provide the best opportunities for the Sami people to continue their traditional livelihood practices of reindeer herding, salmon fishing, berry gathering, hunting, and harvesting.

Food sovereignty and food security as theoretical frameworks

The concept of food ethics has been used to explain the general features of the value and importance of food in Sami culture. In this section, the theoretical frameworks of food security and food sovereignty will be used to stress the connection between food systems, policies, justice, and wellbeing in society.

The term 'food security' was coined during reconstruction after World War II. At that time, it was applied with the aim of making nations food secure. The meaning of the phrase 'making nations food secure' is not an easy one to explain because it has taken on different connotations over time. Initially it was used to discuss issues of food supply and availability. It meant ensuring that people would have enough food for a long period of time at the national and international scale (FAO 2003). Gradually, the term has expanded and been applied to cities, communities, households, and individuals (Pinstrup-Andersen 2009; Conway 2012). Due to ecological, social, and political changes, the term 'food security' has been used in various ways and for a range of functions. Therefore, Indigenous Peoples, including the Sami, may have an understanding of 'food security' that differs to that of the settlers. Herein lies the significance of 'indigenising' the discourse around food security (discussed later in this chapter). At the World Food Summit in 1996, the Food and Agriculture Organisation of the United Nations (FAO) defined the term in this way: food security exists 'when all people, at all times, have physical and economic access to sufficient, safe and nutritious food that meets their dietary needs and food preferences for an active and healthy life' (FAO 1996). Earlier in 1994, the United Nations Development Program had already identified food security as one significant aspect of human security (UNDP 1994). Therefore, the FAO (1996) definition is significant as it acknowledges food as a human right, as stated in Article 11 of the United Nations' International Covenant on Economic, Social, and Cultural Rights (ICESCR 1966, 7).

However, in this context, a 'lack of food security is largely understood as an "access" issue' (Murdock and Noll 2015, 328). Nonetheless, this is not always the case; an example is the contamination of Columbia River salmon. Like many Sami communities, a number of Native American tribes, including the Warm Springs Tribe, the Yakama Nation, the Umatilla Confederated Tribes, and the Nez Perce Tribe, have based their livelihoods on salmon fishing from the Columbia River in the Pacific Northwest (USA) for many generations. However, in the late 1950s, an increase in farming, timber industries, and mining activities played a role in the contamination of the Columbia River, as described by the Indigenous scholar Lori Lambert (2008).

As a consequence, large amounts of toxins contaminated the salmon in the river, making it unfit for human consumption. To mitigate this predicament, the federal government offered food replacements to the Native American tribes in the form of processed canned fish. However, they continued to practice traditional fishing activities and to consume fish from the polluted Columbia River, disregarding governmental health warnings.

It is a good strategy to meet the dietary needs of Indigenous Peoples by ensuring adequate distribution of and access to safe, nutritional food, as per the 1996 FAO definition of food security. However, in this case, the governmental solution was put forward without consulting Native Americans about their food. Therefore, this solution failed to acknowledge the deep connection between traditional salmon fishing and Indigenous identity. This scenario shows how an 'access-focused' solution can easily fail to acknowledge the cultural and social value of salmon for Indigenous People such as the Sami or the Native Americans of the Pacific Northwest. Replacing fished salmon with canned fish assumes a culturally universal interpretation of food practices that, obviously, proved unsatisfactory in this case. It is now clear that both the Native tribes and the Sami people possess a special food culture which is deeply connected to their way of life and to their natural environment, including the non-human world of animals, plants, and inanimate nature (Legat et al. 1995). These food cultures are therefore deeply connected to their Indigenous identity.

In response to disillusionment with food security, understood as the dominant global discourse on food policy, some scholars claim that, when based only on economic growth and market strategies, food security may not be the most suitable political tool for eradicating poverty and inequality (Pimbert 2008; Schanbacher 2010; Wittman et al. 2010). To create a fairer food system, they rather support the idea of food sovereignty, which is based on values such as sustainability, environmental impacts, equal access to lands, and the protection of traditional livelihood activities. Food sovereignty discourses are part of the debate on food justice, a field that examines how injustices and inequalities in the current food system are patterns connected to gender, class, and race (Glennie and Alkon 2018, 7). The term 'food sovereignty' was established internationally in 1996 during the FAO World Food Summit by La Via Campesina, a self-managed organisation working with matters of food, justice, and identity (Patel 2009). Over the years, the term 'food sovereignty' has attracted varied interpretations. According to community activists, the concept indicates 'how people and communities should have sovereignty over their food systems' (Werkheiser 2016, 378). Food sovereignty advocates for the right of local communities to practice full control over their food. Sami communities, for example, should be free to identify their own food policies that do not necessary follow market mechanisms and mass production. Thus, food sovereignty focuses on the needs of local people and highlights the significance and implications of local food production (Menser 2008; Schanbacher 2010). From a food sovereignty perspective, the

Sami people have the right to control the Sapmi natural environment, which is the source of most of their traditional food.

Ultimately, the frameworks of food ethics, food security, and food sovereignty emphasise the relationship between the Sami peoples' livelihood activities and their dignity, self-determination, and identity as the Sami Indigenous community, even though the discourses around these topics remain in the settler's realm and language.

Moving beyond non-Indigenous frameworks: the connection between self-determination and Sami traditional livelihood practices

The anthropologist Elaine Gerber suggests that 'food and eating serve as important and symbolically laden sites through which people contest their identity and mark their independence' (Gerber 2014, 516). The value of food as a symbol of self-determination and identity for the Sami people is crucial and deserves further investigation to move this debate into a context that considers Indigenous perspectives. From this perspective, the substitution of salmon with canned fish, in the Columbia River's case, reinforces a structure of subordination (Plumwood 2000; Cohoon 2014) of Indigenous People. Under the framework of food sovereignty, this substitution seriously hinders the Native tribes' freedom and self-determination as Indigenous People. Equally, from a food ethics perspective, controlling the Sami's food harvesting, gathering, and production affects their self-determination as Indigenous People who want to preserve their traditional way of living. Yet, in recent decades, national governments of regions the Sami people inhabit have established regulations that limit the salmon fishing and reindeer herding practices of Sami communities. These restrictions on their livelihood practices eliminate the Sami's traditional means of support, strip them of their autonomy to make free choices about their traditional livelihoods, deny their right to self-determination, and force them to depend on external market-based supplies for their daily subsistence.

In struggles over food sovereignty, the connections between questions of food and other matters such as land use and political, economic, ecological, and social issues emerge quite clearly (Werkheiser 2016, 378). In food security discourses within Indigenous communities – also called Indigenous food security – all these aspects are not only connected but also tightly interrelated. For instance, in Indigenous communities, ecological problems can have repercussions at multiple levels: cultural, nutritional, political, and economic. In a more practical sense, issues such as pollution, climate crises, and the overuse of natural resources can directly affect fisheries; the lives and food supplies of reindeer, including lichens; and the food supplies of other wildlife in the European High North (Kyllönen et al. 2006, 696–697). As a consequence, Sami people have suffered food insecurity, defined as insufficient access to and availability of food for an active and healthy life

(FAO 2015). Moreover, we can infer that food insecurity and environmental instability have the potential to shake the delicate political equilibrium of this region and disrupt the Sami's political autonomy on several fronts. Food scarcity in Sami lands presents serious challenges to the Indigenous population, especially by severely limiting their social and physical freedom. The Sami people are not able to exercise free choice and are pressured into buying food in supermarkets to compensate for their food insecurity. This scenario directly threatens Indigenous People's self-determination, dislocating them from food self-sufficiency, food independence, and forcing them to depend on food markets. Therefore, to empower Sami people, significant attempts should be made to decrease their dependence on global food markets and revitalise Indigenous Sami food traditions. Scholars such as Delormier et al. (2017) and Coté (2016) have suggested similar solutions to help other Indigenous communities regain their autonomy and sovereignty over their food systems.

According to Kyle Powys Whyte (2016), a Potawatomi Indigenous scholar and activist, food motivates Indigenous Peoples' collective capacities. The term 'collective capacities' refers to a complex ecological system, or ecology, of interrelations among humans, non-human animals, plants, spiritual creatures, and landscapes based on their own origins, cultural narratives, religion, ways of life, political and economic systems. The Australian feminist philosopher Val Plumwood, for instance, calls Indigenous Peoples' wisdom and respect for food 'sacred eating' (Plumwood 2000, 298–303). Applying Whyte's notion of 'ecology' to Sami communities highlights the interconnections between their livelihood practices, wise resource use, traditional ecological knowledge, and worldview. This is also linked to past, present, and future generations; to natural cycles of life and death; and to animate and inanimate elements of nature (Helander-Renvall 2016; Casi 2019). This view is shared by scholars who depict the value of food as a hub or force 'pulling certain people, nonhumans, and ecosystems together in ways that promote collective action' (Werkheiser 2016, 383; Whyte 2016, 10).

Using the dominance of their own culture, settlers try to transform Indigenous territories in their own homelands economically, politically, and culturally. Industrialisation, resource extraction, deforestation, and forced integration are all part of this process (Whyte 2016, 12–13). Whyte (2016), in agreement with anthropologist Tate A. Lefevre (2015) and other Indigenous, feminist, and environmental justice scholars (LaDuke 1993; Calhoun et al. 2007; Walker et al. 2013), calls this process 'settler colonialism', a structure of oppression by the dominant civilisation. In this gradual process, the settlers' overbearing attitude targets the natural environment for resource extraction and land use, polluting the air, soil, and water and destroying aquatic and terrestrial habitats that Indigenous Peoples use for food production (Agyeman et al. 2010). Consequently, the cultures of local Indigenous communities are erased as the adaptation capacities of their ecologies are removed. This limits traditional Indigenous ways of life and eradicates their plans, possibilities, and hopes for the future (Walkers et al. 2013). From this perspective, the Sami people living according to their traditional Indigenous way of life experience intersectional coercions. This means that they experience deeper simultaneous social pressures than non-Sami residents in the same area. The term 'intersectionality' was first used by the civil rights advocate Kimberlé Williams Crenshaw to describe the coercions women of colour experience under the dominant culture (1989). The term describes the multiple concurrent factors that oppress black women in society; considering these pressures separately does not illustrate the full situation. Since then, the notion of intersectionality has expanded to other several social categories and fields of study. Hence, from an intersectional perspective, the Sami people face more interlinked, simultaneous social pressures than non-Sami in the same region. Some of those social pressures are due to race, class, and religion. However, in the specific case of Sami people, other pressures apply as well. Sami face oppression due to long-standing colonial attitudes, human rights violations done to those living according to Sami culture, racism and unequal treatment, forced integration, destruction of their social systems, disruption of intergenerational knowledge transmission, dispossession of their homelands, historic traumas, etc. In addition, colonial attitudes continue to affect the Indigenous Sami food systems on a greater scale than in the past (Sarche and Spicer 2008). To present a more complete picture, an intersectional viewpoint has led Indigenous activists to reformulate the links among nutrition, colonialism, and certain diseases (Whyte 2016).

'Decolonising' food security and 'Indigenising' food sovereignty

Food ethics, food security, and food sovereignty were useful frameworks for highlighting the relationships among food systems, livelihood practices, and identity in Sami communities. However, in order to bring those discourses closer to Indigenous matters, I suggest, in line with Indigenous scholars (Coté 2016; Greaves 2016), to engage in processes of 'Indigenising' food sovereignty and 'decolonising' food security when applying these frameworks to Sami communities.

When interviewing Sami living in the Norwegian High North, the political scientist Wilfrid Greaves (2016) found that Sami identified few significant factors that define security. In addition to traditional food and Indigenous ways of life, the Sami identified other priorities as follows: protection of the natural environment, autonomy, and self-determination within the settler Nordic states, and establishment of Sami identity through cultural practices (Greaves 2016, 473). Greaves points out that different communities define security – which also includes food security – in various ways depending on their history and social and political development. Although the Sami people identified several dangerous threats to their livelihood, they deliberately chose not to describe these priorities as security issues (Greaves 2016, 462). In their view, language about 'security' only refers to hard security and military interventions by nation-states (Greaves 2016, 472–473). This term therefore remains in the realm of the dominant society.

To develop comprehensive food security policies in Sami communities, it would be helpful if the Nordic states employed a more inclusive, respectful approach that heeds the core needs of Indigenous Peoples as well as their own viewpoints. Along these lines, Greaves emphasises the need for Nordic states to decolonise their ideas of security from their political agendas by allowing affected Sami populations to define security from their own Indigenous perspectives (Greaves 2016, 463). In addition, Indigenous communities should be given appropriate tools to fight food insecurity within their homelands. Important 'tools' would include safeguarding the protection of Sami lands from pollution and extractive companies and fostering the healing of previously contaminated lands. This process should stem from a decolonialising practice in which the Nordic states acknowledge the overriding value of traditional Sami livelihood practices. This will also support the Nordic states' negative duty to not perpetrate 'unjust harm on another' (Orend 2002) and therefore to not limit the survival of the Sami's 'collective capacities' (Whyte 2016). In this context, the notion of 'decolonisation' will be used to empower the Sami people, as Indigenous scholar Angela Cavender Wilson suggests (2004). In this chapter, 'decolonising' does not mean rejecting all Western theories but rather 'centring [Indigenous] concerns and worldviews and then coming to know and understand theory and research from [Indigenous] perspectives and for [Indigenous] purposes' (Tuhiwai Smith 1999, 39).

'Indigenising' food sovereignty goes a long way toward clarifying and building a broader understanding of the colonial aspects embedded in the idea of food sovereignty. In line with Coté, the verb 'indigenise' means regenerate the relationships between Indigenous People and their lands by revitalising their ecological knowledge and food systems (Coté 2016, 1). In specific, indigenising food sovereignty, Coté notes, 'means reframing it within Indigenous People's struggles for autonomy, self-sufficiency, and self-determination rather than within assertions of domination, control, and authority over ancestral homelands' (2016, 9). Words such as 'authority' and 'domination' do not belong to Indigenous worldview but rather to the settlers' viewpoint; hence, they should be avoided when talking about Indigenous food sovereignty.

Therefore, in the context of the legal right to self-determination, indigenising food sovereignty in Sami communities means giving the Sami people the power to define their own food systems, as mentioned in the Declaration of Nyéléni concerning food sovereignty in Indigenous communities (Via Campesina 2007). In these contexts, the revitalisation of Sami traditional food systems through ecologically sound methods based on traditional principles plays a pivotal role in strengthening Sami self-determination, which, in turn, would help decrease their dependence on the market food system. Moreover, the promotion of Sami principles and traditional teachings about land use should play a prominent role in the design of Sami food security policies for present and future generations. In this sense, efforts to 'indigenise' food sovereignty would strongly encourage the recovery of traditional Sami food practices (Coté 2016; Whyte 2016).

Conclusions

Food has a fundamental relationship with other aspects of human life. This chapter has explored the connections among the Sami food system, Sami traditional livelihood practices, such as reindeer herding and salmon fishing, Sami identity and self-determination as Indigenous People. These relationships have been considered from a philosophical perspective under the frameworks of food ethics, food sovereignty, and food security. In the era of globalisation, more effective food security strategies are needed to address food insecurity challenges in Indigenous communities. Theories of food sovereignty play an important role in bringing cultural identity to the foreground of the discourse around food, without being too idealistic (Thompson 2015, 75–76). From the perspective of food ethics, strengthening the relationships among food, identity, and culture can emphasise the need to develop a more respectful attitude toward food matters. This chapter has also shown that traditional Sami food choices and livelihood activities are connected to the natural environment and are part of Indigenous Peoples' ways of life. They are therefore interlinked with the Sami people's identity and dignity. In addition, in Sami communities, food plays a role in community bonding, functions as an intergenerational knowledge transmitter, and helps enhance cooperation. As such, traditional Sami food habits are not simply a matter of consumerist preferences for one food over another. They belong to the category of 'overriding values', values that are too essential to be ignored or discarded.

In conclusion, to empower Sami Indigenous People, I suggest engaging in processes that 'indigenise' food sovereignty and 'decolonise' food security. In Sami communities, from a practical standpoint, this means letting the Sami people define food security and food sovereignty and determine their own food policies, avoiding terms such as 'domination' and 'authority', which do not belong to their worldview. Moreover, when developing and implementing food security policies affecting the Sami community, it would be of great benefit in the long run to take a more inclusive approach in order to develop an understanding of the Sami peoples' perspective on food security.

References

Agyeman, J., P. Cole, R. Haluza-Delay, and P. O'Riley. 2010. *Speaking for ourselves: Environmental justice in Canada*. Vancouver: University of British Columbia Press.

Åhrén, M. 2016. Saami–Nordic co-operation. In Mats-Olov Olsson et al. (eds.), *Encyclopedia of the Barents Region, Vol. 2*, 281. Oslo: Pax. https://www.cfans.umn.edu/sites/cfans.umn.edu/files/WhitePaperFinalVersion2011.pdf.

Barnhill, A., M. Budolfson, and T. Doggett. (eds.). 2018. *The Oxford handbook of food ethics.* New York: Oxford University Press. doi:10.1093/oxfordhb/9780199372263.013.39.

Cavender Wilson, A. 2004. Reclaiming our humanity: Decolonization and the recovery of Indigenous knowledge. In D. Abbot Mihesuah and A. Cavender Wilson (eds.), *Indigenizing the academy: Transforming scholarship and empowering communities*, 255-262. Lincoln: University of Nebraska Press.

Calhoun, A., M. Goeman, and M. Tsethlikai. 2007. Achieving gender equity for American Indians. In S. S. Klein et al. (eds.), *Handbook for achieving gender equity through education*, 525–552. New York: Routledge.

Casi, C. 2019. Sami food practices and traditional ecological knowledge. In E. Vinnari and M. Vinnari (eds.), *Sustainable governance and management of food systems: Ethical perspectives*, 207–211. Wageningen, The Netherlands: Wageningen Academic Publishers, https://doi.org/10.3920/978-90-8686-892-6_28.

Cohoon, C. 2014. Ecofeminist food ethics. In P.B. Thompson and D.M. Kaplan (eds.), *Encyclopedia of food and agricultural ethics*, 521-533. Dordrecht: Springer.

Comstock, G. 2013. *Research ethics: A philosophical guide to responsible conduct of research.* Cambridge: Cambridge University Press.

Conway, G. 2012. *One billion hungry: Can we feed the world?* Ithaca, NY: Cornell University Press. doi:10.1111/1477-9552.12029.

Coté, C. 2016. 'Indigenizing' food sovereignty: Revitalizing Indigenous food practices and ecological knowledges in Canada and the United States. *Humanities* 5(3): 57. doi:10.3390/h5030057.

Crenshaw, K. 1989. Demarginalizing the intersection of race and sex: A black feminist critique of antidiscrimination doctrine, feminist theory and antiracist politics. *University of Chicago Legal Forum* 1: 139–167. chicagounbound.uchicago.edu/uclf/vol1989/iss 1/8.

Debucquois, C. 2014. Access to land and the right to food. In P. Thompson and D. Kaplan (eds.), *Encyclopedia of food and agricultural ethics*, 8–12. Dordrecht: Springer.

Delormier, T., K. Horn-Miller, A. McComber, and K. Marquis. 2017. Reclaiming food security in the Mohawk community of Kahnawà:ke through Haudenosaunee responsibilities. *Maternal and Child Nutrition* 13(3): 1–14. doi: 10.1111/mcn.12556.

(FAO) Food and Agriculture Organization of the United Nations. 1996. *Rome declaration on world food security and world food summit plan of action*; World Food Summit 13–17 November 1996, Rome. Retrieved from http://www.fao.org/3/w3613e/w3613e00.htm.

(FAO) Food and Agriculture Organization of the United Nations. 2003. *Trade reforms and food security: Conceptualizing the linkages, commodity policy and projections service commodities and trade division*; Rome, FAO. Retrieved from www.fao.org/docrep/005/y4671e/y4671e06.htm.

(FAO) Food and Agriculture Organization of the United Nations. 2015. *The State of Food Insecurity in the World 2015. Meeting the 2015 international hunger targets: Taking stock of uneven progress*, Rome. Retrieved from http://www.fao.org/publications/sofi/2015/en/.

Gerber, E. 2014. Eating, feeding, and disability. In P.B. Thompson and D.M. Kaplan (eds.), *Encyclopedia of food and agricultural ethics.* Dordrecht: Springer. doi:10.1007/978-94-007-0929-4_40.

Gertz, E. 2017. Proposed Arctic railway would cut through Lapland reindeer habitat: A new railway project would help Finland and Norway take advantage of melting sea ice to ship products from the Arctic Coast. But Indigenous communities fear the rail line would disrupt their reindeer herding traditions. *Arctic Deeply*, part of *News Deeply*, 3 August. Retrieved 5 October 2018 from https://www.newsdeeply.com/arctic/articles/2017/08/03/proposed-arctic-railway-would-cut-through-lapland-reindeer-habitat.

Glennie, C. and A.H. Alkon. 2018. Food justice: Cultivating the field. *Environmental Research Letters* 13(7). doi:10.1088/1748-9326/aac4b2.

Greaves, W. 2016. Arctic (in)security and Indigenous Peoples: Comparing Inuit in Canada and Sámi in Norway. *Security Dialogue* 47(6): 461–480. doi:10.1177/0967010616665957.

Helander-Renvall, E. 2016. Saami – Traditional ways of life and worldview. In Mats-Olov Olsson et al. (eds.), *Encyclopedia of the Barents Region, Vol. 2*, 300–303. Oslo: Pax.

ICESCR. 1966. UN General Assembly, *International Covenant on Economic, Social and Cultural Rights*, 16 December 1966, United Nations Treaty Series, vol. 993, p. 3. Retrieved 14 June 2019 from https://www.ohchr.org/EN/ProfessionalInterest/Pages/CESCR.aspx.

Kant, I. 1983. *Grounding for the metaphysics of morals*. Tr. James Wesley Ellington. Indianapolis: Hackett Pub., 1785.

Kant, I. 1996. Critique of practical reason. In I. Kant *Practical Philosophy*. Tr. and ed. by M. J. Gregor. New York: Cambridge University Press, 1788.

Kyllönen, S., A. Colpaert, H. Heikkinen, M. Jokinen, J. Kumpula, M. Marttunen, K. Muje, and K. Raitio. 2006. Conflict management as a means to the sustainable use of natural resources. *Silva Fennica* 40(4): 687–728.

LaDuke, W. 1993. A society based on conquest cannot be sustained. In R. Hofrichter (ed.), *Toxic struggles: The theory and practice of environmental justice*, 98–106. Philadelphia, PA: New Society Publishers.

Lambert, L. 2008. Salmon and contamination in the Columbia River. *Enduring legacies: Native case studies*. Washington, DC: Evergreen State College. Retrieved from nativecases.evergreen.edu/collection/cases/salmon-and-contamination.

Lefevre, T. A. 2015. Settler colonialism. In J. Jackson (ed.), *Oxford bibliographies in anthropology*, 1–26. Oxford: Oxford University Press.

Legat, A., M. Chocolate, and S.A. Zoe. 1995. *Dogrib land: The importance of knowing. Report prepared for the Dogrib Treaty 11 Council and BHP Diamonds Inc.* Hay River, NT: Dene Cultural Institute.

Menser, M. 2008. Transnational participatory democracy in action: The case of La Via Campesina. *Journal of Social Philosophy* 30(1): 20–41.

Mepham, T. B. 2000. The role of food ethics in food policy. *Proceedings of the Nutrition Society* 59: 609–618. Retrieved from www.cambridge.org/core/journals/proceedings-of-the-nutrition-society/article/role-of-food-ethics-in-food policy/2A65E474D88CDC5BFB97A841A35A0710.

Murdock, E. and S. Noll. 2015. Beyond access: Integrating food security and food sovereignty models for justice. In D. Dumitras, I.M. Jitea, and S. Aerts (eds.), *Know your food: Food ethics and innovation*, 325–332. Wageningen, The Netherlands: Wageningen Academic Publishers. doi.org/10.3920/978–90–8686–813-1_49.

Nilsson, L.M., L. Dahlgren, I. Johansson, M. Brustad, P. Sjölander, and B. Van Guelpen. 2011. Diet and lifestyle of the Sami of southern Lapland in the

1930s–1950s and today. *International Journal of Circumpolar Health* 70(3): 301–318. doi:10.3402/ijch.v70i3.17831.

Nyyssönen, J. 2016. Saami – Political history. In Mats-Olov Olsson et al. (eds.), *Encyclopedia of the Barents Region, Vol. 2*, 283–289. Oslo: Pax.

Orend, B. 2002. *Human rights: Concept and context*. Ontario: Broadview Press.

Patel, R. 2009. Grassroots voices: Food sovereignty. *The Journal of Peasant Studies* 36(3): 663–706.

Pedersen, S. 2016. Saami – Other livelihoods apart from reindeer herding. In Mats-Olov Olsson et al. (eds), *Encyclopaedia of the Barents Region, Vol. 2*, 281–282. Oslo: Pax.

Pedersen, S. 2016. Saami Parliament in the Barents Region. In Mats-Olov Olsson et al. (eds), *Encyclopedia of the Barents Region, Vol. 2*, 282–283. Oslo: Pax.

Pimbert, M. 2008. *Toward food sovereignty: Reclaiming autonomous food systems*. London: International Institute for Environment and Development.

Pinstrup-Anderse, P. 2009. Food security: Definition and measurement. *Food Security* 1: 5–7.

Plumwood, V. 2000. Integrating ethical frameworks for animals, humans, and nature: A critical feminist eco-socialist analysis. *Ethics and the Environment* 5(2): 285–322.

Samer.se. Food, the website about Swedish Sami. Retrieved 12 June 2019 from http://www.samer.se/4591.

Sarche, M. and P. Spicer. 2008. Poverty and health disparities for American Indian and Alaska Native children. *Annals of the New York Academy of Science* 1136: 126–136.

Schanbacher, W. 2010. *The politics of food: The global conflict between food security and food sovereignty*. Oxford: Praeger Security Press.

Scrinis, G. 2008. On the ideology of nutritionism. *Gastronomica. The Journal of Food and Culture* 8: 39–48.

Siipi, H. 2015. Requirements of safety and acceptability in food security definitions. In D. Dumitras, I.M. Jitea, and S. Aerts (eds.), *Know your food: Food ethics and innovation*, 193–197. Wageningen, The Netherlands: Wageningen Academic Publishers. doi: 10.3920/978-90-8686-813-1.

Staalesen, A. 2018. Finland says new Arctic Railway should lead to Kirkenes; the small Norwegian town on the Barents sea coast is the preferred end station in the great infrastructure project. *The Barents Observer*, 09 March. Retrieved 05 October 2018 from https://thebarentsobserver.com/en/arctic/2018/03/finland-says-new-arctic-railway-should-lead-kirkenes.

Stoor, J. P., N. Kaiser, L. Jacobsson, E. Salander Renberg., and A. Silviken. 2015. We are like lemmings: making sense of the cultural meaning(s) of suicide among the Indigenous Sami in Sweden. *International Journal of Circumpolar Health* 74: 27669. doi: 10.3402/ijch.v74.27669.

Thompson, P. B. 2015. *From field to fork: Food ethics for everyone*. Oxford: Oxford University Press.

Thompson, P. B. and MacDonald. 2013. What food is 'good' for you? Toward a pragmatic consideration of multiple values domains. *Journal of Agricultural and Environmental Ethics* 26: 137–163.

Tuhiwai Smith L. 1999. *Decolonizing methodologies: Research and Indigenous Peoples*. London: Zed.

UNDP. 1994. *Human Development Report*. Oxford and New York: Oxford University Press.

UNRIC, United Nation Regional Information Centre for Western Europe. Retrieved 12 June 2019 from https://www.unric.org/en/indigenous-people/27307-the-sami-of-northern-europe--one-people-four-countries.

Vía Campesina. 2007. *Nyéléni Declaration*. Sélingué, Mali: Forum for Food Sovereignty. Retrieved 18 August 2019 from http://nyeleni.org/spip.php?article290.

Walker, R., D. Natcher, and T. Jojola. 2013. *Reclaiming Indigenous planning*. Montreal, Canada: McGill-Queen's Press.

Werkheiser, I. 2016. Individual and community identity in food sovereignty: The possibilities and pitfalls of translating a rural social movement. In M. Rawlinson and C. Ward (eds.), *The Routledge handbook of food ethics*, 377–387. Oxford: Routledge Press.

Whyte, K. P. 2016. Indigenous food sovereignty, renewal and US settler colonialism. In M. Rawlinson and C. Ward (eds.), *The Routledge handbook of food ethics*, 354–365. Oxford: Routledge Press.

Wittman, H., A. Desmarais, and N. Wiebe. 2010. The origins and the potential of food sovereignty. In H. Wittman, A. Desmarais, and N. Wiebe (eds.), *Food Sovereignty: Reconnecting Food, Nature and Community*, 1–14. Oakland, CA: Food First.

Zwart, H. 2000. A short history of food ethics. *Journal of Agricultural and Environmental Ethics* 12: 113–126.

8 Food security management in the Western Russian Arctic zone

Current status and information support issues

Maxim Shishaev, Zhanna Kasparyan and Pavel Lomov

Introduction

Ensuring food security is an important strategic goal for the sustainable development of any territory. However, it is a complex task. It has, firstly, several levels of significance (world, national, regional, social group, household, and individual), and, secondly, it unites many actors, indicators, and factors linked by direct and indirect feedback with immediate and delayed effects. In the Arctic territories, this problem is further complicated by special geographical, economic, cultural-ethnic, historical, and medico-physiological factors. Extreme natural conditions not conducive to the development of agriculture increase the economic costs of supplying the population with food products due to transportation costs. Weak economic infrastructure and low volumes of agricultural production force residents to depend on imported food from other regions and the seasonality of these imports (basic vital goods are provided within 'Severny zavoz' – a set of annual government measures to deliver goods to the faraway northern territories). The region is characterised by special traditions related to food production and consumption. These can be seen in local diets and nutrition, particularly those of Indigenous Peoples, whose traditional lifestyle includes harvesting and storing food for future use. Northerners' metabolisms determine their consumption needs; the kind, notion, and norms of human nutrition in the North differ in many ways to those of other regions.

To effectively monitor and manage the food security of the Arctic territories, it is necessary to identify and evaluate the various factors that affect the wellbeing of people who work and reside in this region. The specific features of the Arctic significantly complicate the development of a unified methodology for assessing the area's food security. In addition, residents of the Arctic territories differ significantly in their composition, way of life, and cultural and historical roots. They therefore have differing and sometimes mutually exclusive needs that define the idea of 'food security'. Observing the parity of these requirements is a rather difficult task for decision-makers

DOI: 10.4324/9781003057758-8

in the field of social policy and security. Solutions should be based on reliable information in the form of high-quality statistical data characterising various important aspects of food security, together with effective methods for their processing and analysis.

This chapter describes the methodological and information basis for food security management in the Russian Western Arctic. It also evaluates the opportunities and limitations of using modern information technologies for this task. It pays special attention to possibilities of evaluating the role of local food in the region's food security. Particularly, it addresses the question of integrity and considers the complexity of the current situation. It examines the availability, quality, and integrity of information concerning food security in the Russian Western Arctic, the penetration of IT in the region, the possibilities and limitations of using ontologies in this region, and evaluates the IT readiness of the regional food security management system.

The food security framework in Northern Russia

Food security in Russian discourse

The concept of food security in Russian practice has not yet been given a clear scientific justification or legal definition (Uskova et al. 2014). In addition, the term 'food security' is often replaced by the term 'food independence', which leads to confusion in terminology, criteria, and indicators. In Russian, the term 'food security' includes the idea of 'food safety', further complicating the definition of the term.

Based on the system approach and taking into account that food security is ultimately a system of political and economic measures that affect individuals, groups, or households as subjects of the existing socio-economic system, we define 'food security' here as a subsystem of national economic security focused on the uninterrupted supply of the population with quality (safe) food in quantities sufficient for (or exceeding) established physiological norms. The mechanisms for ensuring food security are complex; they include managing socio-economic and institutional–legal norms aimed at preventing threats to the population's supply of quality food.

The current regulatory documents of the Russian Federation, primarily the Doctrine of Food Security of the Russian Federation, interpret food security as an economic state in which physical and economic access to food products is guaranteed to all inhabitants (National Security Strategy 2009; Doctrine of food security of the RF 2010; Indicators of food security of the RF 2013). However, an exclusively economic view of food security seems insufficient, because such an approach obscures individual and social components and completely ignores qualitative safety indicators, as will be discussed below.

Although the scientific community is actively discussing the theoretical and methodological aspects of the problem of food security in Russia, a unified opinion has not yet been established (Sobol 2010). In addition, there are significant variations in the understanding of the term 'food security' in Russia and in international practice; internationally, greater importance is placed on the issues of personal accessibility and food safety.

Food security framework: criteria, indicators, factors, actors, threats, and instruments

Most scholars agree on several criteria that define food security, based on the Rome Declaration of 1996 (FAO 1996):

1 physical availability of sufficient quantities of safe and nutritious food;
2 economic accessibility of food of appropriate volume and quality for all social groups of the population; and
3 food safety for consumers; that is, the ability to prevent the production, sale, and consumption of low-quality food products that could damage the population's health and the economic independence of the national food system (food independency).

The effective functioning of the food security system should satisfy two more requirements:

4 sustainability; that is, the ability of the national food system to minimise the impact of seasonal, weather-related, and other fluctuations to provide food to the population throughout the country; and
5 effectiveness, which means that the national food system is developing in the mode of extended reproduction, increasing volumes in accordance with the growing needs of the population for food.

If these basic criteria for food security are accepted by the scientific community, the choice of indicators is much more difficult. This is probably due to the various ways of dividing the socio-economic subsystem of food security, which can be investigated from the perspective of different subject areas (economics, politics, public health, sociology, etc.). As a result, many different methods and indicators are proposed in the scientific literature (Uskova et al. 2014).

At the same time, the legal acts of the Russian Federation contain bulky – and therefore useless for research – systems of a huge but non-aggregated datasets. The Food Security Doctrine of the Russian Federation contains 33 different main target indicators (and over one hundred secondary ones). These indicators are not combined into an aggregate system and mostly reflect statistical socio-economic indicators. It is not possible to conduct international comparisons using this document. In addition, the Doctrine prioritises food production, paying little attention to issues of accessibility and food quality. For example, the requirements for physical and economic access to food, along with food safety issues, appear at the end of the list, and the criteria are in fact indicators of self-sufficiency that are more relevant to food independence than food security.

The FAO indicator system used for interethnic comparisons and assessing food security differs significantly to that adopted in the Russian Federation. The food security indicators of the Russian Federation do not consider indicators such as the hunger index; the production of food in the value estimate per capita; the proportion of children with retarded growth; or the incidence of anaemia, hypovitaminosis, iodine deficiency, or obesity. To assess the quality of a food security system, it is necessary to consider a complex set of indicators that characterise both the quality of food products and the level of public health.

The FAO indicators are not always appropriate for Russian territories. For example, to assess physical access to food, the FAO uses the density index of hard-surfaced roads and railways. However, in the Arctic territories, where the population is dispersed extremely unevenly and the territories themselves are sparsely populated, roads were built in a tree-like scheme, unlike their network structure in other regions of the Russian Federation, and so the use of this indicator is very controversial.

As yet, no comprehensive scientific justification or indicators of food security have been developed for the Russian Arctic. Nor are there scientifically and statistically substantiated human nutrition norms for this region, except perhaps physician recommendations that have been implemented (among other measures of social policy) in some Arctic territories. For example, a Murmansk city program provides preschool children with fish oil supplements.

The complexity and inconsistency of the conceptual foundations of food security requires the use of new, modern approaches. However, the assessment system recommended by the Russian Federation does not apply at the individual or the household level. It does not consider diet composition and related physiological parameters of health, nor the cost of a minimal food basket and its proportion in household expenditures.

Several indicators are used to assess regional food security (Uskova et al. 2014):

1 natural/absolute indicators of production (consumption) of basic foods and nutrients;
2 volumes of consumption and production of food and individual products related to optimal (recommended) volumes;
3 technical and economic factors indicating the economic and material status of the agroindustrial complex (financial and credit indicators, cultivated areas, livestock, production, and availability of agricultural machinery, fertilisers, etc.);
4 socio-demographic indicators, including population reproduction and the dynamics of labour resources in rural areas; and
5 medical and biological indicators, including incidence of certain diseases.

The system of indicators for assessing regional food security differs fundamentally to the main national and international criteria of food security.

Firstly, the indicators of food independence of individual regions seem to be disputable, and for the Arctic regions, this index will always be very low. Secondly, national indicators assign the task of ensuring food security to national rather than regional authorities. Thirdly, according to the FAO, the volume of the edible stock of grain defines food security at the international level. This approach is unacceptable for assessing the food security of some regions, especially the Arctic, as a grain economy is impossible in this climate. Fourthly, the country's food security is based on specific mechanisms designed to ensure it that do not operate in the Arctic regions (customs tariffs, compensation fees, excises, sales taxes, quotas, etc.). For some Arctic territories, food imports are the primary source of food security (northern delivery).

Along with the indicators defining food security, several other influencing factors should be taken into consideration. Despite a significant body of literature addressing this issue, there is currently no generally accepted methodology (Indicators of food security of the RF 2013; Uskova et al. 2014). Meanwhile, the methodology established by the federal law does not provide an unambiguous integral assessment of the situation and does not take regional specifics into account. At the regional level, methods for defining food security have not been legally formalised in all regions. In the Murmansk region, for example, the food security doctrine is under development.

The most important components of the food security system are the entities that interact with it – actors. The actors involved in food security can be analysed from the standpoint of institutional theory. There are several reasons for this. Firstly, there are many contradictions in the system of actors' relationships, the study of which is a separate scientific task. Secondly, as many authors have pointed out, informal relations in agricultural production directly relate to the task of ensuring food security (Maksimova 2014; Wegren et al. 2016).

The Russian discourse around food security is characterised by a wide variety of different views on the subject. These views can be grouped into four levels: individual, local, regional, and national. At each level, food security is viewed from different perspectives and involves different actors, indicators, and criteria. Different management mechanisms and information sources are available for assessing food security at each level. A summary of the differences in food security on each of these different levels is presented in Table 8.1.

Regional aspects of food security in Russian North considering the example of the Murmansk region

The main provisions of the (not yet adopted) Food Security Doctrine of the Murmansk (2010) region also fail to provide a specific methodology for monitoring food security (Sopov 2010; Parshukov et al. 2017). The general goal is to ensure the population of the Murmansk region stable provision of

Table 3.1 Levels in the Russian discourse on food security

	The individual level	*The municipal level*	*The regional level*	*The federal level*
Actors	- Citizens - Households - Fishermen and hunters - Mushroom pickers - Household subsidiary plots - Indigenous minorities of the north - Unity of citizens (for example, the Society for the Protection of Consumer Rights) - NGOs	- Civil society and private organisations - NGOs - Organisations that monitor food quality	- Regional executive and legislative powers - Regional monitoring bodies - Regional public and non-profit organisations - Regional market actors - Scholars looking at the region	- Federal government - Executive government bodies - Supervising organisations - State, public, and private organisations - Market actors - Policymakers - Researchers - NGOs
Factors	**Endogenous:** - Natural and climatic features of the region - Yield of the year - Availability of territories for traditional crafts and husbandries **Exogenous:** - Socio-economic: purchasing power of the population	- Status and dynamics of local markets: labour, food, agricultural production - Local corruption (for example, seizure of the municipal market by Azerbaijanis, bribery, etc.)	**Endogenous** - Dynamics of the population - Regional level of food production - Dynamics of economic development - Medico-demographic factors: morbidity, prevalence of certain diseases, etc. - Welfare/regional susceptibility to certain infectious diseases - Ecological situation - Import share - Monetary income per capita; consumer basket cost - Living wage - State of farming and food production - Degree of land degradation - Trade turnover - Productivity - Quality of goods - Consumption of food per capita	**Demographic** - Population composition and migration **Economic** - State support of the agroindustrial complex, transport, and logistics - State support for small businesses **Socio-economic** - Inflation rate - Democratisation of society **Political** - Sanctions - State policies regarding foreign economic activity **Environmental** - Natural and climatic conditions

			Exogenous - Socio-political situation - Composition of the food basket - Income per capita - Food independence - Rational norms of food consumption (actual) - Correspondence of the level and structure of the final consumption of food products to physiological norms of rational nutrition - Agribusiness potential volume and structure of exported and imported food - Quality of food	**Scientific and technical** - Scientific and technical developments
Indicators	**Used by the state:** - Economic accessibility - Physical accessibility of food – mainly quantitative socio-economic indicators **Used by actors:** - Quality of food - Saturation of the product market (quantity and variety) - Satisfaction with the state of the market	- The state of the municipality's product market – quality, quantity, stocks, diversity - The degree of the population's satisfaction with the assortment of products - Local medical statistics on morbidity, mortality, and the quality and quantity of food products		- Non-aggregated quantitative indicators characterising the spheres of a consumption, b production and national competitiveness, and c organisation of management
Threats	- Low yield - Decrease in the number of wild animals that can be hunted for food - Extensive economic development of traditional management territories - Decrease in the standard of living - Climate change - Social and political instability - Natural and technogenic cataclysms - Decreases in: - Product quality - Assortment of products - Product availability	- Socio-economic tensions impede the population's sustainable physical or economic access to food - Risk of social or political tensions and mass protest - Negative dynamics of municipal economic development - Road network conditions - Instability and deficit of municipal budgets - Yield of the year - Purchasing power	- Regional climatic and geographical features - Dominance of short-term goals in social and economic policies - Low agroclimatic farming potential - High dependence on imported foodstuffs - Unstable food market conditions - Increased competition in regional markets - Financial instability - Dependence on import - Social tensions	- Macroeconomic risks, (decrease in investment attractiveness of the real domestic sector, decrease in the competitiveness of domestic products, dependence on foreign economic conditions) - Technological risks (including inconsistency of food safety requirements) - Agroecological risks - Foreign trade risks

(Continued)

	The individual level	*The municipal level*	*The regional level*	*The federal level*
Instruments	- Individual (household) economic planning - Family settings and rules - Direct exchange of experience (individual communications) - Public information resources (mass-media and the Internet) - Consumer demand - Involvement of public organisations in consumer rights issues	- Instructions and rules of trade - Product quality control - Activities of the consumer protection society or other NGOs - Veterinary and other official food quality control - The cost of land rent - Cost of rent for premises - Interbudgetary transfers	- Product quality control - Lack of regional legislative mechanisms (currently, a monitoring system has been proposed but not adopted) - State program 'Development of agriculture and regulation of markets agricultural products, raw materials and food for 2014–2020 (30 September 2013, No. 563-PP) - Lack of informational support for managerial decisions	- Relevant regulatory acts that determine economic conditions for the country and its individual sectors
Relevant information sources	- Interpersonal communication - Traditional (signs, rules, terms, etc.) - The Internet - Mass media	- Statistics of controlling organisations - Municipal statistics - The Internet - Mass media (unstructured information)	- Regional statistics - Analytical reports of ministries and departments - The Internet - Mass media - Official regional reports, reviews, papers, and analytics - Sociological surveys of the population (qualitative assessment)	- Mass media - State statistics - Federal legislative acts - Official reports, reviews, papers, and analytics - Sociological surveys of the population (qualitative assessment)

high-quality natural food products that meet normal physiological needs for energy and nutrients in the conditions of the Far North. The Doctrine also outlines how to implement this provision by using unique, local, natural, climatic, hydrobiological, demographic resources and by integrating the agricultural and fisheries complexes in the Murmansk region with participants in the agrifood markets of the Russian Federation and the Nordic countries (Sopov 2010).

Thus, assessing the food security of the Arctic regions requires the development of a different kind of methodology. It should be based on international approaches and take advantage of the open statistics available for the Russian Federation. Nilsson's (Nilsson et al. 2013) approach seems promising.

Indicators of food security in the Arctic territories should meet several requirements. They should be sufficiently small in number, open, informative, coherent, collected and processed regularly, and, most importantly, sufficient for creating a unified information system that can be used to forecast the food needs of the Arctic population. Some of these issues are reflected in Nikitenko and Trofimova (2016).

The main threats to regional food security are:

- Physical and moral depreciation of the material and technical foundation;
- dependence of food production on imported technologies and raw materials;
- reduction of biological resources and ecological deterioration (water);
- emigration of qualified personnel;
- decrease in the size of the rural population and the aging of this population;
- decrease in food self-sufficiency, bankruptcy of agrarian businesses, and reductions in the number of small businesses;
- unfavourable external conditions (sanctions, economic crises, price disparities, etc.);
- decrease in financial support and in access to credit;
- social and economic deterioration; and
- conflicts among Indigenous Peoples, non-Indigenous residents, and new arrivals at all socio-administrative levels and all conflicts between individual actors.

In conclusion, food security is complicated and multi-dimensional. Since the ultimate criteria of food security are the availability, accessibility, and adequacy (quality and volume) of food supplies for individuals and these qualities depend on food security policies, the topic should be considered on several levels, from individual to local and national. In the Russian Western Arctic, there are widely divergent views on food security that consider different sets of factors, actors, and indicators. Official views of food security focus primarily on the socio-economic dimension. However, many disparate factors may indicate food security, and how it should be determined has not been adequately defined.

Information support for food security in North-West Russia

Information support for food security management in Russia on the regional level

Information support for monitoring and managing food security in Russia has received serious attention at the state level. In 2011–2013, the Russian Ministry of Agriculture financed the development of an information system for monitoring and forecasting food security and the development of other state information resources regarding food security in the Russian Federation (Russian Federation 2018). State authorities currently use this system to collect, store, organise, and consolidate data (Ministry of Agriculture of Russia 2018a). Data are collected from the Federal Service of State Statistics (Rosstat), regional executive reports, and the Federal Customs Service of the Russian Federation. The data include food balances, price information, customs, and official statistics (Ministry of Agriculture of Russia 2018b). In total, about 64,000 indicators are monitored, retrospectively. However, the current system has some shortcomings, including low productivity and a weak capability for automated analysis. In addition, the information system and the collected data are available only to employees of relevant public services. Therefore, the creation of a modernised information system with advanced analysis and forecasting capabilities is under discussion (Kostyleva 2016).

The main source of data about indicators of food security is official statistics and statistical reports by regional and federal authorities. Statistics are collected at various intervals and include a wide range of general socio-economic indicators and indicators characterising food consumption. For example, the quarterly sample survey of household budgets collects data about the consumption of food and nutrients. Goods and services purchased for personal consumption are classified as such under the departmental categories Classification of Individual Consumption by Purpose (COICOP).

Along with quarterly and annual surveys, Rosstat conducts specialised statistical observations less frequently; some of these examine indicators of food security. One involves the selective observation of the population's diet. Rosstat surveys this under the framework of statistical observations on socio-economic problems once every five years. Selective diet observations consist of a sample survey that includes representatives of various segments of the population throughout the Russian Federation and covers 45,000 households. The information collected includes health status, reserves and sources of nutrition, and volumes and characteristics of consumed food and products. Processed and raw data (microdata) of selective observations for 2013 are currently available (Rosstat 2018). Another example of regular but low-frequency observations of indicators of food security is the All-Russian Agricultural Census, which is conducted every ten years. For now, the results of the 2016 agricultural census were published. These findings provide a detailed picture of the use of agricultural land, including cultivated crops. Data are organised according to different categories of land use, including peasants (farmers), personal subsidiary, and other individual households (Murmanstat 2016).

Statistical observations in the Russian Federation have a hierarchical structure that reflects the country's administrative structure. Observations are conducted using uniform methods by subjects of the Russian Federation, and the results are aggregated by the relevant federal bodies (Russian Federation 2019). This organisation is reflected in the structure of the statistical data, making it possible to project the country's food security on the regional level. However, on the lower levels – local and individual, and particularly for assessing the role of local products – the use of these data is problematic. Data that are collected frequently (at least once a year) are mainly reduced to generalised economic indicators and provide insufficient detail for such a task. On the other hand, data from selective observations are sufficiently detailed but are not collected frequently enough to monitor dynamics. Moreover, even these data are not detailed enough for a deep assessment of the role of local products in ensuring food security. For example, the classifications used in a sample survey of household budgets combine all meat from Northern species, including reindeer meat, into one category of 'Other meat products'. This makes it difficult to assess threats to food security associated with regional-specific risks. These issues make it practical to use secondary sources of data – the media, quality control bodies, alternative studies by public organisations, and others (Kapustina 2016).

Digitalisation in North-West Russia as a new opportunity for food security information support using intelligent data processing techniques

Today, digital technologies are common all over the world, and the Russian Arctic is no exception. According to the Digital Society Index, 2018, Russia is now among the top ten countries with the most developed digital economies. It is ranked second in the world for citizen involvement in digitalisation processes. Under the national Digital Economy program (Russian Federation 2017), Russia plans to provide 80% of government services in electronic format by 2025. The creation of 50 smart cities is also in the planning stage, along with the launch of ten industry digital platforms and the transition of ten enterprises to high-tech formats. If the digitalisation of public administration and industry in Russia is largely a subject of long-range planning, then the spread of digital technologies at the personal level takes place here and now. More than 80% of Russian citizens personally use digital technologies. Russia ranks sixth in the world in number of Internet users; over 71% of the

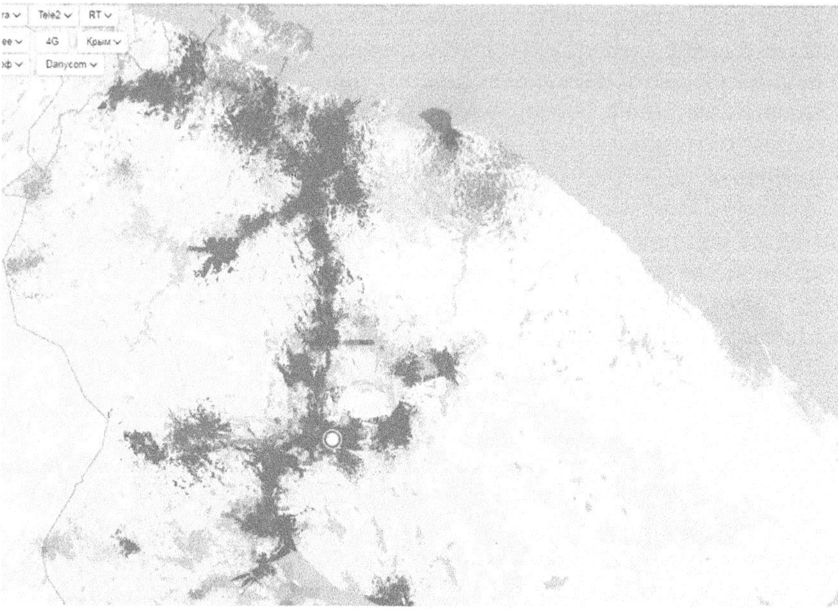

Figure 8.1 Coverage of mobile networks in Murmansk region.
Source: Karta (2019).

total population is online (Internet Users by Country, 2016). Residents of Northern Russia have also seen the penetration of digital technologies. In the Murmansk region, Internet penetration is even higher than the Russian average and reaches 90% (HSE 2017; Figure 8.1). The telecommunications infrastructure of the Russian Arctic has developed rapidly in recent years. A number of projects to develop terrestrial and satellite telecommunications involve partners from different countries (Communication in extreme conditions 2016; Commission for Arctic Development 2017; Rostelecom 2018). Third- and fourth-generation mobile network coverage in the region is growing and already covers the territories inhabited by Indigenous Peoples, as shown in Figure 8.1 (Rostelecom 2019).

Obtaining high-quality data on indicators of food security is key to ensuring adequate management of food security. In the Russian Arctic, data on the role of local food in ensuring food security are very limited. On the other hand, the spread of digital technologies and the Internet have led to a rapid increase in digitally available data. By some estimates (see IDC 2010), data stored on the Web will reach 35 zettabytes by 2020. The digitalisation of all aspects of life has created the prerequisites for the use of digital data in various forms and from various sources (document storage, social media, etc.). Information from such sources that directly or indirectly characterises food security could be used to monitor and manage it. Digital data could serve as an addition and in some cases as an alternative to traditional statistical research. This is especially important for assessing individual food security (including the role of local food), since official statistics do not always address relevant factors. Therefore, these factors can be assessed only through costly field research or by using a virtual alternative. The FAO (2018a) stated that evaluating food security on the local or individual level requires additional information not included in official statistics. Such information should be gathered by the direct household surveys, such as interviews with actors or focus group discussions. Implementing such investigations in a traditional 'offline' manner is a quite expensive and time-consuming task. Online interviews and discussions, as well as automatic analyses of digital data, could offer a much more cost-effective alternative.

Digital data may be used in many ways. Many studies (Christensen and O'Sullivan 2015; Mertens et al. 2015; FAO 2018a) discuss the possibilities of using social network analysis (SNA) methods to study various aspects of food systems: the geographic structure of food markets, risk analyses, and the influence of community network structures on key properties of food systems. SNA can also be used to assess the effects of alternative policies on the configuration of food systems as a whole (FAO 2018a). Similar approaches can be used with virtual social network services; this simplifies and reduces the cost of analysis. The FAO (2018a) has described methods for using SNA to assess regional food security. While that paper considers real-life SNA, the same approach could be used with online social networks (SNS). To assess food security on the basis of SNS data a challenging problem of establishing the correspondence between virtual and real social objects should be solved. A number of studies have addressed this (Goga 2015; Smirnov et al. 2016; Hao et al. 2018). One common approach involves a combined analysis of SNS content and meta-information associated with SNS messages. The aim of such an analysis is to identify markers in virtual objects that point to real geographical and social properties. Extracting real-world geographical attributes of virtual socio-objects is crucial to analysing regional food security using virtual data, particularly in the Arctic. SNS content analysis is a special kind of text analysis. SNS messages are characterised by short texts, unique grammatical structures, specific lexicons, etc. However, in addition to technical problems, this use of SNS data faces serious limitations associated with the use of personal data. This issue is at the forefront of current discussions on the topic and is the subject of legal regulations (EU Regulation 2016/679 of 27 April 2016, GDPR 2018; Federal Law 'On Personal Data' N 152-FZ 2018).

Along with analysing the structural properties of food systems (which is the focus of SNA), digital data can also be used to analyse other aspects of food security. For example, Widener and Li (2014) used information technology to measure the prevalence of healthy and unhealthy food across the contiguous United States. The study used geolocated Twitter data processed using sentiment analysis. Neznanov and Parinov (2015) extracted social data via SNS (from the Russian part of LiveJournal); a similar method could also be used to examine food security.

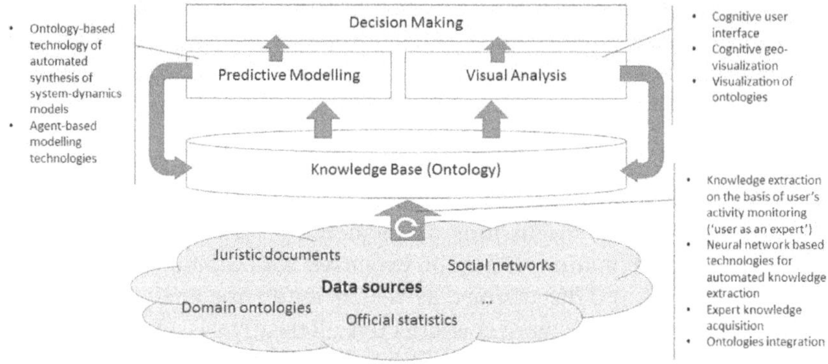

Figure 8.2 Structure of an ontology-based information system for food security management.

The availability of adequate source data defines the possibilities of using sophisticated intelligent IT to monitor and manage food security; such methods could include predictive modelling and cognitive visual analysis. Figure 8.2 provides an example of a possible system implementing such techniques. To use modern, sophisticated IT to analyse food security, we need not data but information and knowledge. As the FAO points out, 'food systems are complex non-linear, multidimensional and heterogeneous networks of social, economic, institutional, and environmental relations evolving over space and time' (2018a, p. 3). The FAO report (2018b) stresses the importance of understanding all the complex interrelationships of factors, indicators, and components of modern food systems to ensure sustainable food security. Understanding implies clear organisation of data, contextualisation of information, and accumulation of experience operating such systems. A common modern solution to this problem is the use of formalised knowledge in the form of ontologies. The possibilities of using ontologies to collect, organise, and analyse information about food security in the Russian Arctic are discussed in the next section.

Ontology as a formal representation of knowledge in the field of food security

Thomas Gruber provides one of the most famous definitions of ontology, calling it an 'explicit specification of a conceptualisation' (Gruber 1993, p. 220), which assumes representation the system of mental concepts in some language. Later, Borst (Borst 1997, p. 12) and Studer (Studer et al. 1998, p. 173) extended this definition by adding the requirements of a shared character of conceptualisation and formal specification. Therefore, 'an ontology is a formal, explicit specification of a shared conceptualisation'.

A conceptualisation is a representation of objects that are assumed to exist in some area of interest and the relationships among them. A shared conceptualisation is one that some group of agents (humans, organisations, information systems, etc.) accept. A formal specification is a description of the conceptualisation using a formal language such as Resource Description Framework (RDF); Schreiber and Raimond 2014), Web Ontology Language (OWL; Grau et al. 2008) or the Simple Knowledge Organisation System (SKOS; Baker et al. 2013). In RDF, an ontology can be described as a set of triplets in the form: 'subject–property–object'. It is suitable for creating semantic networks of terms. An RDF schema (RDFS) extends the descriptive capabilities of RDF by providing the syntax to separate classes and instances and to define properties and class hierarchies. OWL represents an ontology as a formal theory consisting of logical expressions (axioms) about objects in the subject domain. This offers extensive logical inference capabilities and ensures the formal consistency of the ontology. In contrast to OWL, SKOS follows the principle of minimal ontological commitment. It represents different types of knowledge organisation systems, including vocabularies, thesauri, and classification systems. Its concepts and relations are loosely specified, using thesaurus-style relationships like 'broader' rather than the logically formalised relationships used in OWL.

In the field of information support for food security, ontologies can be applied to a wide range of problems that require nontrivial logical inferences. Due to the technological separation of the inference mechanisms and logical rules (knowledge), ontologies are most effective where the applied logic is dynamic. This is particularly characteristic of the tasks involved in assessing individual food security. Along with predictive modelling and interactive visual analysis, ontologies may be used for analysing and synthesising dynamic supply chains; analysing the impact of new products on the qualitative composition of diets; and identifying and analysing implicit factors in food security, particularly the impact of new normative legal acts on food security indicators.

To illustrate the possibilities of logical inference over OWL ontologies, we will consider its application for identifying factors that affect food production (Figure 8.2). Let the OWL ontology include the set of axioms listed in Table 8.2.

Here classes: Meat(), GameMeat(), Activity(), and FactorAffectingGameMeatProduction() are given. GameMeat() is a subclass of Meat(). Class Activity() has some instance with name 'Hunting'. Class FactorAffecting-GameMeatProduction() is specified using a logical expression that defines the conditions under which instances become members of that class. Thus, creating the event 'AnnouncementOfHuntingProhibition' in the ontology and defining the 'influences-on' relationship with 'Hunting' places this announcement in the class 'FactorsAffectingGameMeatProduction'. Similarly, logical inference can be used to solve other tasks ensuring food security.

Table 8.2 Example of an OWL ontology applied to factors affecting food production

Axiom	*Description*
GameMeat(X) → Meat(X)	Game meat is a type of meat
has-source (X, 'Hunting') AND Activity('Hunting') → GameMeat(X)	If X is a product of hunting, and hunting is an activity then X is a game meat.
FactorAffectingGameMeatProduction(X) ↔ Event(X) AND influences-on(X, Y) AND has-source (Z, Y) AND GameMeat(Z)	If X happened, and X influences on Y, and Z is a product of Y, and Z is a game meat then X is a factor influencing game meat production.

The development of ontologies involves a number of problems. The primary problem is that ontology developers must be highly qualified subject domain specialists and must also have additional specific knowledge and skills in field of languages for ontology representation. As a result, many existing ontologies are low quality, containing logical inconsistencies and incomplete concepts and relationships (not all relevant concepts and relationships are represented). Poor quality can also manifest in too-minimal domain definitions; the use of different approaches to formalisation; and a low level of formal semantics, which inference engines use to infer new knowledge. Using multiple ontologies in an operation (especially low-quality ones) requires additional effort from knowledge engineers and subject domain specialists to overcome the structural and semantic heterogeneity of different ontologies. A common way to improving the quality of ontologies being developed is the use of upper-level ontologies that describe abstract concepts and their connections and can be used in all subject areas. Another approach is to use ontology design patterns that describe typical ontology fragments (Gangemi and Presutti 2007).

Some existing open ontological resources related to the production and consumption of food are:

AGROVOC is a multilingual agricultural thesaurus that covers all areas of interest to the Food and Agriculture Organisation (FAO) of the United Nations, such as food, nutrition, agriculture, forestry, fisheries, scientific and common names of animals and plants, the environment, biological concepts, techniques of plant cultivation, etc.

The **Global Agricultural Concept Scheme** (GACS) maps the 10,000 most used concepts from the CAB Thesaurus (developed by Centre for Agriculture and Biosciences International) and the NAL Thesaurus (produced by the National Agricultural Library, the USDA, and the Inter-American Institute for Cooperation on Agriculture).

The **UNESCO thesaurus** is not directly connected to food security but is widely used all around the world. It is a controlled, structured list of terms used in subject analysis and the retrieval of documents and publications in the fields of education, culture, natural sciences, social and human sciences, communication, and information. It is continuously enriched and updated, and its multidisciplinary terminology reflects the evolution of the UNESCO's programs and activities. It contains 7,000 terms in English and in Russian and 8,600 terms in French and in Spanish.

Applying these ontological resources to food security issues presents some problems, both technical issues and problems with the organisation of their content. One technical problem is the absence of a programming API of web resources that represent these thesauruses. This means they cannot be used in the online mode as a component of their own information systems. Instead, users must create and regularly update their own subsystem or microservice to process requests for the contents of thesauruses. The large volumes of these thesauri make it difficult for domain specialists to work with their downloaded versions since they require substantial computing resources on personal computers. It is therefore preferable to create a separate web service for working with their content. Finally, the lack of supported free program libraries for working with the thesauri complicates the use of the SKOS model to represent the concept systems of the listed ontologies. In addition, the joint use of SKOS thesauri and OWL ontologies, which are the most powerful tools for automated reasoning, in a single information system requires harmonisation due to the differences in these models. For example, it is necessary to determine whether to translate an SKOS concept as an OWL class or instance.

In addition to the technical problems of using these ontological resources, some additional challenges complicate their use for assessing the role of local food in food security in the Russian Arctic. Firstly, the small number of relationships and simple concept definitions from the point of view of formal semantics, as well as the existence of 'multiparent' concepts, make it difficult to directly use these resources in data processing scenarios requiring logical inference. Secondly, the sets of concepts presented in these thesauri do not include plant-based food products specific to the Arctic regions. For example, they do not contain forest species of berries and mushrooms. Therefore, additional work is needed to expand the conceptual systems to account for the traditional dietary habits of people living in the Arctic zone. Thirdly, the structures of the conceptual systems presented in the thesauri contain fragments that can be regarded as inaccurate or contradictory. For example, in the AGROVOC thesaurus, the concept 'Venison' is designated as a narrower for the concept 'Game meat', but the concept 'Reindeer' as a source of venison is absent. Similarly, the concept 'Ducks' has as its parent (relation 'broader') the concepts 'Poultry' and 'Birds'. At the same time, it has different relationship – 'Include in' – with the concept 'Waterfowl', even though all these concepts could be considered parents for 'Ducks'.

In sum, ontologies can be an effective tool for automated solutions for non-trivial information support tasks related to food security management, including assessing the role of local products in food security. On the other hand, existing ontological resources can only be used as dictionaries for creating ontologies. Appropriate ontologies for food security management should include additional relationships between the concepts and logical restrictions appropriate to specific tasks related to ensuring food security in the considered region. To be used as a tool for comprehensive food security analysis, an ontology must reflect the deep subject relationships of the domain. For example, Sen's entitlement theory relationships (Sen 1976) describing the causes of famines may be defined in ontology axioms.

Conclusion

The Doctrine of Food Security of the Russian Federation defines food security as providing every citizen with food of sufficient quantity and quality. The country's food security management system reproduces Russia's administrative structure and includes federal, regional, and local (municipal) levels. Food security systems at different levels use significantly different indicators and factors and involve different actors and links among actors. Thus, food security is a complex multidimensional problem that must be considered from different perspectives. At the federal level, this area is quite well developed: There is a regulatory framework, criteria, and indicators of food security are defined, and a monitoring system based on official statistics is generally formed and operates. The most problematic aspects of the federal food security system are the tendency to consider food security as an exclusively economic topic, the identification of security with independence, and the imperfection of the system of indicators. This makes it difficult to assess food security from the standpoint of individual security, the declared ultimate criterion. At the regional and lower levels, the situation is significantly worse. Regional and especially local regulatory frameworks are poorly or incompletely developed. As a result, many aspects of food security specific to regions, municipalities, and households remain 'under the radar' of state monitoring and regulation. This includes the role of local products in regional food security.

The widespread use of digital technologies and advanced tools for automated intelligent data processing offer new opportunities for collecting and analysing information for the purpose of monitoring and managing food security. Digital coverage of the Russian North is quite good: The Internet is widely available, almost 100% of the population use digital devices, and there is a state digitalisation program promoting the provision of state services in digital form. Furthermore, powerful, modern intelligent IT are capable of automating many functions involved in collecting, processing, and analysing information for food security management in general and assessing the role of local products in particular. At the same time, there is a gap in this area: a loose conceptual base, the absence of solid databanks, and the lack of Russian-language ontologies on relevant issues. Despite the large volume of regularly collected statistical information, the quality of the available statistical data is insufficient to assess the role of local products and assess individual food security. Thus, Russia has sufficient technological readiness to improve food security with the help of modern IT, but its IT readiness at the subject level is weak.

One way to eliminate this gap could be to reflect problems of food security in state digitalisation programs. Key tasks include developing a nationally unified conceptual and methodological apparatus for analysing food security that is also harmonised with global practice and adapting modern intelligent IT to the tasks of monitoring and analysing food security. The practical steps towards this adaptation could be the creation of high-quality ontologies of the subject area, defining the used system of concepts, and the internal logic of food security management tasks to cover all levels, including regional, food security management.

References

Baker T., S. Bechhofer, A. Isaac, A. Miles, G. Schreiber, and E. Summers. 2013. Key choices in the design of Simple Knowledge Organisation System (SKOS). *Journal of Web Semantics* 20: 35–49. doi: 10.1016/j.websem.2013.05.001.

Borst, W. 1997. Construction of engineering ontologies. PhD thesis, Institute for Telematica and Information Technology, University of Twente, Enschede, The Netherlands.

Christensen, L. O. and R. O'Sullivan. 2015. Using social network analysis to measure changes in regional food systems collaboration: A methodological framework. *Journal of Agriculture, Food Systems, and Community Development* 5(3): 113–129. doi: 10.5304/jafscd.2015.053.013.

Commission for Arctic Development. 2017, October 18. The Russian Federation and Finland discussed the project of laying a fibre-optic communication line along the Northern Sea Route. Retrieved 27 February 2018 from https://arctic.gov.ru/News/1959c8dc-31b4-e711-80d7-00155d006312?nodeId=4f828d76-8f58-e511-8259-e82aea5c46bbandpage=1andpageSize=10.

Communication in extreme conditions. 2016. Interview with Evgeny Buydinov, Deputy Director General for Innovative Development of the Federal State Unitary Enterprise 'Space Communication' (GP KS). *Electrosvyaz*, 18 May. Retrieved 27 February 2018 from http://www.elsv.ru/svyaz-v-ekstremalnyh-usloviyah/.

Digital Society Index. 2018. Retrieved 27 February 2018 from https://www.dentsuaegisnetwork.com/news-releases/dentsu-aegis-launches-digital-society-index-2018.

Doctrine of Food Security of the RF. 2010. Decree of the president of the Russian Federation of 30 January 2010 No. 120: On approving the doctrine of food security of the Russian Federation. Retrieved 1 September 2018 from http://www.kremlin.ru/acts/bank/30563.

EU Regulation 2016/679 of 27 April 2016, GDPR. Retrieved 16 November 2018 from https://publications.europa.eu/en/publication-detail/-/publication/3e485e15-11bd-11e6-ba9a-01aa75ed71a1.

FAO. 1996. Rome declaration on world food security. Retrieved 1 September 2018 from http://www.fao.org/docrep/003/W3613E/W3613E00.HTM.

FAO. 2018a. Social network analysis for territorial assessment and mapping of Food Security and Nutrition Systems (FSNS). A methodological approach. Retrieved 6 September 2018 from http://www.fao.org/3/I8751EN/i8751en.pdf.

FAO. 2018b. The state of food security and nutrition around the world. Retrieved 12 July 2018 from http://www.fao.org/state-of-food-security-nutrition/en/.

Federal Law of the Russian Federation: On Personal Data, N 152-FZ. Retrieved 16 November 2018 from http://www.consultant.ru/document/cons_doc_LAW_61801.

Gangemi, A. and V. Presutti. 2007. Ontology design for interaction in a reasonable enterprise. In P. Rittgen (Ed.), *Handbook of ontologies for business interaction* (pp. 48–68). Hershey, PA: IGI Global.

Goga, O. 2015. Matching user accounts across online social networks: Methods and applications. *Université Pierre et Marie Curie* – Paris VI, 2014. Retrieved 21 July 2018 from https://tel.archives-ouvertes.fr/tel-01165052.

Grau, B., I. Horrocks, B. Motik, B. Parsia, P. Patel-Schneider, and U. Sattler. 2008. OWL 2: The next step for OWL. *Journal of Web Semantics* 6(4): 309–322.

Gruber, T. R. 1993. A translation approach to portable ontologies. *Knowledge Acquisition* 5(2): 199–220.

Hao et al. 2018. When social computing meets soft computing: Opportunities and insights. *Human-centric Computing and Information Sciences* 8, 8. Retrieved 21 July 2018 from doi: 10.1186/s13673-018-0131-z.

HSE. 2017. National Research University Higher School of Economics. *Indicators of the Digital Economy: 2017.* Retrieved 6 August 2018 from https://www.hse.ru/primarydata/ice2017.

IDC. 2010. The digital universe decade – Are you ready? Retrieved 6 September 2018 from https://www.emc.com/collateral/analyst-reports/idc-digital-universe-201005.pdf.

Indicators of food security of the RF. 2013. Order of the Government of the Russian Federation of 18 November 2013 N 2138-r: *On the approval of the list of indicators in the field of ensuring food security of the Russian Federation.* Retrieved 1 September 2018 from http://www.consultant.ru/document/cons_doc_LAW_154706/?frame=1.

Internet Users by Country. 2016. Retrieved 6 September 2018 from https://www.internetlivestats.com/internet-users-by-country/

Kapustina, I. 2016. The organisation of monitoring in the sphere of food securityю. *Symbol of Science* 8(1): 107–111.

Karta. 2019. Karta pokrytiya MTS, Megafon, Yota, Tele2, Beeline, Rostelekom, Sberbank, Tin'koff, TTK, SkyLink LTE [Coverage map MTS, Megafon, Yota, Tele2, Beeline, Rostelecom, Sberbank, Tinkoff, TTK, SkyLink LTE]. Retrieved 6 September 2019 from https://4g-faq.ru/karty-pokrytiya/.

Kostyleva, T. (2016, March 21). The Ministry of Agriculture is not going to develop an IS in the field of ensuring food security 'from scratch'. *Digital Russia.* Retrieved 6 September 2018 from http://d-russia.ru/minselxoz-ne-budet-razrabatyvat-is-v-sfere-obespecheniya-prodovolstvennoj-bezopasnosti-s-nulya.html.

Maksimova, T.P. 2014. Potential of informal institutions as the determinant of national food security: Theoretical aspects. In K.A. Khubiev (ed.), *International scientific conference "The economic system of modern Russia: ways and goals of development", November 19, 2014: Abstracts of the conference participants.* Retrieved

10 August 2018 from https://www.econ.msu.ru/ext/lib/Category/x48/x91/18577/file/Razdel%2013.pdf.

Mertens, F., M. Fillion, J. Saint-Charles, P. Mongeau, R. Távora, C. José Sousa Passos, and D. Mergler. 2015. The role of strong-tie social networks in mediating food security of fish resources by a traditional riverine community in the Brazilian Amazon. *Ecology and Society* 20(3): 18. Retrieved 27 August 2018 from doi: 10.5751/ES-07483-200318.

Ministry of Agriculture of Russia. 2018a. List of information systems of the Ministry of Agriculture of Russia. Retrieved 5 September 2018 from http://mcx.ru/analytics/infosystems/.

Ministry of Agriculture of Russia, 2018b. On the creation of the state automated information system in the field of providing food security of the Russian Federation. Retrieved 6 September 2018 from http://minsvyaz.ru/uploaded/files/prezentatsiya-po-voprosu-v-minselhoz.pdf.

Murmanstat, 2016. All-Russian Agricultural Census 2016. Retrieved 26 August 2018 from http://murmanskstat.gks.ru/wps/wcm/connect/rosstat_ts/murmanskstat/ru/census_and_researching/census/national_census_2016/.

National Security Strategy. 2009. Decree of the President of the Russian Federation of 12 May 2009 No. 537: On the National Security Strategy of the Russian Federation until 2020. Retrieved 1 September 2018 from http://www.kremlin.ru/acts/bank/29277.

Neznanov A.A. and A.A. Parinov. 2015. Analysing social networks services using formal concept analysis research toolbox. In R. Missaoui, S. Kuznetsov, and S. Obiedkov (eds.), *CEUR Workshop Proceedings. Proceedings of the International Workshop on Social Network Analysis using Formal Concept Analysis (SNAFCA 2015)* (pp. 43–54), Malaga.

Nikitenko, M. and I. Trofimova. 2016. Food security in the Arctic zone of the Russian Federation. *Society: Politics, Economics, Law* 9: 33–37.

Nilsson, L.M., J. Berner, A. Dudarev, G. Mulvad, J. Øyvind Odland, A. Parkinson, A. Rautio, C. Tikhonov, and B. Evengård. 2013. Indicators of food and water security in an Arctic Health context – results from an international workshop discussion. *International Journal of Circumpolar Health* 72: 1. Retrieved 1 September 2018 from https://www.tandfonline.com/doi/full/10.3402/ijch.v72i0.21530

Parshukov, D.V., D.V. Khodos, N.I. Pyzhikova, E.V. Pyhanova, and E.Y. Vlasova. 2017. Monitoring and evaluation of food security in the Russian Federation. *International Journal of Applied Business and Economic Research* 15: 195–207.

Rosstat, 2018. Results of federal statistical observations on socio-demographic problems. Retrieved 6 September 2018 from http://www.gks.ru/free_doc/new_site/inspection/itog_inspect1.htm.

Rostelecom. 2018. Underwater fiber-optic line of Kamchatka – Sakhalin – Magadan. Retrieved 27 February 2018 from https://www.rostelecom.ru/projects/FarEast_FOCL/.

Rostelecom. 2019. Coverage maps for MTS, Megafon, Yota, Tele2, Rostelecom, SkyLink LTE. 3G, 4G, 2G and cellular. Retrieved 23 September 2019 from https://4g-faq.ru/karty-pokrytiya/.

Russian Federation. 2017. The digital economy of the Russian Federation, ordinance dated 28 July 2017 No. 1632-r. Retrieved 26 August 2018 from http://static.government.ru/media/files/9gFM4FHj4PsB79I5v7yLVuPgu4bvR7M0.pdf.

Russian Federation. 2018. State tenders. Retrieved 5 September 2018 from http://zakupki.gov.ru/epz/main/public/home.html.

Russian Federation. 2019. Rosstat development strategy until 2024. Retrieved 24 September 2019 from https://gks.ru/storage/mediabank/Strategy.pdf.

Schreiber, G. and Y. Raimond. 2014. RDF 1.1 Primer, W3C Working Group Note, 25 February 2014. Retrieved 2 June 2014 from http://www.w3.org/TR/rdf11-primer/.

Sen, A. 1976. Famines as failures of exchange entitlements. *Economic and Political Weekly* I(31–33): 1273–1280.

Smirnov I. et al. 2016. In search of lost profiles: The reliability of 'VKontakte' data and its importance in educational research, *Voprosy Obrazovaniya* 4: 106–122.

Sobol, T.S. 2010. The theoretical and methodological aspect of Russia's food security in modern conditions. Retrieved 21 July 2018 from http://www.edit.muh.ru/content/mag/trudy/12_2010/11.pdf.

Sopov. 2010. Food security of the Murmansk region: Past, present, and future. Retrieved 1 September 2018 from http://www.b-port.com/analytics/item/38862.html.

Studer, R., R. Benjamins, and D. Fensel, 1998. Knowledge engineering: Principles and methods. *Data andand Knowledge Engineering* 25(1–2): 161–198.

Uskova, R.J., A.N. Selimenkov, A.N. Anischenko, and A.N. Chekavinsky. 2014. *Prodovol'stvennaya bezopasnost' regiona* [Food security of the region]. Vologda, Russia: ISEHRT RAN.

Wegren, S., F. Nilssen, and C. Elvestad. 2016. The impact of Russian food security policy on the performance of the food system. *Eurasian Geography and Economics* 57(6): 671–699.

Widener, M.J. and W. Li. 2014. Using geolocated Twitter data to monitor the prevalence of healthy and unhealthy food references across the US. *Applied Geography* 54: 189–197.

9 Arctic food crisis management[1]

Christer Pursiainen

What if a serious food crisis were to hit the Arctic? How might it manifest? Is a crisis management system in place that could handle it?

This chapter starts by justifying the chosen research question. It continues with a concise presentation of the phases of crisis management, which provide the structure for further discussion. To illustrate the current state of Arctic crisis management, three possible food crisis scenarios are considered: radiological fallout and contamination, animal disease epidemics, and the impact of oil spills on fisheries. While the third scenario is more generic, the first two focus on the Finnish Arctic (Lapland), which is assumed to be representative, at least of the Arctic regions of the other Nordic countries.

It is concluded that there is a great deal of sectorial knowledge about these possible crisis scenarios and that numerous crisis management practices are in place. Nationally, these crisis management systems are sophisticated, and prevention, preparedness, and response strategies in particular appear to be well planned and organised. International and regional arrangements complement the national level and are essential to managing crises, which often cross national boundaries. However, as major Arctic food crises are – fortunately – relatively rare, there is a lack of historical data that would provide a firm basis for risk assessment. Furthermore, climate change introduces some unexpected dynamics.

What seems be missing is a holistic picture of potential food crises in the Arctic. Any major food security crisis in the area would not affect only one sector; it would impact many health-related, social, economic, environmental, and political dimensions. Moreover, subnational, national, transboundary, and international cooperation is challenging. To this end, a more holistic perspective would be beneficial for managing these types of crises.

What is Arctic food security?

The entire populated Arctic belongs to developed countries with generally high living standards, even if the Arctic area itself is somewhat marginalised. The Arctic is characterised by a harsh, cold climate; a unique

DOI: 10.4324/9781003057758-9

ecosystem and above-average environmental vulnerability; long distances and limited accessibility; and the presence of Indigenous Peoples.

This is, of course, an oversimplification as the Arctic is highly heterogeneous. It simultaneously contains some of the most developed technological population centres with access to quick transport to the economic hubs of the world, as well as remote areas with small populations that rely on traditional ways of life. Similarly, climatic conditions vary considerably within the Arctic area, from permanently ice-covered seas to open sea waters and from extreme cold to rather stable non-extreme temperatures.

Yet the Arctic is often considered from a broader perspective. Its potential for future mineral resource exploitation and as a global transportation route (due to climate change) makes it an area of global interest, in its own right.

The widely accepted definition of food security derives from the World Food Summit's 1996 Rome Declaration (WFS 1996, point 1): 'Food security exists when all people, at all times, have physical and economic access to sufficient, safe and nutritious food to meet their dietary needs and food preferences for an active and healthy life'. A report commissioned by the Arctic Council (Nilsson et al. 2013) characterises the term 'Arctic food security' as a cross-cutting issue comprising social, economic, environmental, political, and cultural dimensions. These include such concerns as food and water-borne diseases; the increasing incidence of lifestyle diseases; the high cost of healthy foods; contamination; changes to ecosystems that impede access to food; high fuel costs; and the loss of traditional knowledge.

Due to the presence of Indigenous Peoples in the region, Arctic food sovereignty is also much discussed. Food sovereignty emphasises the right of local people to control food production and distribution, instead of multinational corporations (e.g., Rudolph and McLachlan 2013; Sage 2014).

The definitions above are clearly very broad. They are related to the more generic concept of human security, which is allegedly traceable to use by the UN in the early 1990s. Due to the vagueness of the term, it has sometimes been claimed that human security is too expansive to be operationalised (e.g., Paris 2001). In terms of food security, almost everything related to food therefore becomes 'securitised'. A phenomenon that is originally seen as a non-politicised issue can first become politicised and consequently a legitimate object of public discussion and policy-making, and after that the issue can be securitised, that is, it is presented as a security threat which may require extraordinary measures (e.g., Balzacq 2015).

This chapter examines the three above-mentioned crisis scenarios related to Arctic food security. While several other types of risks could be suggested, these three scenarios provide a framework for discussing some elements of food security in more detailed and concrete terms than a consideration of generic concerns would. The main question addressed here is: How would three specific food crisis scenarios be managed should they emerge?

The crisis management cycle

What is crisis management? A crisis can be defined in a variety of ways depending on the discipline, field, or context. However, at least three key criteria were identified as early as the 1960s (Hermann 1963). First, a threat is posed to the essential goals or values of an actor. Second, there is limited time for decision-making, due to an approaching deadline and/or the increasing costs of inactivity. Third, there are numerous unpredictable events or uncertainties connected to the situation, making it difficult to form a clear picture of the potential outcomes of decisions.

This broad definition is applicable to almost all types of crises, regardless of whether they are categorised according to their causes (e.g., malicious, non-malicious, man-made, technological, natural, social, economic, complex); their consequences (e.g., social impact, economic impact, human impact, reputation loss); or the degree of the ability to predict and influence them (e.g., Veenema and Woolsey 2012; Bellavita 2006; Gundel 2005). The definition also covers our three crisis scenarios.

The second part of crisis management is management. Management is a set of interacting elements to establish policies and objectives and to achieve those objectives (ISO 2015). This definition originates from the International Organization for Standardization 9000 family of standards, which focuses on the quality of management in general.

In this context, the concept of a crisis management cycle (Pursiainen 2017; cf. Heath 1998; Heck 1991; Rosenthal and Pijnenberg 1991) is useful for theory and practice. If a crisis is a time-limited phenomenon and a kind of deviation from the normal state of affairs, crisis management also includes the normal state before and after the crisis. The cycle frame helps distinguish the different phases of this management system.

The customary divisions of pre-crisis, during crisis and post-crisis are usually further subdivided, with some variations, into more detailed and prescriptive phases. These may include risk assessment, prevention, preparedness, response, recovery, and learning. Effective crisis management presupposes the efficient management of all individual phases. The cycle based on these phases is illustrated in Figure 9.1.

Risk assessment. Trying to prevent or prepare for a crisis becomes meaningless if one does not know what to prevent or prepare for, which risks should be considered, and how they should be prioritised. Therefore, risk assessment (e.g., Pritchard 2015; Yoe 2012; Aven 2008) is a precondition for the subsequent phases of the crisis management cycle.

The field of risk assessment is standardised. Following the ISO 31000 family of standards (ISO 2018; ISO/IEC 2019), risk assessment consists of three tasks: risk identification, risk analysis, and risk evaluation. Each of the risk assessment phases can be approached in a number of quantitative, semi-quantitative, or qualitative ways or combinations thereof. The last evaluation phase defines which risks will be selected for further treatment.

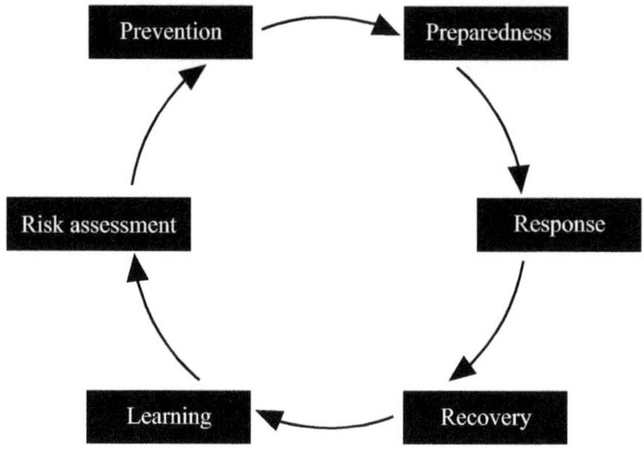

Figure 9.1 The crisis management cycle.

This chapter considers whether any proper risk analyses have been conducted for our three illustrative Arctic food security crises. If so, what do these analyses tell us?

Prevention. In relation to natural, man-made, and technological hazards, the United Nations International Strategy for Disaster Reduction (UNISDR/UNDRR 2009; cf. ISO/PAS 2007) defines prevention as the outright avoidance of the adverse impacts of hazards and related disasters. However, as complete avoidance of losses is not always feasible, the task of prevention often reverts to mitigation. Hence, the terms prevention and mitigation are usually grouped together or even used interchangeably. Other terms, such as risk reduction or risk treatment, are also used to express similar ideas. This chapter will explore whether any risk prevention strategies are in place to avoid or mitigate the possible Arctic food security crises mentioned above.

Preparedness. If serious risks cannot be completely eliminated by preventive (risk treatment) efforts, one should be prepared for them so as to mitigate the consequences of a crisis when it hits. Preparedness generally includes such elements as contingency planning; stockpiling equipment and supplies; in-advance coordination of rescue or repair efforts; planning evacuations; informing the public; and associated training and exercises. Preparedness efforts must be supported by formal institutional, legal, and budgetary capacities (e.g., Haddow, Bullock, and Coppola 2011; UNISDR/UNDRR 2009; Perry and Lindell 2003). This chapter will evaluate the presence and quality of these preparedness elements in relation to the imagined Arctic food security crises.

Response. In the context of large-scale societal disasters, regardless of the root cause, UNISDR/UNDRR (2009) defines a response as the provision of emergency services and public assistance during or immediately after a disaster in order to save lives, reduce health impacts, ensure public safety, and meet the basic subsistence needs of the people affected. It also mentions that disaster response focuses predominantly on 'immediate and short-term needs' and is sometimes called disaster relief. In practice, a crisis demands coordinated responses and proper crisis communication. Response partially overlaps with preparedness, especially with contingency planning. To understand how response works, we must rely on some historical cases. To this end, the present study investigates how food security crisis response in the Arctic has worked in the past.

Recovery. UNISDR/UNDRR (2009) defines recovery as the restoration, and improvement where appropriate, of facilities, livelihoods and living conditions of disaster-affected communities, including efforts to reduce disaster risk factors. It adds that the recovery task of rehabilitation and reconstruction begins soon after the emergency phase has ended and should be based on pre-existing strategies and policies that facilitate clear institutional responsibilities for recovery action and enable public participation. Like response, recovery is best evaluated by considering an historical case. However, successful recovery is, according to the definition above, largely related to 'pre-existing strategies and policies'. This affords us a limited opportunity to evaluate the state of recovery in relation to possible Arctic food security crises.

Learning. Post-crisis learning is often viewed as a three-phase process: lessons identified, lessons learned, and lessons institutionalised (e.g., Pursiainen 2017; Milton 2010). Hence, an organisation has truly learned only when the results of that learning have been institutionalised in the organisation's safety and security culture and behavioural practices. As in the case of response and recovery, we can truly evaluate this phase only in relation to real crises. However, a successful learning process also presupposes a preplanned system to facilitate post-crisis learning. Do we find some evidence of learning in our three Arctic food security cases?

Three crisis scenarios

It is difficult to find a bona fide holistic crisis management strategy for the Arctic, and even more difficult to find a strategy for Arctic food security (for a literature review, see Loring and Gerlach 2015). The document that comes closest to a holistic treatment is the Arctic Council's 2013 report on Arctic food security. However, it merely lists some general concerns as well as previous and ongoing relevant projects. This state of affairs is rooted in the lack of a proper risk assessment, which is, however, on the wish list: 'In the longer term, it will be increasingly important for society to understand ongoing changes in the Arctic and their associated risks, as well as to act on early warning signals for potential thresholds and other types of change' (Nilsson et al. 2013, 115).

On the other hand, climate change has inspired some scholars (e.g., Clement 2019; Wesche and Chan 2010; Meakin and Kurvits 2009) to focus on

Arctic food security in particular. The numerous Intergovernmental Panel on Climate Change (IPCC) reports usually include some literature reviews and data relevant to Arctic food security as well. These kinds of studies generally arrive at the vague conclusion that while climate-related changes threaten some aspects of food security in the Arctic, increased temperatures and decreased sea ice cover may also create new opportunities for local food production. These impacts will be influenced not only by environmental change but also by economic, technological, and political forces. As the negative impacts of these changes are difficult to anticipate, resilience or adaptation strategies should be developed. Some recent studies emphasise the need for more multilevel governance systems and planning to prepare for Arctic food security crises in local communities (e.g., Birkland et al. 2019; Orttung et al. 2019).

Notwithstanding the relatively sparse holistic literature, there is considerable knowledge about specific sectors or scenarios; for some potential crises, there are fully developed crisis management strategies. The three illustrative crisis scenarios considered here – the first two mainly from the Finnish crisis management perspective and the third more generally – all share the generic characteristics of an Arctic food crisis but also differ considerably.

Radiological fallout and contamination

Radiological fallout could expose people to radiation directly; it could also expose them indirectly through contamination of the environment and the food chain. Under the crisis categories described by Boin, McConnell and 't Hart (2008, 41), this is a crisis that would develop instantly and terminate gradually because radiation could remain in the environment for a very long time. While closely related to food security, the issues at stake are also much broader. In addition to human and animal health consequences, the economic, environmental, social, and political (both domestic and international) challenges could be significant.

There are some historical cases of this Arctic scenario, and three of them were very serious. The first occurred before the early 1960s, when Soviet nuclear weapon tests at Novaya Zemlya caused major atmospheric fallout and other contamination in the Arctic area. In addition to the local and regional fallout, small-particle radioactive plumes can travel all over the world in the second layer of the Earth's atmosphere and may not fall for years. In the European North/Arctic, the fallout remained rather small-scale. The Limited Test Ban Treaty signed in 1963 eliminated most of this threat; tests in the atmosphere, underwater, or in outer space (but not underground) were banned, and the Soviet Union was one of the parties to the Treaty.

The second example was in 1978. Cosmos 954, a Soviet reconnaissance satellite driven by a nuclear reactor, scattered radioactive debris over northern Canada when it dropped from the sky due to a malfunction. This prompted a major operation to locate the radioactive pieces of the satellite.

The operation covered a total area of 124,000 square kilometres and (naturally) sparked a diplomatic crisis. Similar unexpected events cannot be completely ruled out. In August 2019, there was a missile testing accident involving an allegedly nuclear-fuelled missile in Arkhangelsk in the Russian Arctic; this reportedly caused at least local radiological fallouts (e.g., *New York Times* 2019).

The third incident is the Chernobyl (in the then-Soviet Union, now Ukraine) nuclear power plant (NPP) disaster in 1986 (cf. Moeko Minagawa's analysis in Chapter 3 of this book). While the total contamination/fallout from Chernobyl consisted of 20 different nuclides (Pöllänen et al. n.d., 195), the most dangerous was Cesium-137 due to its long half-life. Paradoxically, while only a small part was released into Finnish Lapland compared to the southern parts of the country (for a global contamination map, see Izrael n.d.), its consequences were more severe and long-lasting in the North due to the food chain: lichen → reindeer → human beings. Lichen is the staple diet of free-ranging reindeer, and reindeer in turn comprise a typical source of livelihood for the local population. Lichen is also extraordinarily vulnerable to long-term contamination by radiological fallout. In Northern Norway and Sweden, where the fallout was more severe than in Finland, increased radiation levels were measured in reindeer meat decades after the disaster.

The most obvious potential source of this type of crisis remains a radiological release from an NPP, and the closer the proximity, the more severe the consequences are likely to be. In this vein, the Finnish (Lapland) crisis management system is well worth examining, alongside comparable information from Sweden and Norway.

Risk assessment. After the Chernobyl disaster, a large-scale collection of lichen and reindeer meat samples was conducted in Finland. Testing showed that Chernobyl had considerably less severe consequences than the fallout from nuclear weapons tests a couple of decades earlier. Yet Chernobyl directly affected food security as a result of contaminated reindeer meat. The reindeer herders, most of whom are Indigenous Sami people, reported that they might consume 0.5 kg of reindeer meat per day; the average Finn eats this amount in an entire year. Therefore, it was concluded that reindeer herders received considerably higher radiation doses than the Finnish population on average (Rissanen and Rahola 1990).

In terms of risk assessment, Chernobyl undoubtedly prompted more concerted efforts toward radiation-related food crisis management in Finland, especially in the areas of prevention and preparedness. By the mid-1990s, detailed analyses of the consequences of nuclear or radiological contamination for water safety (STUK 1995), the food industry (Kantala 2005), and dairy products (STUK 2011) were published.

The root cause of this type of food crisis would be an NPP reactor failure. Currently, most risk assessments in this field are conducted on NPPs or NPP reactors, and most conclude that the risk is micro-minimal; the theoretical probability of a severe accident at a Finnish NPP, for instance, is once every 10,000 or 100,000 years, based on sophisticated quantitative methods (e.g., STUK 2013, p. 43). However, in fact, three severe NPP disasters have occurred within just a few decades, in the US (Three Mile Island 1979), Russia (Chernobyl 1986), and Japan (Fukushima 2011).

Perhaps this reality check, engineering calculations notwithstanding, is why, in a list of 21 identified and prioritised risks, the Finnish *National Risk Analysis* includes a 'severe nuclear accident in Finland or in our vicinity' (Ministry of the Interior 2019, 2016). It is obvious that if Chernobyl, located in Ukraine, created notable fallout in Finland, then NPPs in or closer to Finland would be capable of causing much more damage.

Finland currently has two NPPs with four reactor units. Other nuclear facilities in Finland's immediate neighbourhood include two NPPs in Russia (Kola and Sosnovyj Bor; Russia currently has ten operational NPPs with thirty-six operational reactors) and Forsmark NPP in Sweden (Sweden has a total of three operational NPPs and ten reactors). Both Russia and Finland are building new reactors as well.

Located in the Arctic proper, the Russian Kola NPP is often named as a potential risk. The Finnish *National Risk Analysis* from 2015 (Ministry of the Interior 2016) alludes to it, arguing that, although the safety arrangements of Russian power plants have improved over time, they still incur greater risks of a severe accident than NPPs in Finland. The risk analysis further states that the consequences of such an accident may be serious and then calls for major efforts from society to prevent and limit the ensuing damage. An NPP accident could also require long-term actions, such as decontaminating the inhabited environment, restoring normal living conditions, and ascertaining the safety of foodstuffs and drinking water. Full recovery could take decades. For some reason, the updated *National Risk Analysis* from 2018 (Ministry of the Interior 2019) does not include this kind of Russia-related critical estimation, although the conditions have not changed.

Despite the minimal theoretical probability of an NPP accident, Sweden and Norway have also considered the possibility of radiological fallout from an NPP and identified such an event as one of the risk priorities in their national risk assessments (MSB 2016; DSB 2014), paying specific attention to food security. In the Swedish *National Risk Analysis*, there is a section headed 'Emissions of hazardous substances (CBRNE)', that is, chemical, biological, radiological, nuclear, or explosive substances. The section illustrates a NPP disaster exercise set at Oskarshamn NPP in southern Sweden. The impact assessment concludes that the effects of such a scenario on food production would be limited. However, the scenario creates major difficulties for individual farmers as the fallout area is deemed to be unusable for farming for years to come. As the forest would also be affected, restrictions on wild game meat would be required, and residents would be advised to avoid eating berries or mushrooms from the area. The EU's interim thresholds on goods with potential radioactive contamination would come into force immediately after such an incident;

> this means that anyone who puts food products on the market must be able to guarantee that they do not exceed these thresholds. Exports from Sweden to the rest of Europe and the world will probably decrease, irrespective of whether these thresholds can be met, because people will be concerned about eating foods that come from an area affected by a fallout.
>
> (MSB 2016, 74)

Norway's *National Risk Assessment* similarly includes an NPP fallout scenario, although the country itself has no NPPs. The document states that the likelihood of a serious nuclear event affecting Norway is low:

> An assessment has been made of the likelihood of an accident at a [...] plant, with large emissions that affect Norway. This is expected to occur once every 5,000 years, which gives a likelihood of 0.02% that it will occur in the course of a year.

However, if such an accident did occur, the impact could be severe: 'Radioactive contamination causes exposure to ionising radiation, either directly or through ingesting contaminated foods or breathing in contaminated air'. As for food security, the assessment points out that 'nature, the environment, and food production will be hit hard, and the slaughter of animals, destruction of milk, etc., could become necessary. Action will be required for several decades' (DSB 2014, 129–135).

Prevention. The only way to prevent a food crisis caused by radiological contamination from an NPP is to control the root cause, namely, an NPP malfunction. The mechanism to prevent an NPP accident in Finland involves very strict regulations on the technological redundancy of safety systems and organisational safety culture. The national regulatory body is the Radiation and Nuclear Safety Authority (STUK), which controls and supervises compliance with national and international regulations.

Of course, STUK's authority is limited to Finnish NPPs. Even if STUK's professionalism and resources can be taken as a given – although Fukushima was a stark reminder that even a country with highly developed nuclear safety, such as Japan, can fail miserably – the issue of other countries' NPPs remains. In other words, prevention measures cannot easily be extended outside the country. Being a party to EU nuclear safety regulations (Nuclear Safety Directive, etc.) and International Atomic Energy Agency (IAEA) safety standards adds to the national arrangements. After the Fukushima disaster, the EU, together with national regulatory authorities, organised stress tests for NPPs to identify deficiencies in prevention and to enhance safety. To date, there have been no NPP disasters in the EU.

A special case is Russia, which is not a member of the EU but is a member of the IAEA. Russia's safety culture arouses concern in the Finnish public at times due to whistle-blower reports about misconduct. True, Finland and other Nordic countries have tried to extend their prevention strategies to Russia. For a couple of decades, they have, in cooperation with Russian authorities, partially funded bilateral and multilateral programmes aimed at enhancing the safety of Russian NPPs in neighbouring areas. However, it is difficult to evaluate the success of this cooperation in terms of prevention.

No specific strategies to prevent an (Arctic) food security crisis related to radiological contamination can be identified, and, indeed, it would be difficult to figure out what such strategies might consist of aside from generic work to improve nuclear and radiation safety.

Preparedness. If an NPP disaster and subsequent food crisis cannot be prevented, then preparedness is key. According to Finnish government, 'Finland is prepared for accidents at nuclear power plants. [...] The general requirements for preparedness for nuclear and radiological emergencies are laid down by acts and decrees' (Ministry of the Interior 2016, 32).

While individual nation-states have national preparedness strategies, the Nordic countries' competent authorities in the field of nuclear and radiation safety have agreed on joint guidelines for protective measures in the early and intermediate phases of a nuclear or radiological emergency (National Institute of Radiation Protection et al. 2014). This document includes a section entitled 'Protective measures for foodstuffs, drinking water, and animal feed'; these measures rely heavily on EU-level regulation. The issue largely concerns the maximum permitted levels of different radionuclides in food and water. The document stipulates that if the permitted levels are exceeded, those products will not be used. If residents are in danger of receiving radiation doses via other pathways, these maximum levels are lowered so that the total dose does not exceed the permitted levels. In the case of natural food products, such as fish, plants, game, and mushrooms, the response phase usually includes an official recommendation to avoid these foodstuffs entirely.

Early warning is essential to preparedness. Chernobyl also served to accelerate the development of an early warning measurement system in Finland. Currently, the whole country is covered by 260 gamma radiation monitors that measure radiation levels in real time. The findings of these measurements are also available to the public (STUK n.d.). An airborne radiation measurement system has also recently been established. This is in accordance with the coordinated EU system as well as the system maintained by the Council of the Baltic Sea States (CBSS) network of nuclear and radiation safety authorities, of which Russia is also a full member. In 2016, STUK also updated its agreement with Rosatom, the Russian authority in the field, on the communication and early warning system in case of radiological incidents.

Response. In complex crises where several actors are involved, the two most important tasks are, first, coordination and leadership and, second, communication with the public. In Finland's response to the Chernobyl accident, both of these elements were awry (see the case study in Forsberg et al. 2004). The extended radiation level was measured in Finland, but, according to the rules at that time, it was too low to press the red button. The issue was made public only when the radiation had reached Sweden; then Swedish authorities publicised the information. For a while, it was unclear where the radiation had originated, as the Soviet Union covered up the disaster. When the source was revealed, it immediately precipitated a political crisis in Finland because Finnish leadership was deemed to have deliberately withheld information from the public. This bad image was exacerbated by bad reporting. When the Finnish public were first informed about the disaster, STUK radiation experts who had no experience in crisis communication at that time were invited to speak about Becquerels and Sieverts. Meanwhile, the confused, concerned population were unsure whether they should let their children play in the yard. Considerable time elapsed before leaders at the highest political level stepped in. No specific attention to the Arctic can be identified at this stage, not least because the immediate fallout doses were measured in the south of Finland.

Today, a crisis of this type would probably be handled more efficiently. In almost all developed countries, especially if they have NPPs, rather clear-cut rules are in place in case of emergencies. In Finland, for instance, a constantly updated regulation defines the roles and responsibilities of multiple actors, including the nuclear and radiation safety authority, the government, and different ministries and agencies in the fields of emergency management, health, environment, and food security. The National Supervisory Authority for Health and Wellbeing (VALVIRA) and the Finnish Food Authority (formerly EVIRA) can both issue mandatory orders about the use of foodstuffs in these kinds of crises. So, ostensibly, the decision-making system would work, even if a crisis necessitated deviation from the crisis response plans. Indeed, contingency planning should always focus on planning how to improvise (Stern 2013).

Yet Fukushima served as a salutary wake-up call in many areas of response, including communication. An NPP disaster over 7,500 kilometres away from Helsinki, with no immediate impact on human or animal health in Finland, crashed the STUK web server because 15,000 citizens suddenly wanted information about what had happened and what its impact would be.

Recovery. Real recovery can only be evaluated *post factum*, although successful recovery presupposes advance planning. If radiological fallout and contamination were to occur across a massive area, recovery could not entail total decontamination. In our Arctic food crisis scenario, recovery primarily relates to the livelihood activity of reindeer herding. To this end, recovery would involve measuring the radioactivity level of reindeer meat to decide whether it could be used domestically or exported.

At the time of the Chernobyl disaster, there was no recovery strategy. All decisions were made *ad hoc*. This meant that a variety of solutions were implemented in the affected area. Immediately after Chernobyl, Finland did not establish an acceptable Bq/kg radiation dose level for reindeer meat. Sweden and Norway did, although they set different acceptable dose levels. Just a few years later, the EU, which Finland and Sweden (and Norway through the EEA) joined in the mid-1990s, harmonised the situation by establishing joint dose levels as a precondition for food that is exported in the single market.

Learning. Much has been learned over the years, especially since Chernobyl. The Finnish system seems to be a self-corrective one in which previous experience is translated into new knowledge, and there is a clear effort to institutionalise the political, organisational, and technological lessons learned from previous shortcomings.

Animal disease epidemics

Serious animal diseases are those that spread particularly easily, usually through animal products, animal transportation, grazing livestock, instruments, animal keepers, freely moving pets or wild animals, or insects. If not properly managed, these risks may morph into epidemics. Currently, European countries classify animal diseases in a variety of ways, but the European Union is expected to harmonise the disease classification until 2021. A-class diseases are the most transmittable and serious.

In Finland, specifically in the northern part of the country, several diseases have potentially serious and widespread consequences and should be monitored. Each of these diseases is regulated somewhat differently depending on its classification. These potentially serious include (to mention but a few) foot-and-mouth disease (FMD), which may infect cattle and reindeer; chronic wasting disease (CWD), which has been found in reindeer and moose; the Orf virus, which can be found in sheep, goats, and reindeer; the African swine fever virus (ASFV), which may be transmitted to farmed pigs from wild boar or through human activity abroad; avian influenza (bird flu), which may be transmitted to poultry (and is highly unlikely in the Arctic); and infectious hematopoietic necrosis (IHN), which is a disease that affects fish (mainly salmon). Some of these diseases may be transmitted to humans and cause specific health problems. In any case, animal diseases, even if mostly harmless to humans, may severely damage the reputations of farms or companies, require long-term regulative bans on selling and exporting, or necessitate the mass slaughter of animals.

Obviously, climate change cannot be overlooked in this field, either. For some time now, studies have argued that warmer temperatures could improve the survivability of diseases and reduce development times and extend transmission times for marine fish diseases, terrestrial parasitic diseases, bacterial diseases, and insect-borne diseases in Arctic fauna. These may be caused by disease agents that already exist in the environment, but they are likely to be favoured by climate warming and assisted by insects that thrive in warmer conditions and transmit diseases. Or new diseases could spread due to agents that enter the Arctic through host species that are able to establish new ranges because of climate change; these agents may spread to Arctic species that are more vulnerable because of a lack of natural immunity (e.g., Parkinson and Butler 2005; Bradley et al. 2005).

An even more serious imagine-the-unimaginable threat scenario is related to the thawing of the permafrost due to climate change. In the past, many infected animals were buried in the soil, including those with diseases such as anthrax and smallpox. For some time, scholars (e.g., Revich and Podolnaya 2011) have been warning that in Yakutia, Eastern Siberia in particular, permafrost melting may lead to the release of viable anthrax organisms in areas where an anthrax outbreak occurred in the past and infected animals were buried. In 2016, this scenario took place, leading to the infection of more than 2,000 animals, at least one human death, and dozens of hospitalised patients (e.g., Doucleff 2016; Nilsen 2016). Finland can again serve as a primary example of crisis management in this area, with particular attention paid to the Arctic conditions where appropriate.

Risk assessment. The Finnish *National Risk Assessment* from 2015 (Ministry of the Interior 2016) does not mention animal diseases. However, the *Security Strategy for Society* from 2017 (Turvallisuusneuvosto 2017) has a short section on biological threats that includes such diseases. Consequently, the updated version of the *National Risk Assessment* from 2018 (Ministry of the Interior 2019) also includes a short and rather superficial chapter to that effect. There is no holistic official risk assessment of animal diseases or, more generically, food security that focuses on the Arctic or on Finnish Lapland.

The competent food security authorities are more specific in this respect. To assess the risk of an animal outbreak, each potential disease must be examined separately. Thus, one can find, for instance, a risk profile of ASFV transmission routes to Finland (EVIRA 2011). An example which clearly stands as a best practice model of animal disease risk assessment is the Finnish Food Authority (the then EVIRA) 2011 comprehensive quantitative assessment on the spread of FMD (Lyytikäinen et al. 2011). As the last FMD outbreak in Finland was in 1959, the historical data were either not very useful or unavailable. True, the 2001 epidemic that mainly affected the UK, along with the subsequent epidemics in 2007 and 2011 in some other European countries, led to a considerable increase in data and research literature. However, these data are not easily applicable to other countries and conditions.

The Finnish 2011 risk assessment therefore uses a Monte Carlo simulation, a rather sophisticated quantitative risk assessment method. The analysis starts with a randomly selected infected farm and then iterates the possible courses of contagion until no farm is infected. To be effective, this exercise must use detailed data on livestock farm locations, density, direct and indirect contact with other farms, and production structures. These data are supplemented with some assumptions about the protective measures that would be put in place after the first case of infection had been identified. Overall, the risk assessment contains more than 100,000 simulation iterations.

The simulations show that the potential for the spread of FMD in Finland is low. One-third of the outbreaks would be sporadic in one farm and would not cause further contagion, lasting on average 3.5 weeks until the farm was disinfected, after which the restrictive measures would be lifted. The most typical outcome, occurring in more than 60% of the iterations, would result in an outbreak affecting an average of five farms and lasting approximately five weeks. Even in the worst-case scenario, the outbreak lasted only ten weeks and affected twenty-nine infected farms; no simulated outbreak spiralled out of control.

As for the Arctic, although the situation would depend on the production structure of the farms, the FMD risk analysis shows that the northernmost regions would fare even better. This is because the number and density of animal production farms is considerably lower and their disease transmission contacts with other farms are likewise reduced.

However, this risk assessment does not consider reindeer, so Lapland, the Arctic area of Finland, poses some specific challenges. First, several farms in this region engage in mixed cattle and reindeer farming, which increases the chances of animal disease transmission. Second, reindeer are mostly free-ranging and organised into cooperatives with several reindeer keepers. In early spring, they are herded together because of the lack of natural food (lichen) due to snow cover. In the autumn, they are herded together again for counting and selected slaughtering. At other times, should they be infected, it would be very difficult to find and isolate the infected from the non-infected.

Prevention. Disease prevention in food animals in Finland relies on risk treatment that refers to the assumption of an imported disease and its development in Finland through livestock and poultry farming malpractice. A fundamental element of prevention is, of course, a functioning authority system. As in most Arctic countries, food security in Finland is organised hierarchically. The highest authority is the respective ministry, in this case, the Ministry of Agriculture and Forestry. Under it works the Finnish Food Authority, which supervises several Regional State Administration Agencies. They, in turn, work closely with local authorities and with official municipal veterinarians and inspectors, especially in issues affecting food security. The roles and duties of each level are defined in legislation. Cross-sectorial coordination and cooperation between other sectoral authorities is implemented when needed (Lounela 2018).

The main instrument of prevention is combined EU and national regulations, as well as guidelines and guidance on importing animals and animal products, the biosecurity of animal holdings, the isolation of infected animals if needed, and sometimes vaccination (Lounela 2018). However, vaccination is generally not a solution. For example, in the case of FMD, routine vaccination would cause more harm than good, at least economically. The situation would be the same if the disease were detected in neighbouring countries or on a Finnish farm. The results of the simulations of the spread of the disease, coupled with an economic analysis of different scenarios, do not support emergency vaccination policies in Finland; vaccination would have a negligible effect on the spread of the disease even in worst-case outbreaks (Lyytikäinen et al. 2011).

Preparedness. As it is impossible to prevent all animal diseases, the solution is to be ready to identify them and to create the capacities and capabilities to respond to an outbreak. This involves contingency planning, emergency stocks, training and exercises, education and information about the symptoms, a rapid communication system, checking of suspected cases, the isolation of animals, long-term monitoring and random samples, and risk-based testing of imported animals and animals close to the border (Lounela 2018). The preparedness system also includes early warning, which entails well-regulated responsibilities for farmers and authorities. In an outbreak, specifically trained veterinarians would conduct an epidemiologic inquiry that would include identifying contact farms (farms that have had direct or indirect contact with the suspected farm or that may have acquired the infection from the same source) (Lyytikäinen et al. 2018).

Response. If an animal disease is identified, several routine tasks ensue. These may include slaughtering and destroying animals; clearing, cleaning, and disinfecting animal shelters; and destroying or handling manure and other infectious material. Overlapping with these measures, any further spread of the disease must be prevented. This means diagnosing and mapping the origin of the disease; examining contact farms; isolating the infected area; establishing protection and surveillance zones around the infected farm(s), usually for some kilometres; restricting the movement of animals and animal products in specified zones; and, for some diseases, vaccinating animals (Lounela 2018).

Recovery. Livestock farmers may struggle to recover their livelihood if many animals must be slaughtered or if serious harm is done to the farmer's reputation. One recovery measure, therefore, is financial support for affected animal owners. In Finland, owners are entitled to compensation for slaughtered animals and lost property. Compensation for the loss of production is discretionary and may cover up to 75% of the losses incurred.

Learning. The competent Finnish authorities would need to make systematic, institutionalised efforts to draw lessons from each disease case. After processing a suspected disease or eradicating a disease case, the success of the measures will be routinely assessed, and contingency plans and guidelines updated. This is often followed by arranging additional training or education.

Oil spills and fishery collapse

Fishery is a natural part of the Arctic population's economic activity. Oil spills may lead to a crisis in this area. The only major disaster in the (sub-) Arctic area to date was the Exxon Valdez tanker accident in 1989. Herring fishery in the affected area collapsed four years later, allegedly because herring embryos were exposed to crude oil (Incardona et al. 2015).

The probability of major oil spills in Arctic seas is inevitably increasing alongside climate change, which enables resource exploration and extraction activities and has opened new transport routes in the area. Major offshore oil spill disasters can occur for three reasons: failures on offshore platforms, offshore pipeline ruptures, and accidents involving tankers or freight or passenger ships.

Risk assessment. Oil spill risk assessment in the Arctic is a rather well-researched issue though it usually examines the consequences rather than the causes (e.g., Aune et al. 2018; Nevalainen, Helle, and Vanhatalo 2017). It has been concluded (Wenning et al. 2018) that six risk/consequence assessment methods currently exist for the Arctic in particular; all are aimed at informing decision-makers about better crisis management strategies should a major oil spill occur.

Yet compared to many other sea areas, Arctic oil spill risk assessment suffers from a lack of data, both on historical oil spills and basic ecological background data. For this reason, apart from the Exxon Valdez *post factum* assessment (e.g., French-McCay 2004), scenario-based laboratory tests, models, and simulations have been used to compensate for the lack of historical data. This is no mean feat, however, as many variables, such as location, weather, season, oil type, spill size, affected species, short- and long-term consequences, can affect the impact of a spill. Challenges specific to the Arctic generally include the environment's sensitivity (mainly seen in the long recovery time); long distances from existing response infrastructure; and ice-covered waters, which make oil less accessible (NTNU 2018; Bambulyak et al. 2014; Eger n.d.). On the other hand, the situation is dynamic. Recent studies (e.g., Nordam et al. 2017) suggest that future oil spills in the Arctic in a warming climate may cover larger areas and cause increased shoreline exposure due to reduced ice coverage.

It is also difficult to assess the impact of an oil spill on fish stock or individual fish species and, indirectly, on the fishing industry. Some researchers argue that instead of a traditional species-specific risk assessment, ecological risk assessments should focus on food chains to better reflect the long-term impacts of a spill (Nevalainen, Helle, and Vanhatalo 2017, 92–93, 2018). Studies of the impact of oil spills on fish populations that use simulations (e.g., Carroll et al. 2018) have concluded that commercial fish species (such as cod) would be affected through the impact on fish roe and larvae, while the adult population would retain full reproductive potential and a sufficient number of juveniles would survive to replenish the population.

This, however, would cause recruitment losses for mature cod stock for a certain length of time.

Prevention and preparedness. In this scenario, prevention and preparedness are often intertwined. The most important tool is regulation that combines organisational and technological performance criteria (see Wenning et al. 2018; Lloyd 2012). In addition to national legislation, international, and regional agreements (see Steigelman 2017) are particularly important as many maritime activities take place outside national waters. These include, most notably, the International Maritime Organisation's (IMO) *International Code for Ships Operating in Polar Waters* (the *Polar Code*, see IMO 2017), which sets mandatory prevention and preparedness measures for commercial shipping activities.

The Arctic Council has also prepared operating guidelines for oil spills (see Arctic Council n.d.), including a clear-cut system of mutual assistance among the eight Arctic states in the case of an accident. One local example is Svalbard's (Norway) risk assessment (Sysselmannen Svalbard 2013, 13–15), which summarises national and local regulations, the roles and responsibilities of the authorities and shipping companies, and preparedness-related resources and capabilities that apply in the case of an oil spill.

Obviously, oil exploration and shipping companies' self-interest encourages them to build the best possible safety systems for the equipment and machinery they own or operate. The challenge, however, lies in balancing profit-seeking with investing resources in the protection or resilience of the systems. This balance is likely to be imperfect, as the discussion below on liability reveals.

Response. The response to an oil spill is particularly difficult in the Arctic due to its harsh conditions and remoteness. However, at least four response strategies for mitigating oil-spill exposure are adaptable to Arctic conditions. The first is a mechanical response, in which equipment are used to contain and recover oil on the sea's surface. A second option is to deploy chemical agents and additives on the sea's surface and subsurface to change the physical and chemical properties of the oil. Third, spilled oil on the sea's surface can be burned in a controlled way. A final option is natural degradation: taking no action apart from monitoring the movement of the oil spill (Wenning et al. 2018; cf. Aune et al. 2018).

Each of these response methods has its advantages and limitations, and their effectiveness depends on the time of year and environmental factors unique to the Arctic environment. Perhaps more importantly, a real crisis, with its attendant idiosyncrasies, does not always allow sufficient time to identify the optimal strategy. However, the quicker and the more successful the response is, the less fish stocks would be affected. Nevertheless, as the risk simulations show, issues such as location, weather, season, oil type, and affected species may vary considerably and affect the outcome for fisheries.

Recovery. Oil spill recovery is costly and often contested among stakeholders. It begs several questions: Who will bear the responsibility? Does tort liability legislation, which includes a variety of insurance and recovery fund schemes, function well enough to support proactive recovery in this rather complicated international area? (See, e.g., Steigelman 2017; Johnstone 2015; Keil 2015; Lloyd 2012).

Fishery recovery after an oil spill has several dimensions. The recovery of the fish population and recovery of fishery are related but still distinct issues. Some modelling and simulation exercises suggest that, even in the event of a major oil spill in the Arctic, the commercial fish stock would not easily drop below a sustainable limit (although the Exxon Valdez herring data seem to suggest otherwise). The adult fish population usually survives by moving elsewhere. Therefore, the recovery of the fish population largely depends on the continued level of fishing. Even if the oil spill caused the loss of 100% of recruits in a single year for, say, cod, eliminating the contribution of the affected year's juveniles, the population would recover to 99% of the default value after ten years. However, a 10% reduction in fishing efforts for eight years after the spill would allow the stock to recover in eight years (Carroll et al. 2018; cf. Langangen et al. 2017). This possible solution leads to more questions: Who should be responsible for regulating these issues, and who would compensate fishers for possible losses in this transboundary context?

Learning. Although Exxon Valdez has been the only major Arctic oil spill to date, it yielded some data that have been utilised for academic risk assessment in data-based analyses and simulations. In more practical terms, 'many lessons were learned from this accident, including the important one that some cleaning methods used to remove the oil led to greater environmental damage than the oil itself' (Eger n.d.). The lack of post-crisis data means that additional well-designed models and simulations are the way to increase knowledge in this area. In some sense, this serves to merge the learning and risk assessment phases.

Conclusions

This chapter has considered three possible Arctic food crisis scenarios: radiological fallout and contamination, animal disease epidemics, and the impact of oil spills on fisheries. The results of this analysis are summarised in Table 9.1. Without repeating the details, at least five general arguments presented in this chapter can be highlighted.

First, while the Arctic region has unique characteristics (long distances, harsh climate, lack of disaster data, etc.) that should be considered in crisis management strategies, it should not be treated as an isolated case. General crisis management theory provides a good structure that can be tailored to the Arctic.

Second, when discussing 'Arctic food security', a more detailed analysis shows that different crisis scenarios may differ significantly. Indeed, a working crisis management system must concentrate on particular sectorial scenarios.

Table 9.1 Summary of the characteristics of Arctic food crisis management in three scenarios

Crisis scenario	*Specific Arctic pros and cons*	*Risk assessment*	*Prevention*	*Preparedness*	*Response*	*Recovery*	*Learning*
Radiological fallout and contamination (re. food security) Finland (and Nordic countries) as an example	*Positive:* Remote area, relatively small population, less direct human contamination. *Negative:* Closeness of several NPPs. Relatively large impact on food chain, and indirectly on human health through reindeer meat consumption.	Quantitative reactor-specific risk assessments, showing minimal risks. Prioritised in national risk assessment (conflicting with the reactor-specific technological assessments).	National legislation and regulation. EU- and international (IAEA)-level regulation and standards. Limited cross-border prevention approach, besides the obvious cross-border characteristics of the scenario.	Good national-level systems of planning, training, and early warning. Regional and international assistance systems in place.	Good national-level systems in terms of roles, responsibility, and communication. Regional and international assistance systems in place.	Nationally established and an EU-level harmonised system concerning the dose rates in food products. National and probably EU-level programmes to financially assist recovery process.	Increasing post-crisis knowledge and respective institutional solutions, related to all phases of crisis management.
Animal disease epidemics Finland as an example	*Positive:* Low density of farms leading to less infection transmissions. *Negative:* Free reindeer grazing leading to larger possibility of being infected by wild animals; more difficult to detect and isolate infected animals.	Disease-specific and well-prepared risk assessments on some diseases exist, showing non-major risks. Lack of background data on materialised relevant epidemics. Prioritised in national risk assessment. No specific regional Arctic risk assessments.	National legislation and regulation. EU-level regulation and standards. Good monitoring system.	Good national-level systems of planning, training, and early warning.	Good national system in terms of roles re. responsibility and communication between authorities.	National partial compensation system (75%) for animal farmers for their economic losses.	Systematised post-event learning after any disease detection and spread.

(*Continued*)

Crisis scenario	*Specific Arctic pros and cons*	*Risk assessment*	*Prevention*	*Preparedness*	*Response*	*Recovery*	*Learning*
Oil spills and fishery	*Positive:* Currently relatively limited tanker and other traffic, though increasing rapidly. *Negative:* The environmental sensitiveness (mainly the long recovery time), long distances from existing response infrastructure, and ice-covered waters as the oil is less accessible in such conditions.	Increasing scientific risk assessment literature on Arctic oil spills, focusing overwhelmingly on consequence side at the cost of causes. Lack of background data on materialised, relevant oil spills, compensated for by developed modelling and simulation.	National legal regimes in all Arctic countries. International (IMO, etc.) legal regimes.	National legal regimes in all Arctic countries. International (IMO, etc.) legal regimes. Regional and international assistance systems.	While the situational factors are crucial (location, weather circumstances, season, oil type, spill size, affected species, etc.), there are at least four response methods for oil spills that can be used in the problematic Arctic conditions.	Liability and compensation systems exist but are contested. Fish population recovery crucially depends on how much fishery is limited during the recovery time.	The sparse post-disaster data hinders learning, but is compensated for by modelling and simulation.

Note: All online links checked in August 2019.

Third, all the scenarios described here touch upon several dimensions of society. Hence, crisis management must involve simultaneous cross-sectorial vision, and institutional solutions and analyses must be coordinated.

Fourth, the crisis management system of any (food) crisis is organised hierarchically. In a crisis, local, national, regional, and sometimes international crisis management approaches would need to be combined into a coordinated toolkit in the spirit of a multilevel governance approach.

Finally, while much has been done, there is still some way to go in addressing all these challenges. If anything, this analysis has shown that the existing crisis management cycle could be structured for a more holistic approach that addresses the unique features of the Arctic region as well as national, institutional, and sectoral idiosyncrasies. This would support cross-sectorial learning, highlight potential bottlenecks, and indicate appropriate investment priorities in Arctic food crisis management.

Note

1 The author is greatly indebted to several people interviewed for this chapter for their valuable inside information and opinions, especially to those in the Finnish crisis management systems for the scenarios presented in the chapter, who donated considerable time to discuss these issues: Maarit Muikku and Kari Peräjärvi (Radiation and Nuclear Safety Authority in Finland); Hanna Lounela (Finnish Food Authority); Pirkko Pirinen (Lapland's Regional State Administration); and Taito Vainio (Finnish Ministry of the Interior). Finally, the author is grateful to the anonymous reviewer for useful comments. All the information, interpretations, and conclusions, as well as related errors, naturally remain the sole responsibility of the author.

References

Arctic Council (n.d.). Retrieved from https://arctic-council.org/index.php/en/about-us/working-groups/eppr.

Aune, M., A.S. Aniceto, M. Biuw, M. Daased, S. Falk-Petersen, E. Leue, C.A.M. Ottesend, K. Sagerup, and L. Camus. 2018. Seasonal ecology in ice-covered Arctic seas – considerations for spill response decision making. *Marine Environmental Research* 141: 275–288.

Aven, T. 2015. *Risk analysis*. Second edition. Chichester: John Wiley and Sons.

Balzacq, T. (ed.) (2015). *Securitization theory: How security problems emerge and dissolve*. London: Routledge.

Bambulyak, A., R.U.F. von Bock und Polach, S. Ehlers, and A. Sydnes. 2014. Challenges with oil spill risk assessment in Arctic regions: Shipping along the Northern Sea route. Proceedings of the ASME 2014 33rd International Conference on Ocean, Offshore and Arctic Engineering OMAE2014, June 8–13, San Francisco, California, USA.

Bellavita, C. 2006. Changing homeland security: Shape patterns, not programs. *Homeland Security Affairs* 2(3). Retrieved from http://www.hsaj.org.

Birkland, T.A., K. Schwaeble, M. Grabowski, M. Lowe, T. Sharkey, and W. Wallace. 2019. *Emergency management, climate change, and complex governance in the Arctic.* EasyChair Preprint 1170. Retrieved from https://easychair.org/publications/preprint_download/ppKk.

Boin, A., McConnell, A. and 't Hart, P. 2008. *Governing After Crisis: The Politics of Investigation, Accountability and Learning.* Cambridge: Cambridge University Press.

Bradley, M.J., S.J. Kutz, E. Jenkins, and T.M. O'Hara. 2005. The potential impact of climate change on infectious diseases of Arctic fauna. *International Journal of Circumpolar Health* 64(5): 468–477.

Carroll, J.L., F. Vikebø, D. Howell, O.J. Broch, R. Nepstad, S. Augustine, G.M. Skeie, R. Bast, and J. Juselius. 2018. Assessing impacts of simulated oil spills on the Northeast Arctic cod fishery. *Marine Pollution Bulletin* 126: 63–73.

Clement, J. 2019. From crisis to resilience in the Arctic. *Arctic Today*, 20 June. Retrieved from https://www.arctictoday.com/from-crisis-to-resilience-in-the-arctic/.

Doucleff, M. 2016. Anthrax outbreak in Russia thought to be result of thawing permafrost. *NPR*, August 3. Retrieved from https://www.npr.org/sections/goatsandsoda/2016/08/03/488400947/anthrax-outbreak-in-russia-thought-to-be-result-of-thawing-permafrost.

DSB. 2014. *National Risk Analysis 2014.* Oslo, Norway: The Norwegian Directorate for Civil Protection.

Eger, K.M. n.d. *Effects of oil spills in Arctic waters.* Retrieved from http://www.arctis-search.com/Effects+of+Oil+Spills+in+Arctic+Waters.

EVIRA 2011. *Afrikkalaisen sikaruton mahdollisia maahantuloreittejä – riskiprofiili* [Possible entry routes for African swine fever – risk profile]. EVIRA, April. Retrieved from https://www.ruokavirasto.fi/globalassets/tietoa-meista/julkaisut/julkaisusarjat/julkaisuja/elaimet/afrikkalaisen-sikaruton-mahdollisia-maahan tuloteitteja---riskiprofiili.pdf.

Forsberg, T., C. Pursiainen, R. Lintonen, and P. Visuri (eds.). 2004. *Suomi ja kriisit. Vaaran vuosista terrori-iskuihin* [Finland and crises. From 'the Years of Danger' to terrorist attacks] Helsinki: Gaudeamus.

French-McCay, D.P. 2004. Oil spill impact modelling: Development and validation. *Environmental Toxicology and Chemistry* 23(10): 2441–2456.

Gundel, S. 2005. Towards a new typology of crises. *Journal of Contingencies and Crisis Management* 13(3): 106–115.

Haddow, G.D., J.A. Bullock, and D.P. Coppola. 2011. *Introduction to emergency management.* Fifth edition. Amsterdam: Elsevier.

Heath, R. 1998. Dealing with the complete crisis: The crisis management shell structure. *Safety Science* 30: 139–150.

Heck, J.P. 1991. Comments on 'the Zeebrugge ferry disaster.' In U. Rosenthal and B. Pijnenburg (eds.), *Crisis management and decision making*, 75–79. Dordrecht: Kluwer.

Hermann, C.F. 1963. Some consequences of crises which limit the viability of organisations. *Administrative Science Quarterly* 8: 61–82.

IMO 2017. Polar code. Retrieved from http://www.imo.org/en/MediaCentre/HotTopics/polar/Pages/default.aspx.

Incardona, J.P., M.G. Carls, L. Holland, T.L. Linbo, D.H. Baldwin, M.S. Myers, K.A. Peck, M. Taga, S.D. Rice, and N.L. Scholz. 2015. Very low embryonic crude oil exposures cause lasting cardiac defects in salmon and herring. *Scientific Reports*, September 8, 5. Retrieved from https://www.nature.com/articles/srep13499.

ISO 2015. *Quality management systems – Fundamentals and vocabulary*. ISO 9000:2015.

ISO 2018. *Risk management – Guidelines*. ISO 31000:2018.

ISO/IEC 2019. *Risk management — Risk assessment techniques*. Edition 2.0. IEC 31010.ISO/PAS 2007. *Societal security – Guideline for incident preparedness and operational continuity management*. ISO/PAS 22399:2007.

Izrael, Yu. A. n.d. *The atlas of caesium-137 contamination of Europe after the Chernobyl accident*. Retrieved from https://inis.iaea.org/collection/NCLCollectionStore/_Public/31/056/31056824.pdf.

Johnstone, R.L. 2015. *Offshore oil and gas development in the arctic under international law: Risk and responsibility*. Leiden: Brill.

Kantala, T. 2005. *Elintarviketeollisuuslaitosten ja niiden ympäristön puhdistustoimenpiteet säteilytilanteessa* [Measures for purifying food industry facilities and their surroundings in a radiation hazard situation]. STUK-A212, Helsinki.

Keil, K. 2015. Spreading oil, spreading conflict? Institutions regulating Arctic oil and gas activities. *The International Spectator* 50(1): 85–110.

Langangen, Ø., E. Olsen, L.C. Stige, J. Ohlberger, N.A. Yaragin, F.B. Vikebø, B. Bogstad, N.C. Stenseth, and D.Ø. Hjermann. 2017. The effects of oil spills on marine fish: Implications of spatial variation in natural mortality. *Marine Pollution Bulletin* 119:102–109.

Lloyd 2012. *Arctic risk liability in the Arctic, from Lloyd's Report 2012: Arctic opening – Opportunity and risk in the High North*. Retrieved from http://www.arctissearch.com/Arctic+Risk+Liability+in+the+Arctic.

Loring, P.A. and S.G. Gerlach. 2015. Searching for progress on food security in the North American North: A research synthesis and meta-analysis of the peer-reviewed literature. *Arctic* 68(3): 380–392.

Lounela, H. 2018. *Varautuminen helposti leviäviin eläintauteihin* [Preparedness for easily transmitted animal diseases]. Elintarvikevirasto (EVIRA). Presentation 14 March 2018.

Lyytikäinen, T., J. Niemi, L. Sahlström, T. Virtanen, and H. Lehtonen. 2011. *The spread of foot-and-mouth disease (FMD) within Finland and emergency vaccination in case of an epidemic outbreak*. Evira Research Reports 1/2011.

Meakin, S. and Kutvits, T. 2009. *Assessing the impacts of climate change in food security in the Canadian Arctic*. Arendal, Norway: GRID-Arendal.

Milton, N. 2010. *The Lessons Learned Handbook: Practical approaches to learning from experience*. Oxford: Chandos.

Ministry of the Interior. 2016. *Finland's national risk assessment 2015*, Internal Security, Helsinki, Finland: Ministry of the Interior Publication 4/2016.

Ministry of the Interior. 2019. *National risk assessment 2018*. Helsinki, Finland: Internal security, Publications of the Ministry of the Interior 2019:9.

MSB. 2016. *A summary of risk areas and scenario analyses 2012–2015*. Stockholm, Sweden: Swedish Civil Contingencies Agency.

National Institute of Radiation Protection et al. 2014. *Protective measures in early and intermediate phases of a nuclear or radiological emergency. Nordic guidelines and recommendations*. Retrieved from https://gr.is/wp-content/uploads/2016/06/Nordic-Flagbook-February-2014.pdf.

Nevalainen, M., I. Helle, and J. Vanhatalo. 2017. Preparing for the unprecedented – Towards quantitative oil risk assessment in the Arctic marine areas. *Marine Pollution Bulletin* 114: 90–101.

Nevalainen, M., I. Helle, and J. Vanhatalo. 2018. Estimating the acute impacts of Arctic marine oil spills using expert elicitation. *Marine Pollution Bulletin* 131: 782 –792.

New York Times. 2019. Russia identifies four radioactive isotopes from nuclear accident. *New York Times*, 26 August. Retrieved from https://www.nytimes.com/2019/08/26/world/europe/russia-radiation-accident-isotopes.html.

Nilsen, T. 2016. Scientist: Yamal anthrax outbreak could just be the beginning. *The Barents Observer*, 7 August. Retrieved from https://thebarentsobserver.com/en/arctic/2016/08/scientist-yamal-anthrax-outbreak-could-just-be-beginning

Nilsson, A.E., L.M. Nilsson, A. Quinlan, and B. Evengård. 2013. Food security in the Arctic: Preliminary reflections from a resilience perspective. In Arctic Council (ed.), *Arctic Resilience Interim Report 2013*, 113–117. Stockholm Environment Institute and Stockholm Resilience Centre.

Nordam, T., D.A.E. Dunnebier, C.J. Beegle-Krause, M. Reed, and D. Slagstad. 2017. Impact of climate change and seasonal trends on the fate of Arctic oil spills. *Ambio* 46(Suppl. 3): S442–S452.

Norwegian University of Science and Technology (NTNU). 2018. *Predicting the fate of oil spills in Arctic sea ice*. NTNU, February 16. Retrieved from https://phys.org/news/2018-02-fate-oil-arctic-sea-ice.html.

Orttung, R.W., J. Powell, J. Fox, and C. Franco. 2019. Strengthening food security near the Arctic Circle: Case study of Fairbanks North Star borough, Alaska. *Sustainability* 11(10): 2722.

Paris, R. 2001. Human security: Paradigm shift or hot air? *International Security* 26(2): 87–102.

Parkinson, A.J. and J.C. Butler. 2005. Potential impacts of climate change on infectious diseases in the Arctic. *International Journal of Circumpolar Health* 64(5): 478–486.

Perry, R.W. and M.K. Lindell. 2003. Preparedness for emergency response: Guidelines for the emergency planning process. *Disasters*, 27(4): 336–350.

Pöllänen, L., S. Ristonmaa, J. Sandberg, and O. Vilkamo. n.d. Varautuminen häiriöihin ja onnettomuuksiin ydinvoimalaitoksilla [Preparedness for disruptions and accidents at nuclear power plants]. *Ydinturvallisuus*, Ch. 5, STUK. Retrieved from https://www.stuk.fi/documents/12547/494524/kirjasarjaV_ydinturvallisuus_5.pdf/038afd47-4ee3-4045-9027-23cd3cb88f89.

Pritchard, C.L. 2015. *Risk management. Concepts and guidance*. Fifth edition. Boca Raton: CRC Press.

Pursiainen, C. 2017. *The crisis management cycle*. London: Routledge.

Revich, B.A. and M.A. Podolnaya. 2011. Thawing of permafrost may disturb historic cattle burial grounds in East Siberia. *Global Health Action* 4(1). Retrieved from https://www.tandfonline.com/doi/full/10.3402/gha.v4i0.8482?src=recsys.

Rissanen, K. and T. Rahola. 1990. 'Porojen pekkerellit' neljä vuotta Chernobylin jälkeen [Reindeer becquerels four years after Chernobyl]. *Poromies* 1: 5–9.

Rosenthal, U. and B. Pijnenburg (eds.). 1991. *Crisis management and decision making*. Dordrecht: Kluwer.

Rudolph, K.R. and S.M. McLachlan. 2013. Seeking Indigenous food sovereignty: Origins of and responses to the food crisis in Northern Manitoba, Canada. *Local Environment. The International Journal of Justice and Sustainability* 18(9): 1079–1098.

Sage, C. 2014. The transition movement and food sovereignty: From local resilience to global engagement in food system transformation. *Journal of Consumer Culture* 14(2): 254–275.

Sisäministeriö. 2016. *Säteilytilanneohje* [Radiation hazard situation guidelines]. Sisäministeriön julkaisu 10/2016. Retrieved from http://julkaisut.valtioneuvosto.fi/handle/10024/74874.

Steigelman, T.E. 2017. Clarifying the liability risk of shipping in the Canadian Arctic. *Ocean and Coastal Law Journal* 22(1). Retrieved from http://digitalcommons.mainelaw.maine.edu/oclj/vol22/iss1/5.

Stern, E. 2013. Preparing: The sixth task of crisis leadership. *Journal of Leadership Studies* 7(3): 51–56.

STUK. (n.d.). Säteilytilanne tänään [Radiation situation today]. Retrieved from https://www.stuk.fi/aiheet/sateily-ymparistossa/sateilytilanne-tanaan.

STUK. 1995. *Radioaktiivisen laskeuman vaikutukset vesihuoltoon* [Effects of radioactive deposition on water supply]. STUK-A122. Helsinki. Retrieved from https://inis.iaea.org/collection/NCLCollectionStore/_Public/27/010/27010538.pdf.

STUK. 2011. *Maitotuotteisiin kohdistuvat vastatoimenpiteet ydinonnettomuustilanteessa* [Countermeasures for dairy products in the event of a nuclear accident]. STUK-A185, Helsinki. Retrieved from http://www.julkari.fi/handle/10024/124473.

STUK. 2013. *Finnish report on nuclear safety. Finnish 6th national report as referred to in Article 5 of the Convention on Nuclear Safety.* Retrieved from https://www.stuk.fi/documents/88204/278602/finnish-national-report-stuk-b164.pdf/e6304dfe-01f1-4cfd-966b-0ae5c268cae0.

Sysselmannen Svalbard. 2013. *Risiko- og sårbarhetsanalyse 2013* [Risk and vulnerability analysis]. *Offentlig version*. Norway: Longyearbyen.

Turvallisuusneuvosto. 2018. *The Security Strategy for Society.* Government Resolution 2.11. [in Finnish] Retrieved from https://turvallisuuskomitea.fi/wp-content/uploads/2018/02/YTS_2017_suomi.pdf.

UNISDR/UNDRR. 2009. *UNISDR terminology on disaster risk reduction.* United Nations International Strategy for Disaster Reduction (UNISDR), Geneva, Switzerland, May 2009. Retrieved from http://www.unisdr.org/we/inform/terminology.

Veenema, T.G. and C. Woolsey. 2012. Essentials of disaster planning. In T.G. Veenema (ed.), *Disaster nursing and emergency preparedness: for chemical, biological, and radiological terrorism and other hazards.* Third Edition, 1–20. New York: Springer.

Wenning, R.J., H. Robinson, M. Bock, M.A. Rempel-Hester, and W. Gardiner. 2018. Current practices and knowledge supporting oil spill risk assessment in the Arctic. *Marine Environmental Research* 141: 289–304.

Wesche, S.D. and H.M. Chan. 2010. Adapting to the impacts of climate change on food security among Inuit in the Western Canadian Arctic. *EcoHealth* 7: 361–373.

WFS. 1996. *World Food Summit 13–17 November 1996. Rome Declaration on Food Security.* Retrieved from http://www.fao.org/docrep/003/w3613e/w3613e00.htm.

Yoe, C. 2012. *Primer on risk analysis. Decision making under uncertainty.* Boca Raton: CRC Press.

10 Food security from a food regimes perspective

Victòria Soldevila Lafon

Introduction

The most widely accepted definition of food security is the one given by the Food and Agriculture Organisation (FAO) at the World Food Summit in 1996. According to this definition, food security exists 'when all people, at all times, have physical and economic access to sufficient, safe, and nutritious food that meets their dietary needs and food preferences for an active and healthy life' (FAO 1996).

The recent publication *The State of Food Security and Nutrition in the World 2019. Safeguarding against Economic Slowdowns and Downturns* demonstrates that the goal of worldwide food security is far from achieved. In 2018, more than 820 million people in the world suffered from chronic undernutrition. This represents 11% of the world's population (FAO, IFAD, UNICEF, WFP, and WHO 2019).

Concerns about food security are increasing in many geographical areas. This chapter analyses food insecurity in the Arctic from the perspective of food regimes. Food regime analysis has a long tradition in agri-food studies; however, little literature addresses the issue of food security from the food regimes perspective (some exceptions are McMichael 2009b; Holt-Gimenez and Shattuck 2011; Cloke 2013; Otero, Pechlaner, and Gürcan 2013; Sage 2013; and Saab 2018). In addition, few case studies have shown the repercussions of each agri-food regime on the food security of a specific country or geographical area (Woertz and Keulertz 2015 examined the Middle East and North Africa; Soldevila, Rosell, and Viladomiu 2015 considered Mauritania).

This chapter is structured as follows. First, the theoretical framework – the concept of a 'food regime' – is explained. Second, the First and Second Food Regimes are briefly described. The current food regime (the Third) is explained in more detail in Section 4. Finally, some considerations on the implementation of the Third Food Regime in the Arctic and its consequences for food security in the area are discussed.

Food regimes

The term 'food regime' was first used in the 1989 paper by Friedmann and McMichael 'Agriculture and the state system: the rise and fall of national

DOI: 10.4324/9781003057758-10

agricultures, 1870 to present'. The food regime approach analyses the role of agriculture and food in different stages of the global capitalist economy (Friedmann and McMichael 1989; McMichael 2009a). Thus, it allows us to understand the inherent dynamics of the global agri-food system and to unravel the power relationships among participants in this system (farmers, states, consumers, transnational companies, etc.). It also enables us to consider the effects and consequences of these relationships.

A food regime is a dynamic concept. Contradictory relationships within a food regime generate a crisis that forces the transition to a new food regime (McMichael 2009a). The characteristics of the first two food regimes are well established and generally recognised (Friedmann and McMichael 1989; McMichael 2009a; Bernstein 2016; González-Esteban 2018; Krausmann and Langthaler 2019, among others). The First Food Regime began around 1870 and ended with World War I. The Second Food Regime was formed after World War II and, according to some authors, ended with the profitability crisis of the late 1960s and 1970s (McMichael 2009a; Pritchard 2009; González-Esteban 2018, among others). In fact, there is intense debate about the current food regime. Are we currently experiencing a crisis or contradiction of the Second Food Regime and the consequent transition towards a Third Food Regime? Or are we already immersed in the Third Food Regime? (see Friedmann 2009; McMichael 2009a; and Bernstein 2016 for a review of the main arguments of this debate). However, food regime transitions are not globally homogeneous; there are significant differences in the consolidation and evolution of food regimes in different countries and regions.

A food regime approach holds much potential for the analysis of food security. A food regime analysis locates food in the development of global capitalism and connects agrarian production with food consumption (Sage 2013). This integrated vision allows us to understand the economic structures and political processes that lead to hunger, malnutrition, and/or inadequate food consumption (Sage 2013). Cloke advocates for a food systems analysis of food regimes in order to understand the paradoxes of hunger and food overproduction, of the food insecurity of food producers (small farmers and peasants), of obesity and undernourishment, and of undernourishment and food waste (2013). Cloke also points out that it is impossible to implement effective food security policies without considering the systems of power and control in global food systems (2013).

The First and Second Food Regimes

This section briefly describes the main characteristics of the First and Second Food Regimes. This chapter will argue that the Third Food Regime, which will be described in the next section, is dominant in most parts of the world today. The chapter will also focus on the characteristics of each food regime in Global North (which could be quite different to their manifestations in the Global South) because the Arctic is integral to the capitalist systems of Northern countries.

The First Food Regime

Until the end of the nineteenth century, there was no global food regime. Although international trade of luxury goods such as spices and sugar had taken place for centuries, high transportation costs and difficulties in preserving food made international trade in food staples and agricultural products unprofitable. The Industrial Revolution (which brought new, faster methods of transportation) and the formation of the colonial empires changed this situation, facilitating colonial trade and the configuration of an agri-food system that connected agriculture to the rest of the economic sectors worldwide.

According to Friedmann and McMichael, the first European imports of wheat and meat from their colonies (mainly the United States, Canada, and Australia) marked the beginning of the First Food Regime (1989).

One of the main ways agriculture contributes to capital accumulation in capitalist systems is by reducing wages. Industrial capitalists were interested in obtaining cheap food. In the middle of the nineteenth century, U.K. Corn Laws were abolished. This opened cereal imports and intensified the international trade of staple foods. Food imports made it possible to maintain low wages. Furthermore, domestic farmers could not compete with cheap imports, and many were forced to quit their farms and seek jobs as industrial workers.

The ongoing urbanisation process that accompanied industrialisation in Western Europe meant that large population groups could only obtain food by buying it with their wages, since they no longer produced their own food. This circumstance led to chronic malnutrition, especially in urban areas, which was partly alleviated by imports of food staples and through the migration of large numbers of poor workers to the new colonies (Vernon 2007).

However, a significant percentage of the European population continued to work in agriculture. At that time, farmers produced final goods. They sold their products in the market, and their inputs came from the farm itself. Most farms were diversified; their main objective was to feed the family and sell any surplus in local markets. Under this model, the household was the unit of food production and consumption.

The Second Food Regime

After the Second World War, decolonisation process began, and the United States emerged as a hegemonic power, albeit in opposition to the Soviet Union. International economic relations expanded, and a set of supranational organisations and agreements (United Nations, International Monetary Fund, General Agreement on Tariffs and Trade, World Bank, FAO, etc.) were created.

Previously, between 1914 and 1945, international trade (including food trade) collapsed because of protectionist policies. In the United States, the

Depression of the 1930s led to a dramatic fall in agricultural prices and the bankruptcy of many farmers. After that, major investments were made to promote cheap meat production using livestock housed in intensive specialised farms and fed with compound feed. The high productivity of the new livestock system and the interests of large American companies that produced feed grains extended the intensive livestock model to Europe (Viladomiu 1985).

Crop farm productivity also increased exponentially because of mechanisation and technological advances. Farms, while still family owned, increasingly began to specialise in one or a few crops. They no longer produced food for a family but for the market.

This converted farmers from producers of final goods into suppliers of intermediate goods for the large agribusinesses that were being established at that time. This shifted power relations in the agri-food system. Farmers became just one link (and often the weakest) in long agri-food chains that included large corporations producing agricultural inputs (e.g., feed, chemicals, seeds), farmers, food processors, and food retailers. Controlling inputs, finances, and marketing channels allowed the industrial links to control food production even without owning land.

In the Second Food Regime, agriculture contributed significantly to capital accumulation in manufacturing sectors. Increasing wages and low food prices allowed increased consumption of manufactured goods. Farms themselves became significant customers for the manufacturing (e.g., feed, fertilisers, machinery) and services sectors (financial services).

The agricultural policies implemented after World War II and the intensification of food production generated a huge increase in food production and food surpluses in some Global North countries. At the end of World War II, some European countries struggled with food self-sufficiency and even famine. The need to guarantee the food supply for increasingly urban societies encouraged nations to protect national agriculture through agrarian policies that, especially in the case of the Common Agricultural Policy,[1] included strong tariff barriers to agricultural products from third countries. The new food production model led to substantial increases in farm productivity in the Global North, causing an impressive rise in farm output while decreasing labour requirements in the agricultural sector.

The uncontrolled growth of agricultural production in the Global North completely changed the specialisation patterns of agricultural trade: Global North countries became net exporters of food by displacing local production in the Global South. The latter converted their production of staple foods adapted to local diets to the production of 'exotic' foods for upper-middle-class diets in the Global North or to fodder plants used for feed production. In urban areas in the Global South, it was easier to obtain imported food than locally produced food. This led to the 'Westernisation' of diets in urban and peri-urban areas in the Global South (see Soldevila, Rosell, and Viladomiu 2015 for the case of Mauritania).

The Third Food Regime

As mentioned above, there is debate on whether we now live under a new Third Food Regime. McMichael calls it the 'corporate food regime' (McMichael 2005, 2012); other authors call the 'neoliberal food regime' (Pechlaner and Otero 2010) or the 'imperial regime' (van der Ploeg 2010). In this new food regime, some of the dysfunctions of the Second Food Regime are greatly aggravated by the aegis of neoliberalism (e.g., the industrialisation of agriculture and the concentration of large agri-food firms) and the appearance of new features such as intrinsic instability and biofuels. The main characteristics of the Third Food Regime are described below.

Global market and the breakdown of domestic agri-food systems

Although domestic markets are still the main destination of food products, domestic prices align with prices and trends in the world market. According to van der Ploeg (2010, 101),

> The previous differentiation of interconnected, locally or regionally centred markets, that to a degree reflected the specificity of relative factor prices at the regional or local level, is being restructured into one global market, increasingly characterised by the same set of price levels and price ratios.

Thus, the world market has become 'the ordering principle for agricultural production and marketing' everywhere (van der Ploeg 2010, 99).

Agri-food production can be offshored and segmented internationally. A new feature of the Third Food Regime is the 'complete exchangeability of large agricultural systems' (van der Ploeg 2010, 101). One example is asparagus; its production moved first from some European countries to Peru and then from Peru to China (van der Ploeg 2010).

Moreover, global agri-food chains are becoming quite common worldwide. The different phases of food production are carried out in different countries and regions. Complex marketing channels connect production and processing regions (some of which are located in food insecure countries in the Global South due to the low salaries and low cost of land) to high-profit markets in the Global North.

The weakening of nation-states and the liberalisation of agri-food trade have led to the total breakdown of domestic agri-food systems. The role of national agri-food systems is no longer self-sufficiency; it is international competitiveness. Thus, domestic agri-food production and processing is adapted to the demands of global markets and not to the needs of domestic populations (e.g., exotic fruit is produced for export rather than produce staple food for local populations in some food-insecure areas of the Global South). The breakdown of national agri-food systems has led to food shortages in countries with significant agrarian potential.

With the mass production of food for a global market, food consumption patterns have become homogenised. Diets are unified, creating dietary reductionism and the disappearance of local food cultures. However, the world market is segmented by income level. Standardised foods are mass produced for the lower middle classes while luxury foods are produced worldwide for consumption by the upper middle classes of the Global North and the elites of the Global South.

Food empires: the hegemonic powers in the Third Food Regime

Due to the weakening of the nation-states, large transnational agri-food corporations become the hegemonic powers in the new regime, with the connivance of international organisations (especially the World Trade Organisation).

There has been a huge increase in the concentration and centralisation of all stages of the food value chain (input suppliers, farms, food processing, and food retail). A few large companies control the agricultural inputs, the processing sector, and food retailing. Van der Ploeg calls these large transnational companies 'food empires' (van der Ploeg 2010).

Among these large companies, retailers and input suppliers are increasing their power in the food chain. Large supermarkets impose price and production conditions on agri-food processing companies and farmers (e.g., Gereffi 1994; Burch and Lawrence 2005; Pelupessy and van Kempen 2005). When overproduction occurs, retailers can exert permanent downward pressure on the prices paid to farmers and agri-food processors. On the other side, their oligopolistic power allows them to exert upward pressure on food prices for consumers.

The prevailing farm technocratic model generates dependence on oligopolistic input suppliers. Chemical (e.g., fertilisers) and feed companies were crucial to the development of the intensive livestock and crop production that characterised the Second Food Regime. In the Third Food Regime, another agricultural input supplier is taking a powerful position in agri-food value chains: biotechnology companies (Holt-Gimenez and Shattuck 2011; Otero and Lapegna 2016). The power of these companies has been strongly reinforced by 'patents on life', which are safeguarded by the World Trade Organisation's *Agreement on Intellectual Property Rights Related to Trade*. According to Otero and Lapegna, genetically modified crops represent the 'sharpest technological expression of the neoliberal food regime' (Otero and Lapegna 2016, 671).

New roles for agriculture: biofuels and speculation

According to its profit-making logic, the current stage of capitalism has assigned two new roles to the agriculture sector. On the one hand, agriculture becomes an energy producer with the emergence of agri-fuels, meaning that many cultivated lands destined for food production are used to cultivate biofuel plants instead. On the other hand, in an economy in which financial goals clearly dominate productive capital, food products become an object of speculation.

'Accumulation by dispossession' and the removal of all non-capitalist forms of food production

Harvey's phrase 'accumulation by dispossession' refers to the 'new' and 'old' forms of primitive accumulation practices implemented in the neoliberal capitalist era. These include the commodification and privatisation of land and, thus, the displacement of peasants, the privatisations of public assets, the use of the credit system, and the suppression of alternative forms of production and consumption. (Harvey 2007). It can be understood as the expansion of capitalism to any opportunity for profit and the removal of all forms of non-capitalist production and consumption.

In agrarian studies, this concept is often referring to land grabbling (Hall 2013) or as depeasantisation due to the modernisation of agriculture (Otero and Lapegna 2016). However, according to McMichael, in the Third Food Regime, price and credit relations are the key mechanism of accumulation by dispossession: Artificially low food prices displace peasant agriculture and the cultures of provision that now represent new opportunities for accumulation (McMichael 2005). In sum, one of the features of the Third Food Regime is the ruthless elimination of all forms of non-capitalist food production, including peasant agriculture, subsistence farming, and any forms of collective production.

Increasing farm intensification and farm bankruptcy

Nevertheless, 'capitalist' farms are also in danger in the Third Food Regime. The modernisation (or industrialisation) of agriculture has been unable to reverse the fall in farmers' income. The productivist model's proposed solution to this is to increase the scale of production through the intensive use of new technologies and inputs; that is, to increase productivity and reduce production costs. However, the subsequent price drops and the greater power of the other agents in the chain still condemn farmers to low profits. They try to compensate (again) by increasing production and their use of technology, and the cycle continues. This is what some authors have called the 'technological treadmill' (Cochrane 1979, cited in Renting, Marsden, and Banks 2003).

Thus, constant increases of scale are intrinsic to this model, and this has increased the dependency of farmers on capital markets (van der Ploeg 2010). However, even some highly intensive, specialised, large-scale farms find it difficult to survive. They require large investments, but they operate in a highly unstable system (see below). Bankruptcy is becoming usual for large, modern farms.

Instability and recurrent agrarian and food crises

Agri-food trade liberalisation and the overturning of agrarian policies have generated remarkable instability, which has been increased still more by food speculation. Even the FAO recognises than the international grain markets were more volatile in the 1990s than in the early 1970s and that agricultural liberalisation could be associated with increased volatility in production and prices (Clay 2003).

Recurrent food crises are clear proof of the huge contradictions in the Third Food Regime. According to van der Ploeg, these food and agrarian crises emerge because of the ongoing industrialisation of agriculture, the emergence of the agri-food global market, and the establishment of food empires (van der Ploeg 2010). These three processes are the foundation of the Third Food Regime; thus, instability and recurrent food crises are intrinsic to this regime.

Environmental and social impacts of the Third Food Regime

'Accumulation by dispossession' and recurrent agrarian crises are resulting in the displacement of farmers and peasants worldwide. This surplus labour faces increasing difficulties in finding employment in industrial sectors (as happened during the Second Food Regime in the Global North). The situation is especially dramatic in the Global South. Most of the time, mass emigration or illegal activities are the only ways to escape from poverty.

The adverse environmental impacts of intensive farming have been richly documented and include land degradation, water and soil pollution, and CO_2 emissions. In addition, biological diversity is decreasing worldwide. This is due in part to the growing use of biotechnology and in part to dietary homogenisation and dietary reductionism.

Alternatives to the Third Food Regime

In response to some of these dysfunctions, new forms of agrarian production are emerging that move away from the dominant model of the Third Regime or at least try to minimise its most negative impacts. Proposals range from the alternative paradigm of Food Sovereignty to far less sweeping measures such as corporate social responsibility or voluntary sustainability standards.

In the Global North, Alternative Food Networks (AFNs) have increased (e.g., Marsden, Banks and Bristow 2000; Renting, Marsden, and Banks 2003; Goodman and Goodman 2009; Maye and Kirwan 2010; Tregear 2011; Galli and Brunori 2013). Broadly defined, AFNs include all forms of production, distribution, and consumption of food other than the dominant productivist, industrialised model. They include organic producers, farmers who sell directly to consumers, handcrafted products, products with denominations of origin or protected geographical indications, slow food, fair trade products, Community-Supported Agriculture, etc. These new and often shorter chains question some of the paradigms of the dominant model. They aim to increase farmers' incomes by means other than intensification, to re-establish links between producers and consumers (reconnection) and between products and the places they are grown (respatialisation) (Marsden, Banks, and Bristow 2000).

These AFNs coexist with the dominant model in the niche markets that have arisen in response to the excesses of and reactions to the productivist model. However, there is significant debate about whether AFNs can significantly alter or replace the Third Regime foundations.

Another approach is based on Food Sovereignty, something Holt-Gimenez calls 'radical' (Holt-Gimenez and Shattuck 2011). This term was coined by Via Campesina and is defined at the 2007 Forum for Food Sovereignty as 'the right of peoples to healthy and culturally appropriate food produced through ecologically sound and sustainable methods and their right to define their own food and agriculture systems' (Via Campesina, 2007).

Arctic food security from the perspective of food regimes

Food regimes in the Arctic

The Arctic is a highly diverse territory, containing varied landscapes, vibrant ecosystems, diverse people, heterogeneous Arctic cultures, and a land divided among eight countries with different historical and political backgrounds. It is therefore difficult to make generalisations about this region. However, this section roughly identifies some general characteristics of the three food regimes in the Arctic. It should be remembered that changes in food production and consumption have had very different impacts on the different regions of the Arctic.

During the First Food Regime, some regions of the Arctic were immersed in colonisation. As Kulchyski points out, colonialism processes in the Arctic occurred later than in other regions (although contacts and a trickle of immigrants from the South started many centuries before). Arctic colonisation did not involve a massive influx of agricultural settlers or frequent use of military violence. Colonialism in the Arctic has been characterised by the domination of the capitalist mode of production over hunter-gatherer models (Kulchyski 2017).

In the nineteenth century, the Arctic was involved in the world economy through whale hunting and fur trade. Some regions where agriculture was possible suffered from early colonisation. Although colonisation in the Arctic did not involve a massive influx of agricultural settlers, some new farmers settled in the southern regions more suitable for farming activities, resulting in the coexistence of two models of food production and consumption – Indigenous models and settler models. The Indigenous model was based on subsistence food activities (hunting, fishing, gathering, and reindeer herding), and the new settler model was based on farming and large-scale fishing activities. As Nilsson points out, the exchanges between these food systems reinforced food security in the region (see Chapter 11 of this book).

This distinction between these two models of food production and consumption became more blurred as the Indigenous population gradually melded with incoming settlers. Some Indigenous People progressively abandoned nomadism and hunting as they became increasingly more involved in farming activities. Agricultural productions such as livestock and grains or potatoes complemented and/or replaced traditional activities. Frequently, these new farming activities (e.g., harvesting hay and fodder and taking care of livestock) fell primarily to women, while Indigenous men continued to engage in traditional activities (herding reindeer, fishing, hunting, etc.). This transition from hunter/gatherer/pastoral systems to ranching and farming started many centuries ago[2] and occurred very slowly, but this process was fostered during the First and Second Food Regimes.

However, few agricultural products can be produced in sufficient quantities for export in the Arctic. The region therefore retained some self-sufficiency and was partially protected from global food trade. It was not fully integrated into the food trade flows of the First Food Regime; this allowed Indigenous Peoples to preserve traditional methods of food production and consumption.

At the beginning of twentieth century, a new phase of colonialism began in the Arctic. This phase involved the exploitation of resources: minerals, oil, and gas (Kulchyski 2017). The militarisation of some Arctic regions increased during the post-war period. The Second Food Regime coincided with this new phase of consolidation of colonial control over the Arctic regions.

This process was paralleled by the goal to concentrate the region's population and economy (Hansen et al. 2013). Many new workers[3] involved in resource development mega-projects and industrialisation processes immigrated to all regions of the circumpolar Arctic, exposing previously isolated Indigenous communities to Western influences. New immigrants also largely ignoring Indigenous communities' land rights (Kulchyski 2017). There were also significant efforts to move the Indigenous population into settlements.[4] Rural areas were progressively abandoned. The gradual move into settlements represented a major transition towards an economy that mixed subsistence activities with market activities in which people used their wages to purchase food in stores (Organ et al. 2014).

Urbanisation and industrialisation also had a marked impact on the traditional activities of the Arctic population. As Anderson points out, in Siberia,[5] the effects of industrial development ranged from the ardent promotion of reindeer breeding over other Aboriginal economic sectors such as hunting and trapping to the complete marginalisation of Aboriginal economic activities in favour of foreign forms of agriculture and industry (Anderson 1991).

In Russia and Northern European countries, the promotion of reindeer breeding implied restructuring it 'into a form that may not match local imperatives' (Anderson 1991). This involved increasing reindeer meat production to feed the growing numbers of imported industrial workers and to support the new industrial economy. Reindeer meat was commodified: it began to be produced not for subsistence but for sale and export and lost its meaning as an economic unit. New technological advances (e.g., the adoption of the snowmobile in herding in the 1960s) completely altered reindeer herding methods (fewer person-hours were required; this favoured the concentration of big herding operations), supporting more market activities even in such a 'traditional' activity (see Harkoma in Chapter 2 of this book). Similarly, other traditional activities were completely commodified.

Governments promoted these changes. In line with the rules of the Second Food Regime, governments favoured national self-sufficiency and financially supported and protected domestic food production from international competition. From the 1950s to the 1970s, policies in Arctic regions encouraged traditional activities to align with modern agriculture and directed the commodification of local foods (Gombay 2005) with discourses of 'national interest' and 'regional development' (Anderson 1991). In some regions of Northern countries, large-scale reindeer herding ('extensive' reindeer herding) was promoted by national policies and supported by the use of snowmobiles and all-terrain vehicles in herding. The growth of the commercial market for reindeer products, public financial support, and high productivity increased the potential profits of reindeer herding, and both Indigenous People and settlers became increasingly involved in the reindeer industry.

Processing and selling local foods seemed to be one way of involving Indigenous People in market activities, but this transition was not easy and is not yet complete because such activities conflict with Indigenous social traditions and cultural beliefs. As Gombay points out, the traditional Inuit obligation to share local foods with others made it difficult to commodify these foods (Gombay 2005).

The most recent phase of economic capitalism (globalisation) augers even more intense exploitation of natural resources and deeper economic colonialism in the Arctic (Kulchyski 2017). The neoliberalist management of the Third Food Regime has led to decreases in state aid to farmers and the liberalisation of food trade. Farming and fishing are becoming marginal, and primary sector employment is below 10% in all Arctic regions (Hansen et al. 2013). Indigenous People are still more involved in food production than the general population, so they are also more affected by these changes.

The current Arctic food system is heterogeneous: fisheries and aquaculture are often large-scale and export-oriented while agriculture is quite marginal, although greenhouse vegetable production is increasing. Meat production tends to be large-scale as well, while some small and medium producers remain. Organic production is important in some regions (e.g., potatoes in Greenland), and there is potential to increase this (Sustainable Development Working Group 2019).

However, food producers in Arctic face problems similar to those of farmers and fishermen worldwide: Small producers struggle to survive in a competitive market and may need to complement their incomes with other sources (e.g., tourism or handicrafts). On the other hand, large producers (large herding conglomerates or large fisheries) are trapped by increasing price and market instability. Farmers and fishermen frequently clash with governments over environmental and safety regulations, land management, or fishing quotas.

In addition to these changes in food production, the Second and Third Food Regimes fostered nutritional transition toward a Western diet. During these regimes, processed food imports increased significantly in the Arctic, and diets among populations living in Arctic regions changed accordingly. Most non-Indigenous people rely exclusively on imported foods, and Indigenous People are changing their dietary habits as well, including more southern foods in their diets (Rasmussen 1999). This poses a threat to traditional foods. According to Kuhnlein, only 10%–36% of adult dietary energy is derived from traditional food in the Canadian Arctic, and it is particularly worrying that younger people consistently consume significantly less traditional food than older residents (Kuhnlein et al. 2004). The nutritional transition has decreased diet quality and negatively impacted health and wellness (e.g., Bogdanova et al., Chapter 4; Bjerregaard, Chapter 5 of this book). However, traditional food is not only crucial for food security; it is also an essential part of Indigenous cultures (e.g., Nilsson et al. 2013, Harkoma, Chapter 2 of this book; Casi, Chapter 7; Nilsson, Chapter 11; Ouma, Chapter 15).

Food security in the Arctic in the context of the Third Food Regime

The contradictions of the Third Food Regime on food security are evident in the Arctic: The Arctic exports food to global markets while some Arctic regions experience food insecurity.

Colonialism is one of the main causes of food insecurity in the Arctic; it has affected Arctic food security in several ways. Nilsson mentions the significant impact of colonialism on Sami food systems; in the Sápmi, traditional food-related land uses are disadvantaged compared to other types of land use (i.e., large-scale agriculture or resource extraction) (Nilsson, Chapter 11 of this book). According to the Council of Canadian Academics, colonialism has stressed Inuit cultural and social structures, and this fallout has diminished their ability to retain food sovereignty (Council of Canadian Academics, cited in Organ et al. 2014).

According to Loring and Gerlach's literature review about food insecurity in the Arctic, four key factors contribute to food insecurity in the Arctic: over-connectedness to the global system, the impacts of contaminants on subsistence foods, the impact of climate change on subsistence practices, and management and governance constraints (Loring and Gerlach 2015). Over-connectedness to the global system implies, first, dependence on expensive, fuel-consuming technologies (e.g., snowmobiles and motor boards) to perform traditional subsistence activities (i.e., reindeer herding and hunting). It also refers to the rising costs of store-bought foods (Loring and Gerlach 2015). Food safety risks are increasing due to the impacts of contaminants and pathogens on fish, marine mammals, caribou, reindeer, and edible plants and wild berries – that is, on the main subsistence foods of Indigenous People (Loring and Gerlach 2015; Minagawa, Chapter 3 of this book).

Climate change is affecting food security in the Arctic in different ways; traditional food storage cellars are affected by warming temperatures, and climate change poses various challenges to hunting, fishing, and gathering (Loring and Gerlach 2015). However, global warming could also open up new opportunities for food production in the Arctic, especially vegetable production (Sustainable Development Working Group 2019).

Loring and Gerlach also identify management and governance constraints (e.g., restrictive land tenure regimes) as a key factor in food insecurity in the Arctic. There is an 'apparent disconnect between statutory protection and agency attention, on the one hand, and continuing decline in food security and diet-related health outcomes, on the other' (Loring and Gerlach 2015, 382). Kondrashev et al. provide examples of inadequate regulations and laws that have undermined food security for people in the Russian Arctic (Kondrashev et al. 2016). But governance issues are not easy in the Arctic. Each one of the eight Arctic countries has its own development strategy for the area. Although there are some similarities, it is clear that their interests in the area are different (see Kondrashev et al. 2016). Loring and Gerlach advocate for greater involvement of the local population through participatory processes and for active support of the rights of local people to pursue food security on their own terms and eventually to obtain food sovereignty (Loring and Gerlach 2015).

These factors (over-connectedness, the impact of climate change, problems with the safety and quality of traditional foods, and governance) reinforce the dynamics of the Third Food Regime in the Arctic. Dependence on expensive, fuel-consuming technologies to engage in traditional activities, food safety risks affecting subsistence foods, and the impacts of climate change are undermining the ability of local populations to access food outside the market. This reinforces the trend of the Third Food Regime to eliminate all non-capitalist methods of food production, including self-provisioning. The market then becomes the only way to access food.

However, high dependence on store-bought food is a serious problem in isolated, remote areas such as some Arctic regions. Food may not be reliably available, and food supplies may be inconsistent. Therefore, as mentioned before, the Third Food Regime is intrinsically unstable.

In addition, some Indigenous communities suffer from economic and social marginalisation. Poor communities can face increasing problems of food access if their access to food is mediated by monopolistic/oligopolistic food markets; rising costs may make food unaffordable for some households. In addition, highly processed store-bought food introduces problems of food safety and nutritional issues. Therefore, the Third Food Regime is threatening all dimensions of food security (food availability, food access, food supply stability and food safety, and the nutritional content of food). This threat is even more severe in vulnerable areas such as the Arctic.

Access to traditional food and the preservation of subsistence practices are essential to food security in the Arctic (Hossain et al., Chapter 6 of this book). To reverse the current situation, traditional methods of food provision must be valorised. This book provides several examples of how traditional foods could be preserved and valorised (e.g., Montalvan, food preservation methods, Chapter 1; Harkoma, reindeer herding, Chapter 2). There are also some proposals to improve food security and move towards Food Sovereignty in the Arctic (see Herrmann et al. in Chapter 14 of this book for a description of community-led experiences in the Canadian Arctic). However, some threats to Arctic food security (e.g., climate change) are far beyond the scope of nation-state regulations; they require global commitment and would be difficult to avoid.

Acknowledgement

The author wishes to thank the Ministerio de Economía, Industria y Competitividad of the Spanish Government for its support through the project ECO2016–79072-P (AEI/FEDER, UE).

Notes

1 The European Union Common Agricultural Policy (CAP) is an agricultural policy that applies to all EU members. Launched in 1962, it is managed and funded at the European level from the resources of the EU budget.

2 In the eighteenth century, some Arctic populations had already begun to keep goats and cows (e.g., the Forest Sami).

3 Today 1.13 million Indigenous People live in the Northern regions of the Arctic States. The percentage of Indigenous People in the population varies greatly across the Arctic: from 98% in Greenland to less than 7% in North Russia (Young and Bjerregaard 2019).

4 For example, the Inuit were encouraged to move into settlements during the 1960s through large-scale government housing construction projects (Boneesteel and Anderson, 2008).

5 Industrialisation and resource extraction were especially intense and aggressive in the Russian Arctic (mainly under Soviet development policies). At present, approximately two-thirds of industrial activity in the entire circumpolar Arctic region occurs in the Russian Arctic (Glomsrod and Aslaksen 2008).

References

Anderson, D.G. 1991. Turning hunters into herders: A critical examination of Soviet development policy among the Evenki of Southeastern Siberia. *Arctic* 44(1): 12–22.

Bernstein, H. 2016. Agrarian political economy and modern world capitalism: The contributions of food regime analysis. *The Journal of Peasant Studies* 43(3): 611–647.

Bonesteel, S. and Anderson, E. (2008). *Canada's relationship with Inuit: A history of policy and program development.* Retrieved 27 November 2019 from: https://www.aadnc-aandc.gc.ca/eng/1100100016900/1100100016908

Burch, D. and G.A. Lawrence. 2005. Supermarket own brands, supply chains and the transformation of the agri-food system. *International Journal of Sociology of Agriculture and Food* 13(1): 1–18.

Clay, E. 2003. Food security: Concepts and measurement. In FAO (ed.), *Trade reforms and food security: conceptualizing the linkages*, 25–34. Rome: FAO.

Cloke, J. 2013. Empires of waste and the food security meme. *Geography Compass* 7(9): 622–636.

FAO 1996. World food summit: Rome declaration on world food security and world food summit plan of action. Retrieved 16 May 2020 from http://www.fao.org/3/w3613e/w3613e00.htm

FAO, IFAD, UNICEF, WFP, and WHO. 2019. *The state of food security and nutrition in the world 2019: Safeguarding against economic slowdowns and downturns.* Rome: FAO.

Friedmann, H. 2009. Moving food regimes forward: Reflections on symposium essays. *Agriculture and Human Values* 26: 355–344.

Friedmann, H. and P. McMichael. 1989. Agriculture and the state system: The rise and fall of national agricultures, 1870 to present. *Sociologia Ruralis* 29(2): 93–117.

Galli, F. and G. Brunori (eds.). 2013. Short food supply chains as drivers of sustainable development. Retrieved 10 October 2018 from http://orgprints.org/28858/1/evidence-document-sfsc-cop.pdf.

Gereffi, G. 1994. The organisation of buyer-driven global commodity chains: How U.S. retailers shape overseas production networks. In G. Gereffi and M. Korzeniewicz (eds.), *Commodity chains and global capitalism*, 95–112. Westport: Praeger.

Glomsrød, S. and I. Aslaksen. 2008. The economy of the North 2008. Retrieved 15 October 2019 from https://oaarchive.arctic-council.org/handle/11374/35.

Gombay, N. 2005. The commoditisation of country foods in Nunavik: A comparative assessment of its development, applications, and significance. *Arctic* 58(2): 115–128.

González-Esteban, Á.L. 2018. Patterns of world wheat trade, 1945–2010: The long hangover from the second food regime. *Journal of Agrarian Change* 18(1): 87–111.

Goodman, D. and M. Goodman. 2009. Alternative food networks. *International Encyclopaedia of Human Geography* 3: 208–220.

Hall, D. 2013. Primitive accumulation, accumulation by dispossession and the global land grab. *Third World Quarterly* 34(9): 1582–1160.

Hansen, K.G., S. Bitsch, and L. Zalkind. 2013. *Urbanisation and the role of housing in the present development process in the Arctic.* Nordregio. Retrieved 13 October 2019 from http://www.diva-portal.org/smash/get/diva2:700282/FULLTEXT01.pdf.

Harvey, D. 2007. *A brief history of neoliberalism.* Oxford: Oxford University Press.

Holt-Gimenez, E. and A. Shattuck. 2011. Food crises, food regimes and food movements: Rumblings of reform or tides of transformation? *Journal of Rural Studies* 38(1): 109–144.

Kondrashev, A., M. Nikitenko, S. Trofimova, I. Trofimova, and L. Gotsko. 2016. The Arctic states' strategies and the northern regions' food security. *Economic Annals-XXI* 162: 32–37.

Krausmann, F. and E. Langthaler. 2019. Food regimes and their trade links: A socio-ecological perspective. *Ecological Economics* 160: 87–95.

Kuhnlein, H.V., O. Receveur, R. Soueida, and G.M. Egeland. 2004. Arctic Indigenous Peoples experience the nutrition transition with changing dietary patterns and obesity. *The Journal of Nutrition* 134(6): 1447–1453.

Kulchyski, P. 2017. Colonisation of the Arctic. Retrieved 12 October 2019 from https://www.climate-policy-watcher.org/canadian-arctic/colonization-of-the-arctic.htm.

Loring, P.A. and S.C. Gerlach. 2015. Searching for progress on food security in the North American North: A research synthesis and meta-analysis of the peer-reviewed literature. *Arctic* 68(3): 380–392.

Marsden, T., J. Banks, and G. Bristow. 2000. Food supply chain approaches: Exploring their role in rural development. *Sociologia Ruralis* 40(4): 424–438.

Maye, D. and J. Kirwan. 2010. Alternative food networks. *Sociology of Agriculture and Food* 20: 383–389.

McMichael, P. 2005. Global development and the corporate food regime. In F. Buttel and P. McMichael (eds.), *New directions in the sociology of global development (Research in Rural Sociology and Development, Vol.11)*, 265–299. Bingley: Emerald Group Publishing Limited.

McMichael, P. 2009a. A food regime genealogy. *Journal of Peasant Studies* 36(1): 139–169.

McMichael, P. 2009b. A food regime analysis of the 'world food crisis.' *Agriculture and Human Values* 26(4): 281.

McMichael, P. 2012. The land grab and corporate food regime restructuring. *The Journal of Peasant Studies* 39(3–4): 681–701.

Nilsson, L.M., G. Destouni, J. Berner, A.A. Dudarev, G. Mulvad, J. Ø. Odland, and B. Evengård. 2013. A call for urgent monitoring of food and water security based on relevant indicators for the Arctic. *Ambio* 42(7): 816–822.

Organ, J., H. Castleden, C. Furgal, T. Sheldon, and C. Hart. 2014. Contemporary programs in support of traditional ways: Inuit perspectives on community freezers as a mechanism to alleviate pressures of wild food access in Nain, Nunatsiavut. *Health and Place* 30: 251–259.

Otero, G. and P. Lapegna. 2016. Transgenic crops in Latin America: Expropriation, negative value and the state. *Journal of Agrarian Change* 16(4): 665–674.

Otero, G., G. Pechlaner, and E.C. Gürcan. 2013. The political economy of 'food security' and trade: Uneven and combined dependency. *Rural Sociology* 78(3): 263–289.

Pechlaner, G. and G. Otero. 2010. The neoliberal food regime: Neoregulation and the new division of labour in North America. *Rural Sociology* 75(2): 179–208.

Pelupessy, W. and L. Van Kempen. 2005. The impact of increased consumer-orientation in global agri-food chains on smallholders in developing countries. *Competition and Change* 9(4): 357–381.

Pritchard, B. 2009. The long hangover from the second food regime: A world-historical interpretation of the collapse of the WTO Doha Round. *Agriculture and Human Values*, 26(4): 297.

Rasmussen, R.O. 1999. Settlement development and the formal, informal and subsistence sector in the Arctic. *Geografisk Tidsskrift* 1: 171–180.

Renting, H., T.K. Marsden, and J. Banks. 2003. Understanding alternative food networks: Exploring the role of short food supply chains in rural development. *Environment and Planning A* 35(3): 393–411.

Saab, A. 2018. An international law approach to food regime theory. *Leiden Journal of International Law* 31(2): 251–265.

Sage, C. 2013. The interconnected challenges for food security from a food regime perspective: Energy, climate and malconsumption. *Journal of Rural Studies* 29(1): 71–80.

Soldevila, V., J. Rosell, and L. Viladomiu. 2015. Repercusiones de los regímenes alimentarios mundiales en la evolución de la seguridad alimentaria: el caso de Mauritania [Impacts of World Food Regimes on the evolution of Food Security: the Mauritanian case]. *Revista española de estudios agrosociales y pesqueros* 242: 13–48.

Sustainable Development Working Group. 2019. The Arctic as a food-producing region. Retrieved 15 October 2019 from https://oaarchive.arctic-council.org/bitstream/handle/11374/2375/AFPR-Final-Report-1.pdf?sequence=1.

Tregear, A. 2011. Progressing knowledge in alternative and local food networks: Critical reflections and a research agenda. *Journal of Rural Studies* 27(4): 419–430.

Van der Ploeg, J.D. 2010. The food crisis, industrialised farming and the imperial regime. *Journal of Agrarian Change* 10(1): 98–106.

Vernon, J. 2007. *Hunger. A modern history.* Cambridge: The Belknap Press.

Via Campesina 2007. Declaration of Nyéléni. Retrieved 25 November 2019 from https://nyeleni.org/spip.php?article290.

Viladomiu L. 1985. La inserción de España en el Complejo Soja-Mundial [The insertion of Spain in the Soy World Complex]. Serie Estudios. Madrid: Instituto de Estudios Agrarios, Pesqueros y Alimentarios.

Woertz, E. and M. Keulertz. 2015. Food trade relations of the Middle East and North Africa with tropical countries. *Food Security* 7(6): 1101–1111.

Young, T.K. and P. Bjerregaard. 2019. Towards estimating the Indigenous population in circumpolar regions. *International Journal of Circumpolar Health* 78(1): 1653749.

PART III

Arctic food security keys to the future

Strategies to build resilient food regimes and to enhance food security and sovereignty

11 Some reflections on Swedish food strategies from a Sami and an Arctic perspective

Lena Maria Nilsson

Introduction

Food security is targeted in the Sustainable Development Goal (SDG), Target 2.1: 'By 2030, end hunger and ensure access by all people, in particular the poor and people in vulnerable situations, including infants, to safe, nutritious and sufficient food all year round'. Food security exists when 'all people, at all times, have physical and economic access to sufficient, safe and nutritious food to meet their dietary needs and food preferences for an active and healthy life' (F.A.O., 1996). The Scandinavian North is generally considered a food secure area; Scandinavian people today do not suffer from starvation or malnutrition. In Northernmost Sweden, well-stocked grocery stores offer delicious food from all over the world at reasonable prices. As part of the general welfare system, nurseries and schools provide free lunches to all children. However, in discussions of the future impacts of globalisation and climate change, interest in food sovereignty has increased, and the food security of the North has been questioned (Nilsson and Evengård, 2015). In fact, the Nordic countries generally have an insufficient level of food self-sufficiency. Only Denmark, where approximately 50% of the land is used for agricultural production, has a net export of nationally produced food (Karlsson et al., 2018), which means that the Danes would likely be able to feed their entire population in case of a trade arrest. In Finland, Sweden, and Norway, this is not the case.

Food sustainability is another important concept. In this context, modern farming, which is rationalised and depends on fertilisers and fuel, is often contrasted with organic farming, which is less intensive and uses circular systems. It can be argued that both systems are sustainable, industrialised because it requires smaller land aerials and a smaller human work force to produce a fixed amount of food, and organic because it uses fewer non-renewable resources. Neither of the systems includes by default an Indigenous perspective.

In 2017, all political parties of Sweden agreed on a National Food Strategy, the first one to cover all levels of the food chain (Lövén and Bucht, 2017; Näringsdepartementet, 2017a). This national strategy was followed by an

DOI: 10.4324/9781003057758-11

intensified discussion of regional, national, and international food strategies. However, notably missing from these discussions was an Indigenous perspective.

One food strategy study, for example, has used a participatory approach and a scenario methodology to design a food vision for the Nordic countries (Karlsson et al., 2018). This study was well designed and conducted, and its findings significantly deepen the discussion on food sovereignty. Karlsson et al. (2018) invited NGOs representing ecological food production/farming in the four Nordic countries to participate in guided discussions with the researchers. These discussions concluded that implementing full-scale organic farming systems in the Nordic countries; decreasing the production and consumption of meat; and increasing the production and consumption of rapeseed, grain legumes, vegetables, and other crops could give the Nordic countries a food sovereignty level of 130% by 2030. This would also enable all the Nordic countries except Norway to achieve over 100% food sovereignty; in Norway, organic agriculture would only result in 51% food sovereignty (Karlsson et al., 2018). In the normative decisions underlying this scenario, all arable land should be used primarily to produce plant-based human food. Animal feed should be derived primarily from food by-products, and 'semi-natural pastures and Norwegian outfield areas (i.e., forest and mountainous pastures, not counted as agricultural land), where grazing can promote biodiversity and annual cropping is unfeasible'. (Karlsson et al., 2018, page 5). Despite the strengths of this study, one important perspective is missing – that of Indigenous Peoples. According to a Sami worldview (such as that of the present author), there is no such thing as an outfield area in Norway. In the Arctic parts of Sweden, Norway, and Finland, outfield areas are places used since time immemorial to fish, graze reindeer, and gather wild berries and herbs. In a personal conversation, the main author, Johan Karlsson, indicated that the Indigenous perspective was not, in fact, completely absent from the discussions preceding the paper. During one scenario workshop, a person involved in reindeer herding highlighted the lack of a Sami perspective in the discussion. However, this view was not further advanced into the final consensus of the workshop.

Wild herbs and berries are not mentioned in the government bill that forms the background to the National Food Strategy action plan, and reindeer and game meat are addressed only briefly. The importance of reindeer herding to Sami culture is stressed, but its potential importance to food sovereignty is minimised (Lövén and Bucht, 2017). Reindeer and game are only mentioned: 'The production of game and reindeer meat can help to create growth in the food supply chain. There should be easier access to markets' (Näringsdepartementet, 2017a). Consequently, in the governmental action plan connected to this bill, not one action supporting reindeer herding is included (Näringsdepartementet, 2017b).

Food security strategies in Northernmost Sweden have changed profoundly over the last 100 years. In addition, the definition of food security in

Sweden has changed from a traditional Sami perspective to an interpretation rooted in the national welfare policy to one based on the modern global market.

This chapter will discuss how traditional knowledge of foods and food systems in Arctic Sweden could be applied to food security policies today, especially as Northern Europe is increasingly affected by climate change. The chapter will review Sweden's 2017 National Food Strategy and reflect on it from a Sami cultural perspective. This reflection will focus on the following questions: What can we learn from pre-modern food strategies in Sápmi? Which traditional food items may still be produced and harvested? And how could traditional Sami knowledge related to food security contribute to future regional food strategies if it were included in the agenda?

The impact of colonialism on food security

The Sami people are the Indigenous People of Northernmost Norway, Sweden, and Finland and the Kola Peninsula of Russia. This region is also called Sápmi (Figure 11.1).

Throughout modern history, these nation-states' actions regarding the Sami's Indigenous rights to land and water have fulfilled the criteria of colonial policies. An example of these policies in Sweden is provided by (Össbo and Lantto, 2011). Therefore, the establishment of these nation-states will hereafter be referred to as the colonisation of Sápmi. This colonisation began in the mid-1700s, intensified during the industrial era, and continues today.

Colonisation has had a significant impact on the food system and food sovereignty of Sápmi. Traditional food-related land use, such as picking berries and herbs, harvesting the inner bark of pine, fishing, hunting, small-scale

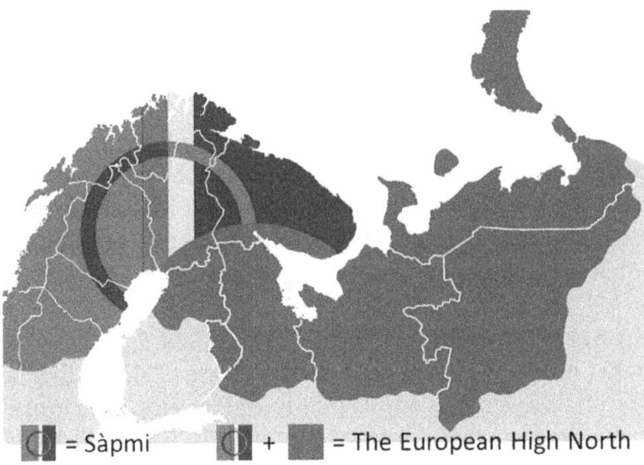

Figure 11.1 The Sapmi region of the European High North (EHN).

farming, and reindeer herding, have been disadvantaged compared to large-scale agricultural, extractive, industrial, recreational, military, and other state-supported land uses. For example, Sami cultural heritage has not been sufficiently considered in environmental impact assessments (2017). Adding the impact of unplanned destructive events such as pollution and soil and water system degradation, the self-sufficiency capacity of traditional food items has severely decreased throughout Sápmi, including Northern Sweden, which is the focus of this chapter.

Food security in the past

It is important not to romanticise or exotify the past. Compared to other parts of the Arctic Boreal region (the taiga), rather advanced agriculture is biologically possible in Sápmi, making this region unique in the circumpolar areas (Mela, 1996). This is also one of the reasons for the area's early colonisation. In pre-modern times, periods of food security in Sápmi were continuously interrupted by periods of food insecurity and starvation due to annual variations in wild and cultivated crops and other unpredictable random events, such as the Swedish–Russian War of 1809 (Kaati et al., 2007) and the heavy snow cover of the 1930s, which significantly reduced local reindeer stocks in Northern Sweden (Lantto, 2000). However, even though Sápmi's past cannot be called food secure, a health survey of the Northern Swedish population in 1929–1931 indicated that residents in the inland areas, inhabited to a large extent by Sami people, had a better nutritional status than the coastal, mainly non-Sami population (Odin, 1937). In the early 1900s, self-sufficiency was still the key element of food security in Northernmost Sweden. Sami food strategies of the early 1900s may therefore deserve some attention in discussions of modern food strategies.

Food security strategies in the early 1900s

Many attempts have been made to distinguish between traditional Sami food culture and traditional Northern Swedish food culture (Ryd, 2015). These attempts have all assumed negligible interactions among the different ethnic groups. However, interviews with elderly Sami reveal the opposite. Food security was achieved through interaction and sharing between local Sami and non-Sami people and between nomadic and settled Sami; all groups contributed their own specialised food items (Nilsson et al., 2011; Ryd, 2015).

The main Sami contributions to this interaction were traditional foods that relied primarily on plant and animal food sources available in the Scandinavian North, such as edible plants and berries, pine inner bark, fish, game, and reindeer meat. Reindeer or goat milk was used for cheese production in some Sami communities in the early 1900s. I have previously described core traditional Sami foods in more detail from a health perspective;

I will not, therefore, address that topic in depth here (Nilsson, 2018). The settled population, who were often of Sami origin or had family connections to the Sami, contributed dairy products such as cow cheese and butter to the food system. The people of Northernmost Sweden were also widely involved in trading food items not locally available or producible such as salt, sugar, coffee, and household cereals other than barley (Fjellström, 1985; Nilsson et al., 2011). Many Sápmi trade routes originated in harbours on the Norwegian coast. These were closer than the central Swedish trade routes to the main settlements, especially considering that the reindeer herding mountain Sami's summer grazing areas were near and sometimes across the Norwegian border. Eastern trade routes with Finland and Russia also contributed significantly to food security, especially in northernmost Sápmi, where trilingual competence in Sami, Finnish, and Scandinavian was not uncommon (Winsa et al., 2000).

Elderly Sami in Jokkmokk, a settlement in the central part of Northern Sweden (Figure 11.1), identify two pillars to their food security strategy. These pillars are knowledge and relationships (Ryd, 2015). Knowledge in this context means traditional knowledge, or *árbediehtu* in Sami. This knowledge includes ecological, economic, nutritional, practical, culinary, and food safety related considerations. The reduction of food waste was a core principle of *árbediehtu*; this included, for example, eating all the edible parts of an animal, using the inedible parts, and sharing available food with others. Therefore, festive events were not defined by fixed dates on a calendar. Rather, festivities were held when someone became rich in certain food items, such as fatty white fish caught from the shoreline (*dra not* in Swedish), a good catch from a hunt, or increased fresh reindeer meat during the pre-winter slaughtering season. A good meal was not defined in culinary terms; a good meal was one that lasted for a long time because it included so much food. A good meal meant enjoying the feeling of satiety and nutritional satisfaction in the company of others (Ryd, 2015).

Towards a market-based food system

World War II challenged Sweden's food system and initiated the shift toward national rather than local food systems. From 1945 to 1990, domestic food production was financially supported and protected from international competition by customs and regulations. Food producers in rural and Arctic regions received financial support from the state so they could continue to produce. The state aimed to maintain around 85% self-sufficiency and to make farming economically comparable with employment as an industrial worker. Maintaining reasonable food prices for consumers was another important issue. These goals were accomplished by decreasing the number of farm units involved in food production and by streamlining agricultural practices by introducing new technical solutions (Flygare and Isacson, 2003). Reindeer herders were similarly redefined as meat producers, and herding was accordingly modernised and motorised (Johnsen and Benjaminsen, 2017). Instead of picking wild berries, families and individuals began to buy berries in stores. Urbanisation and centralisation also occurred during this time. Rural areas in Northernmost Sweden were abandoned for coastal cities and larger cities in central and southern Sweden. The number of people involved in food production decreased. This was to a large extent supported by the Swedish government, due to the high societal costs of domestic food production and the overproduction of food (Flygare and Isacson, 2003). Also, at this time, more exotic food items were imported to Sweden from all around the world, increasing the competition for domestic food. From an international perspective, food prices were relatively high in Sweden during this period.

In 1990, preparatory to applying for European Union (EU) membership, the Swedish agricultural system was liberalised. In other words, the border protections for domestic food products were removed and state financial support for farmers was withdrawn. When Sweden entered the EU in 1995, actions to enhance food security were transferred to a European level. Due to this and other actions following its final entrance into the EU, the Swedish food system gradually transformed from a locally managed, production-oriented one into a neoliberal, market-oriented system. This new system disadvantaged farmers in Arctic Sweden relative to their counterparts in central and southern Sweden (Flygare and Isacson, 2003).

Post-industrial food strategies

Today, Sweden's food self-sufficiency is estimated at 50% by the Federation of Swedish Farmers (LRF, 2018), which is significantly lower than its pre-EU self-sufficiency of approximately 85%. Food prices have been decreasing for a long time, and most Swedes trust that global market food will continue providing them with their daily bread. Consequently, food security and sovereignty have been considered non-issues. However, when Sweden chaired the Arctic Council from 2011 to 2013, the Swedish government highlighted food security as a priority in the context of climate change. An international literature review seeking to identify indicators that could be used to monitor changes in food security in the Arctic revealed a huge knowledge gap, not least in the Nordic countries (Nilsson et al., 2013). Since then, an increasing awareness of the fragility of the food system has arisen, and in June 2017, a National Food Strategy going toward 2030 was approved by all parties in the Swedish Parliament (Näringsdepartementet, 2017a). Though food self-sufficiency was indeed part of the Swedish agenda after World War II, this was the first ever National Food Strategy that explicitly sought to include all parts of the food chain.

Business-oriented food strategies

Three main goals of the 2017 Swedish food strategy are (1) regulatory simplification, (2) efficient market competition, and (3) increased productivity through knowledge and innovation. The goals (Näringsdepartementet, 2017a) state, 'By 2030, food production in Sweden will be competitive. Steadily rising production values and cost efficiency in food production will provide evidence of this'. The phrase 'production values' deserves particular attention. The values described in the strategy are all expressed either in economic terms or as qualitative 'added values' such as environmental friendliness, animal welfare, or culinary excellence. The caloric value of food – that is, the energy content of national food production – seems to be a value of no interest. The Swedish word for nutrients, *näring*, is also used to mean 'business'. In line with this, the National Food Strategy seems to exclusively address the economic aspect of food so that *näring* equals business. On occasion, *näring* is also used to refer to the loss of nutrients to the environment as an unwanted effect of agricultural business. However, the food strategy does not address the population's caloric and nutritional needs beyond a vague aim to increase Swedish food production in general by stimulating the market and to increase the competitiveness of Swedish food on the global market (Näringsdepartementet, 2017a). The National Food Strategy is also consumer-oriented – if the quality of domestic food increases, consumers are expected to prefer Swedish food products and to abandon alternative food items from the global market. This, in turn, is expected to stimulate increased domestic food production.

The National Food Strategy will be followed by an action plan that will be developed through regional food strategies. In some parts of Sweden, these regional food strategies are already completed; in other parts officers are still working on them. Norrbotten County, the northernmost of Sweden's two Arctic counties, has developed a regional food strategy (Stenberg et al., 2016) and strategy for sustainable agriculture (Röös et al., 2016), while Västerbotten County aims to release a strategy in 2019. A pre-study in Västerbotten has identified some background facts to be included in the county's regional food strategy (Jonasson and Persson, 2017). Similar to the National Food Strategy, Norrbotten's regional food strategy and the pre-study in Västerbotten food strategy focus on monetary definitions of success. Thus, the monetary value of the domestic production of different food groups is estimated in millions SEK. In background tables, the production volumes of agricultural products are also described in annual weights (1,000 tons). For foods that are traditionally harvested wild, such as fish, wild berries, and game, it is more difficult to determine precise annual weights; it is also difficult to determine the amount of reindeer meat produced annually in Sweden. These regional food strategies also include secondary food production such as bakeries, restaurants, and delicacies. Secondary food production increases the food's monetary value without increasing its caloric value. From a business point of view, this is, of course, important. From a food sovereignty perspective, it is also somewhat important, such as when a primary food leaves the area and returns in a refined form at a net cost to local consumers. This is the case for a large proportion of the wild berries harvested in Arctic Sweden (Jonasson and Persson, 2017).

The nutritional value of primary domestic food production in Arctic Sweden

This section discusses the nutritional value of the domestic food production of Arctic Sweden. Information available from the regional documents mentioned above has been consolidated here using the data from most recent year in the reports (Stenberg et al., 2016; Jonasson and Persson, 2017). This varies between 2014 and 2016. Where data on food weights are absent, an estimate has been calculated from the production value. For example, wild-caught fish stock in Västerbotten is described as an annual value of 9 million SEK. The report also states that 13,500 tons of farmed fish equals a product value of 487 million SEK. This gives a rate of 36,000 SEK/ton fish. Therefore, the wild-caught fish stock was estimated at 9 million/36,000, that is, 250 tons. Food stocks were recalculated into individual shares by dividing the stock by 516,000 (the population of Arctic Sweden) and by 365 days (one year). It was further divided into caloric values based on published tables of standard portion sizes and caloric contents of common Swedish food items (Sahlén, 1987); one specific item was used as a model for these calculations. For example, salmon was used as the model for fish and bilberries for wild berries. These calculations are shown in Table 11.1.

As shown in Table 11.1, today, the total caloric food production of Arctic Sweden could provide a mean caloric input of 937 Kcal/person/day, which is far below the limit of starvation. The most important Arctic food producer from a caloric point of view is the milk industry, which contributes to more than 50% of the region's total caloric production. In Table 11.1, dairy production is expressed as milk. In reality, however, a large portion of domestic milk is transformed into cheese, especially *Västerbotten*, a Swedish Parmesan. Since Sweden's entrance into the EU, production of *Västerbotten* has increased by 300%, increasing the production value of domestic milk. Another item worth highlighting is cereals. Oat and barley can be cultivated in Arctic Sweden. Today, locally produced cereal is only used for animal feed, so Table 11.1 indicates that cereals contribute no calories/person. Another important plant food is potatoes. According to the regional food strategies, Arctic Sweden is well suited for potato production and could produce far more than it currently does. In addition, it is estimated that only 5% of available wild berries are harvested. A seven-fold increase in wild berry harvesting would provide 500 grams of berries for each person per day (2014). Theoretically this could be achievable.

It is important to stress that the results shown in Table 11.1 should by no means be interpreted as exact values. They are estimates based on previous estimates, and interannual variations in harvests and the growth of the food sources (plant and animal) are not taken into account. However, even if these data have a confidence interval of close to 100%, the current caloric input of Arctic Sweden's domestic food production cannot be considered a strong foundation for food security. The aim with Table 11.1 is not to

Table 11.1 Estimated energy available from domestic primary food production per person and day in Arctic Sweden[a] based on official statistics (Stenberg, 2016, Jonasson, 2017 & Röös, 2016).

	Total weight (1000 ton)	*Weight/person/ day (gram)*	*Type item*	*Energy (Kcal)*
Plant food				
Potatoes	13	68	potatoes	54
Garden products (including flowers)	15	81	broccoli	29
Wild berries, mainly bilberries	13	69	bilberries	40
Cereals (today only cultivated for animal feed)	0	0	-	0
Dairy products				
Milk (a lot is made into Västerbotten cheese)	160	850	3% fat	510
Egg	2.0	11	egg	17
Fish (75% is cultivated)	18	96	salmon	175
Meat				
Agricultural (mainly beef)	9.6	51	beef	79
Game (mainly moose)	3.2	17	moose	17
Traditional (reindeer)	2.0	11	reindeer	16
SUM				937[b]

a Arctic Sweden includes two counties: Västerbotten and Norrbotten.
b Caloric needs for a normal-weight adult with a sedentary job: 2000–2500 Kcal/day.

provide exact figures but return administrative focus to the central values of the Sami food strategies of the past: a product-oriented system that focuses on the nutritional and caloric value of food. In this way, Table 11.1 will serve as an illustration for further discussions.

Future food strategies for a sustainable Sápmi/ Northern Sweden

It is much easier to point out the weaknesses of a strategy than to suggest a better solution. As the effects of climate change become more and more severe, as we enter 'the era of energy transition, combined with emission cuts' (Järvensivu et al., 2018), effective food strategies are key. In a recent background document ordered by the United Nations for the Global Sustainable Development Report 2019, a group of independent Finnish scientists demonstrate how and why market-oriented solutions will fail to adequately address future global challenges. They conclude, 'In the modern global economy, states are the only actors that have the legitimacy and capacity to fund and organise large-scale transitions' (Järvensivu et al., 2018, page 6). Looking exclusively at the food system, the conclusion is likely to be the same. Food cannot be treated like any other commodity on the market. Food is more.

It is necessary for life. It is necessary for culture. It is necessary for humankind. Thus, governments must act. In Sweden, this means that two governments must be involved: the Swedish government, which represents all Swedish citizens, and the Swedish Sami Parliament, which represents Sweden's Indigenous Sami population. Any food strategy, especially one targeting Sápmi and Arctic Sweden, that excludes one of these governments will probably fail from a practical as well as a human rights perspective.

To return to our initial question: To what extent is it possible to apply traditional knowledge of foods and the food systems of Arctic Sweden to modern food security policies? There is no simple answer. Still, as a first step, let us try to identify some relevant lessons in pre-modern Sami food strategies. Then, let us discuss how these lessons might be applied to national and regional food strategies.

Lesson 1: Satiety is the overarching goal

Market-oriented solutions, such as those proposed by the Swedish National Food Strategy, may indeed manage to increase production values via competition. Market solutions may foster the development of new delicacies. However, consumers will get the same amount of energy from an expensive potato or a delicacy made with reindeer beef as they will from cheaper, less refined versions of these foods. In the traditional Sami food system, satiety was the overarching goal. This could be clarified in modern food policy documents; the focus should shift from a market orientation toward a production orientation. A good example of how this could manifest is the Nordic scenario workshop study described in the introduction to this chapter (Karlsson et al., 2018). In addition to modelling a self-sufficient organic farming system (including a transition toward increased intake of plant-based foods), a similar study could also analyse the impact of such a model in terms of the Nordic Nutrition Recommendations. From a macronutrient perspective (the energy balance between carbohydrates, protein, and fat), the model developed by Karlsson et al. (2018) is satisfying; that is, the distribution of macronutrients aligns with the Nordic Nutrition Recommendations (Karlsson et al., 2018). A similar scenario model could be done in an Arctic Swedish context; such a study should include dialogue with Sami food producers and tradition carriers. This would represent an important step toward defining a sustainable food sovereignty strategy applicable to Northernmost Sweden.

Lesson 2: Increase the consumption of traditional, uncultivated, plant-based foods

Today, Arctic Sweden relies on very few food production areas, especially for plant-based foods, as shown in Table 11.1. All plant-based foods cultivated in the North originate in southern latitudes; the only traditional plant-based food item included in the current Arctic regional strategy documents (spring

2019) is wild berries. According to the (limited) calculations presented in Table 11.1, wild berries contribute roughly 22% of plant-based food production in Northernmost Sweden today if counted in weight and 33% if counted in caloric value. A tenfold increase in the harvest of wild berries – which would only use around 50% of the available wild berry stock – would represent a significant contribution to domestic food supply and would decrease dependence on imported tropical fruits. Combining this with increased use of common herbs such as fireweed (*Chamaenerion angustifolium*), sorrel (*Rumex acetosa*), and alpine sow-thistle (*Cicerbita alpine*) would further increase the potential self-sufficiency level of Arctic Sweden. Of these herbs, fireweed deserves special attention. It is abundant all over Sápmi and richer in protein than other traditional herbs; with 4.7% protein, it is almost comparable to peas, which have 5.3% (Nutritionvalue, 2019). Protein-rich herbs will likely gain importance in a climate-related societal transition toward more plant-based foods (Poore and Nemecek, 2018). Fireweed is also an important source of vitamin C, and plants growing at higher latitudes contain more vitamin C (Qvarnström, 2006). In Sami cuisine, fireweed leaves are used with coagulated milk in a dish called *abrek-kombo*, which resembles the more widely known Sami dish *gompa* (Fjellström, 1985). Fireweed can also be prepared as a substitute for asparagus when leaves and flowers near the stem are harvested before midsummer and before the plant has grown taller than 15 centimetres. When it blooms, the leaves and flowers can be used as substitutes for lettuce or spinach. Later in the summer or in autumn, the starchy white part of the stem can be used to check the consistency of soups (according to the author's own experience). More studies evaluating the value of traditional Sami and other wild herbs to food strategy in Arctic Sweden are warranted.

Lesson 3: Decrease the demand for imported food

This lesson is similar to lesson 2 but expands on it. Popular staples such as coffee, tea, cocoa, and sugar cannot be grown in Arctic Sweden. Though a certain amount of trade may be desirable also in the future, decreasing dependence on these colonial staples is another way to increase the region's food self-sufficiency. Fireweed, for example, can be used as a substitute for Chinese tea; the early green leaves are harvested in mid-June to early July (before the purple flowers have begun to bloom) and then fermented and dried. Sugar can be avoided by a return to old preservation methods, such as drying blueberries and herbs and conserving lingonberries and cloudberries in water (described in more detail in (Nilsson, 2018). In this method, clean, fresh berries are covered with fresh water that has been boiled and then cooled; they are then stored in closed bottles in a cool area (Figure 11.2). This method works because of the berries' high content of natural benzoic acid. Increasing the demand for local food is also significant to limiting long-distance transportation in the future. Today, fruits and vegetables are often transported by plane to keep them fresh. A decreased demand for imported food staples, combined by

Figure 11.2 Soaking in water is a traditional way to preserve lingonberries. Today, this staple is usually stored in glass bottles and cans.

an increased demand for local wild and cultivated berries and plants, would thus significantly decrease the carbon dependence of the future food system.

Lesson 4: Use the full potential of animal-based foods

For animal-based foods, the main lesson from the traditional Sami food system is the need to use the full potential of animals after slaughter. Blood and intestinal fat from reindeer and other game are nutritious traditional foods that are rarely found in dishes in Arctic Sweden today. An organic farming system like the one suggested by the scenario workshop described earlier (Karlsson et al., 2018) would provide adequate macronutrients but insufficient micronutrients. That is, according to the calculations shown in this paper, six of the twenty micronutrients listed in the Nordic Nutrition Recommendations would not be produced in sufficient amounts for the population. Thus, residents would need to take supplements to obtain adequate amounts of vitamins A and D, riboflavin, iodine, iron, and selenium (Karlsson et al., 2018). This shortcoming is probably due at least partially to the workshop's normative decision to decrease the amount of animal-based food but still accept more food waste in this sector than in plant-based human food production. Animal fat and intestines are rich in vitamins A and D and in iron. Fish and intestinal foods are rich in selenium and iodine. Therefore, a repeat of this kind of model in a Northern Swedish context would need to include alternative strategies for maximising the nutritional value of animal-based food. Food waste that is not used for human food could also be used as a raw material to produce micronutrient supplements rather than being recirculated as biogas. This is particularly important when society's goal is to decrease the use of and need for animal-based foods.

Lesson 5. Avoid feeding animals with potential human food

An important method for reducing food waste is to define strategies for who eats what and to clearly distinguish human food production from animal feed production. Today, almost all cereals produced in Arctic Sweden are used solely for animal feed. In early 1900s, cereals were grown mainly for human consumption. In Northernmost Sweden, grass, sedges, straw, reindeer lichen, and other herbs and leftovers that ruminants cannot easily digest were prepared for animal consumption by adding salt and boiling the mix into a brew called *döjda*, a loose soup prepared on a stovetop directly in the barn. According to folklore, the governor of Norrbotten County tried to stop this practice in the early 1900s because he thought *döjda* was unhygienic. The governor granted financial support to a poor settler in the parish of Arvidsjaur, Norrbotten on the condition that the settler would immediately stop feeding *döjda* to his cattle. The settler refused, and no support was provided to him (Bergqvist et al., 1908). My mother was born in 1939 and grew up in Malå, near Arvidsjaur. She reports that all the settlers in her family's neighbourhood had stoves in their barns and regularly prepared *döjda* for their cattle on these stoves. In the 1950s, *döjda* was commonly flavoured with sweet molasses from the south; again, my mother reports that the cattle enjoyed this sweetened *döjda*. Karlsson et al.'s (2018) Nordic scenario workshop study proposes that animals should be fed by by-products of human food production (Karlsson et al., 2018). Merging parallel modern practices with traditional knowledge about the preparation and use of *döjda* could be a topic of future research on food security strategies for Arctic Sweden. If successful, such actions would likely partially or completely eliminate the 'food versus feed' competition between humans and domestic animals.

Lesson 6. Promote knowledge and relationships

Building sustainability through knowledge and relationships is an important lesson to learn from the past. In this context, knowledge must include new scientific solutions as well as traditional knowledge, or *árbediehtu*. At universities, researchers continuously produce new scientific knowledge and transfer this knowledge to their students. Some of those students may become researchers who can further problematise, expand, or abandon a given intellectual challenge. The transfer of *árbediehtu*, however, follows another path. A prerequisite for *árbediehtu* transfer is acting or working together, side by side. The learners follow the footsteps of the knowledge carrier, observing, imitating, and developing their own skills (Nutti, 2018). To improve Arctic Sweden's food sovereignty, some skills must be more effectively transferred to future generations. These include gathering skills for harvesting wild berries and plants and horticultural skills for increasing self-sufficiency on a family level. In the early 1900s, school holidays often aligned with harvests in Arctic Sweden; the autumn 'berry picking holiday' was followed by the 'potato harvesting holiday'. Schoolchildren also often contributed to the school's food supply by picking berries as an outdoor school activity. These berries were later served to them for school lunch. Similar practices could be reinstated today. To encourage and teach the population (and especially young people) to gather wild berries and herbs, these activities should be strategically promoted in schools and other suitable arenas.

Relationships are also an important element in this kind of knowledge transfer. In Sweden today, the relationships between Swedish citizens and the Sami people must be strengthened on a governmental as well as societal and individual level. Today, the Swedish educational system largely ignores teaching about Sami society (Svonni, 2015). Teaching future generations to appreciate the traditional survival knowledge shared between Sami and non-Sami settlers throughout history will be a core element of knowledge-based strategies for food sovereignty in the Sápmi of the future, which will be affected by climate change and continuously changing.

Conclusion

To summarise, future sustainable food strategies for Sápmi/Northern Sweden should be outlined and agreed upon in collaboration by the Swedish government and the Swedish Sami Parliament. Such strategies should be product-oriented, ecological, in balance with nature, and based primarily on plants and animals adapted to an Arctic or sub-Arctic boreal climate zone. They should specify caloric values and nutrient values and include a traditional, local perspective, and strategies for waste-reduction. They should also promote knowledge sharing from different knowledge systems and strengthen the relationships among all inhabitants in the area.

Disclosure statement

No potential conflict of interest has been reported by the author.

Acknowledgements

I would like to thank Lars Ericson, agricultural expert at the County Counsel of Västerbotten, for sharing valuable knowledge on food policy development, for recommending literature, and for the interesting discussions and feedback.

References

Bergqvist, O., Svenonius, F. & Bauer, J. (1908). *Lappland: det stora svenska framtidslandet: en skildring i ord och bild af dess natur och folk [In English: Lapland: The great Swedish future country: A depiction in words and pictures of its nature and people].* Stockholm: C.A.V. Lundholm.

F.A.O. (1996). *Rome declaration on world food security: World food summit plan of action* [Online]. Available: http://www.fao.org/3/w3613e/w3613e00.htm [Accessed November 27 2019].

Fjellström, P. (1985). *Samernas samhälle i tradition och nutid [In English: The Sami society in tradition and today]*. Värnamo: Fälths tryckeri.

Flygare, I. & Isacson, M. (2003). Jordbrukspolitiken - ett klassiskt drama [In English: The agricultural policy, a classic drama]. In: Flygare, I., Isacson, M., Nordiska, M. & Stiftelsen, L. (eds.) *Det svenska jordbrukets historia. Band 5, Jordbruket i välfärdssamhället: 1945–2000 [In English: The history of Swedish agriculture, volume 5, Agriculture in the welfare society: 1945–2000]*. Stockholm: Natur och kultur/LT i samarbete med Nordiska museet och Stiftelsen Lagersberg. (227–257).

Järvensivu, P., Toivanen, T., Vadén, T., Lähde, V., Majava, A. & Eronen, J. T. (2018). Governance of economic transition. Background paper to the Global Sustainable Development Report 2019 drafted by a group of independent scientists to a chapter on "*Transformation: The Economy*".

Johnsen, K. I. & Benjaminsen, T. A. (2017). The art of governing and everyday resistance: "Rationalization" of Sámi reindeer husbandry in Norway since the 1970s. *Acta Boreal.*, 34, 1–25.

Jonasson, L. & Persson, S. (2017). Skafferiet mellan kust och fjäll - Livsmedelsproduktion i Västerbottens län [In English: The pantry between coast and mountain - Food production in Västerbotten County]. The County Counsel of Västerbotten, a pre-study report, *not published!*

Kaati, G., Bygren, L. O., Pembrey, M. & Sjostrom, M. (2007). Transgenerational response to nutrition, early life circumstances and longevity. *Eur J Hum Genet*, 15, 784–790.

Karlsson, J. O., Carlsson, G., Lindberg, M., Sjunnestrand, T. & Roos, E. (2018). Designing a future food vision for the Nordics through a participatory modeling approach. *Agron. Sustain. Dev.*, 38, 10.

Lantto, P. (2000). *Tiden börjar på nytt: en analys av samernas etnopolitiska mobilisering i Sverige 1900–1950 [In English: Time begins a new: an analysis of the Sami ethnopolitical mobilization in Sweden 1900–1950]*. Dissertation. Umeå University.

Lövén, S. & Bucht, S.-E. (2017). En livsmedelsstrategi för Sverige – fler jobb och hållbar tillväxt i hela landet [In English: A food strategy for Sweden - more jobs and sustainable growth throughout the country]. Stockholm: Regeringens proposition 2016/17:104, Näringsdepartementet.

LRF. (2018). *Självförsörjning, by LRF, The Federation of Swedish Farmers* [Online]. Available: https://www.lrf.se/politikochpaverkan/foretagarvillkor-och-konkurrenskraft/nationell-livsmedelsstrategi/sjalvforsorjning/ [Accessed November 27 2019].

Mela, T. J. N. (1996). Northern agriculture: Constraints and responses to global climate change. *Agr. Food Sci. Finland*, 5, 229–234.

Näringsdepartementet (2017a). A national food strategy for Sweden – more jobs and sustainable growth throughout the country. Stockholm, Sweden: Government Offices of Sweden, Ministry of Entrepirise and Innovation. Available: https://www.government.se/498282/contentassets/16ef73aaa6f74faab86ade5ef239b659/livsmedelsstrategin_kortversion_eng.pdf.

Näringsdepartementet. (2017b). Regeringens handlingsplan för livsmedelsstrategin [In English: The governments action plan for the food strategy]. Government bill 2016/17:104. Stockholm: Government Offices of Sweden, Ministry of Entrepirise and Innovation.

Nilsson, L. M. (2018). Food, Nutrition, and Health in Sápmi. In: Veslemøy Andersen, E. B. G. W. (ed.) *Nutritional and health aspects of food in Nordic countries*. 1st ed. London: Elsevier. (179–195).

Nilsson, L. M., Dahlgren, L., Johansson, I., Brustad, M., Sjölander, P. & Van Guelpen, B. (2011). Diet and lifestyle of the Sami of southern Lapland in the 1930s–1950s and today. *Int. J. Circumpolar Health*, 70, 301–318.

Nilsson, L. M., Destouni, G., Berner, J., Dudarev, A. A., Mulvad, G., Odland, J. O., Parkinson, A., Tikhonov, C., Rautio, A. & Evengård, B. (2013). A call for urgent monitoring of food and water security based on relevant indicators for the Arctic. *Ambio*, 42, 816–822.

Nilsson, L. M. & Evengård, B. (2015). Food security or food sovereignty: What is the main issue in the Arctic? In: Birgitta Evengård, J. N. L. Ø. P. (ed.), *The new Arctic*. 1st ed. Switzerland: Springer. (213–223).

Nutritionvalue. (2019). *Find nutritional value of a product - Data from USDA National Nutrient Database* [Online]. Nutritionvalue.org. Available: https://www.nutritionvalue.org/Fireweed%2C_raw%2C_leaves_nutritional_value.html [Accessed February 2 2019].

Nutti, Y. J. (2018). Decolonizing Indigenous teaching: Renewing actions through a Critical Utopian Action Research framework. *Action Res.*, 16, 82–104.

Odin, M. (1937). Sjukdomar och sjukdomsfrekvens i övre Norrland särskilt med hänsyn till födans sammansättning [In English: Diseases and the frequency of diseases in northernmost Sweden, with special reference to the composition of the diet]. In: Hellström, N., Odin, M. & Sverige, M. (eds.), *En socialhygienisk undersökning i Västerbottens och Norrbottens län utförd med stöd av Kungl. medicinalstyrelsen under åren 1929–1931. [In English: An investigation into questions of social hygiene in the counties of Västerbotten and Norrbotten, Sweden 1929–1931]*. Lund: Håkan Ohlssons boktryckeri. (1–224).

Össbo, Å. & Lantto, P. (2011). Colonial tutelage and industrial colonialism: Reindeer husbandry and early 20th-century hydroelectric development in Sweden. *Scandinavian Journal of History*, 36, 324–348.

Poore, J. & Nemecek, T. (2018). Reducing food's environmental impacts through producers and consumers. *Science (New York, N.Y.)*, 360, 987.

Qvarnström, E. (2006). *"De tycka emellertid av gammal vana att det smakar gott, och tro dessutom att det är bra för hälsan": samiskt växtutnyttjande från 1600-talet fram till ca 1950 [In English: "They do, however, think of old habits that it tastes good, and also believe that it is good for health": Sami plant utilization from the 17th century until about 1950]*. Dissertation, Sveriges lantbruksuniversitet, Umeå.

Röös, E., Sandin, P. & Brunius, C. (2016). Hållbart jordbruk i Norrbotten – rapport från ett scenarioarbete [In English: Sustainable agriculture in Norrbotten - report from a scenario modelling]. Uppsala.

Ryd, L. (2015). *Urfödan: om självhushållets mat hos folk i Lappland [In English: The original food: About self-sustained food among the people of Lapland]*. Skellefteå, Ord & visor.

Sahlén, U. (1987). *Vad är det i maten? näringsämnen per portion [In English: What do the food contain? nutrients per portion]*. Västerås: ICA-förlaget.

Stenberg, M., Österberg, S.-E., Flinkfeldt, B. & Nyman, H. (2016). Norrbottens Livsmedelsstrategi Nära Mat [In English: The food strategy of Norrbotten, local food], County of Norrbotten. Luleå, Sweden. Available: http://media.xn--nramatbua.nu/2015/09/Norrbottens-livsmedelsstrategi-Nara-Mat_rapport-1.pdf.

Svonni, C. (2015). At the margin of educational policy: Sámi/Indigenous Peoples in the Swedish national curriculum 2011. *Creative Education*, 6, 898–906.

Winsa, B., Baldauf, R. B. & Kaplan, R. B. (2000). *Language planning in Nepal, Taiwan and Sweden*. Clevedon: Multilingual Matters, 118.

12 Bridging food security gaps in the European High North through the Internet of Food

Dele Raheem, Borja Ramis Ferrer, and Jose L. Martinez Lastra

Introduction

Today's interconnected global food system involves a complex supply chain that is prone to risks due to regulatory barriers, disruptions because of economic instability, variations in consumer demand, and the effect of such variations on food production and sustainable development (Sowinski 2012). To achieve sustainability, reduce waste, and ensure the efficient use of resources, changes are required in food supply chains, starting with local suppliers and processors (Derqui et al. 2016). The global food system is wasteful and inefficient, and gaps in access to food have created food insecurity in many parts of the world, including the European High North (EHN) (Godfray et al. 2010; FAO 2017). These gaps comprise food security and also food safety, such as food waste, the failure to track the origin of foods, and compromised safety and quality in processed food products.

Food is a multi-dimensional expression of culture, identity, and community (Dey et al. 2019). Food security issues in the EHN share common elements with other, more frequently studied circumpolar zones, such as the Canadian Arctic and Alaska (Ford 2006; Chan et al. 2006; 2 Lambden et al. 2007; Natcher 2018). These elements include colonialism, the exploitation of natural resources, the Westernisation of food systems, and the disappearance of Indigenous self-sufficiency (Duhaime and Godmaire 2000). Food security for the EHN has nutritional and sociocultural aspects. Indigenous perceptions of livelihood security in the region are grounded in Indigenous socio-cultural traditions and the special relationship of Indigenous People to ancestral territories and resources (Nuttall et al. 2005). Food and its procurement and consumption are often linked to culture and identity, as well as to social, economic, and political organisation (UNHCR 2010).

The Internet of Food (IoF) has been identified (Poppe et al. 2015; Sundmaeker et al. 2016; Dooley et al. 2018) as a novel solution which may help bridge knowledge gaps in food traceability by making pre- and post-processing from farm to table more transparent. It will foster co-creation by food businesses by encouraging the use of digitisation to promote Indigenous and local socio-cultural traditions (Sundmaeker et al. 2016; Koistinen et al. 2017).

DOI: 10.4324/9781003057758-12

Some of this co-creation can be implemented through social engagement with consumers, responding to consumers demands (e.g., in more informative packaging labels), and possibly with smartphone apps related to the food system. However, it is important that the link between food and culture is not lost in the digital disruption of this region's food system. Digitalisation can capture the narratives behind local food processing. Smart approaches can also contribute to sustainable rural development (Prause and Boevsky 2015), as well as assuring customers of the authenticity of organic food and helping detect food fraud. The existing literature on food security in the EHN was reviewed to identify gaps in the region.

This chapter addresses current gaps related to food security in the EHN.[1] Bridging these gaps will require appropriate governmental policies. Digitalisation and mobile communication are regulated by three national policies in the EHN: those of Finland, Sweden, and Norway. It is necessary to identify the role of big data on consumer food choices and to define the relationships among climate change, digitalisation, and consumer food choices. After identifying the food security gaps in the EHN, this chapter describes how the IoF could bridge these gaps. It then analyses the impact of climate change and of food choices in the EHN in the overall food system of the region. Then the chapter calls for food businesses in the EHN to engage in co-creation using the IoF. Finally, the chapter presents the future outlook for the IoF.

Food security gaps in the EHN

The four main pillars of food security are food availability, accessibility, utilisation, and stability (FAO 2008). However, more research is needed on the nutritional, sociocultural, and economic needs of communities so that previous definitions of food security can be expanded to encompass not only the availability of food but also elements of food quality (Chan et al. 2006; Loring and Gerlach 2009; Beaumier and Ford 2010). Food regimes, that is, the role of agriculture and food in different stages of the world-capitalist economy, are examined by Soldevila Lafon in Chapter 10 of this book.

Food quality and access are particularly relevant in the EHN and other Arctic communities; in these areas, traditional foods have cultural and dietary significance (Nuttall 1992). Wild berries, such as *Rubus chamaemorus* (cloudberry), *Vaccinium uliginosum* (bilberry), *Vaccinium vitis-idaea* (lingonberry), *Rubus idaeus* (raspberry), and *Vaccinium oxycoccus* (cranberry), are widely gathered in the EHN. During the berry growing season, they are eaten fresh and preserved for later use (Cormier and Raheem, 2018). Family gardens provide fresh potatoes, strawberries, blackcurrants, onions, and root crops in many regions across the EHN. Mushrooms and herbs can easily be gathered from the forest and often form part of traditional meals in the EHN. Reindeer meat is the most prevalent and important meat produced in the EHN. Other native animals such as elk, rabbits, fowl, and waterfowl are eaten less in the EHN because of the low numbers of wild game of these species (Müller-Wille 2008). However, the impacts of climate change on marine and terrestrial ecological dynamics are threatening access to these traditional foods in the EHN and the entire Arctic region (Ford et al. 2014). Most reports on the current state of adaptation in the Arctic are from North America. Only a few adaptations have been documented in the European or Russian Arctic, and most studies of these regions focus on business, the economy, or infrastructure (Prowse et al. 2009; Ford et al. 2014; Loboda 2014). There are research gaps on the capacity of Arctic communities to adapt to climate change or on the nature of their vulnerability to it (i.e., who and what are vulnerable?). Anticipatory adaptation planning will probably be limited until these gaps are identified and addressed (Ford 2009; Ford and Pearce 2010). Thus, there is a need to identify adaptation strategies and examine their effectiveness in reducing the vulnerability of traditional food systems to climate change. The costs and benefits, including broader non-climatic benefits, of such strategies must also be assessed.

An important way to preserve the culture and tradition associated with local, traditional foods is by adding values to these foods. Digitalisation can boost value addition in the EHN, enhance consumer experiences, and help interest younger people in traditional foods. This could empower local communities by creating jobs, thus ultimately strengthening food sovereignty in the region (Hautamäki et al. 2017; Kuokkanen et al. 2018). Increased food production and processing in the EHN would in turn improve the social and economic conditions of local communities. Small and medium-sized enterprises (SMEs) are drivers for innovation and proactivity, and they will play an important role in local economies and in the competitiveness of the region. The potential of digital technology to create a unique brand for Arctic local foods can be strengthened through cross-border cooperation that investigates the possibilities of shared practices amongst different business enterprises (Natcher 2018).

However, the countries of the EHN (Finland, Sweden, and Norway) have different national policies regarding human activities that are likely to affect food security and food safety (e.g., mining, forestry). Moreover, the region is characterised by long distances between communities and lower populations than the southern regions of these nations. EU policies, such as the common agricultural policy (CAP), do not apply to Norway. Thus, the region will need to harmonise (especially with Norway, which is not an EU member) a common EHN strategy for the smart manufacturing of traditional foods.

The traditional food culture portrayed in the Indigenous Youth, Food Knowledge & Arctic Change (EALLU) cookbook (ICRH 2018) can be combined with scientific knowledge of reindeer husbandry to ensure new ways of developing the economy of reindeer herding and the nutrition of the thirteen Indigenous Peoples of the Arctic (Nenets, Sami, Chukchi, Koryak, Dolgan, Evenki, Even, Yukagir, Dukha, Inuit, Aleut, Gwichin, and Athabaskan).

A combination of traditional and scientific technologic knowledge about food harvesting, storage, distribution, and preservation will support sustainability in many parts of the world including the circumpolar region (FAO, 2013). The West Nordic countries[2] are working together to support artisans in the food sector, they help artisans share knowledge and experience by providing access to consulting (Valsdottir 2016). This is carried out under the Nordbio Innovation Project,[3] which supports over 50 food producers.

Processing food from farm to table in ways that match consumer trends can be supported by systems that connect all participants. Technologies such as digital watermarks and enhanced barcodes allow products to carry information and help enhance the user experience even after the customer leaves the store (Wara and Dugga 2014).

Digital technology helps satisfy consumer safety concerns and minimise recalls of packaged foods. Food manufacturers can also use cloud-based technologies for quality management. Cloud-based technology will enable consumer packaged goods to improve supply chain operating platforms by connecting producers directly with retailers. The low population density in the EHN and the ease of managing the EHN low amount of data make digitalisation an attractive way to bridge food security gaps.

Using the IoF to bridge food security gaps

In the first industrial revolution, water and steam mechanised production. In the second, electrical energy enabled mass production, and in the third, electronics and information technology automated production. The fourth industrial revolution builds on the third revolution, which is the digital revolution that has been taking place since the middle of the last century (Sentryo 2017).

The explosion of Information and Communication Technologies (ICTs), along with its attending high-speed computation and mobile connectivity, has unleashed a global conversation about needs, values, and aspirations around food and food systems (WEF 2018). One emerging technology trend is the effort to realise the Internet of Everything (IoE), which would connect all resources in the value chain, collecting and processing information as goods are produced (Ramis Ferrer and Martinez Lastra 2017; Zheng et al. 2018). The collected data are stored in large repositories that, in turn, must be accessed by data processing engines that may be used for retrieving or even producing implicit information. This requires drawing on several fields of study, including knowledge representation and reasoning, Big Data, and analytics. In the IoE, industrial equipment and a product's lifecycle can be monitored, enabling humans and machines to react to malfunctions and improve processes (Ramis Ferrer et al. 2014; Iarovyi et al. 2016).

This leads to the IoF, which is an emerging area that focuses on the digital aspects, technical innovations, and new data layers around food. These digital additions could change the global food system (Fälström and Jörgensen 2015). Like the IoE, the IoF could be implemented to make food traceable, transparent, and trustworthy and to empower consumers to obtain personalised food that caters to individual food, diet, and health choices (Lange 2017). This could be accomplished with the information that are continuously gathered in a product's lifecycle. However, compared to other services, the food industry has been slow to take advantage of Internet technology. Implementing the IoF in the near future will significantly impact several aspects of the food system. For example, there are technological initiatives to implement *farm to fork* (F2F), which would allow consumers to trace foods from production to consumption and access information from the entire food lifecycle.

Increasing interest in redistributed manufacturing in the EHN and other circumpolar regions will require food businesses to adapt to digitisation and tap into the benefits of the superhighway IoF. Big Data and digital technology will be integral to consumers' everyday lives very soon; devices that can monitor health and highlight the role of nutrition in diet to promote health are already common. The trade-offs of sharing data can help minimise the environmental impact of food processing in the local economy. Therefore, using digital technology to predict consumer trends can make food businesses more efficient, sustainable, and transparent.

Another aspect of food safety relates to the use of pesticides, chemical preservatives, and food fraud and tampering. The widespread implications of these food safety gaps are evident from recent food scandals such as Sudan dye, salmonella found in eggs, horse meat sold as beef, and fipronil found in eggs.

In the future, consumers will need to make informed choices about what to purchase and eat. This means they will need to obtain trustworthy information from food producers. The IoF digitalisation of food production can enable this. For example, in precision agriculture, wireless sensor networks (WSNs) using low-cost sensors measure the soil moisture, plant biomass, and local climate conditions. This makes it possible to improve the facilities where food is grown and/or produced (Wang et al. 2006). Fields and even livestock are set to become sources of high-quality, real-time biophysical data. The intelligence and autonomous behaviour of digital solutions will be game changers; sensors in the soil and the air can conduct surveillance, manage resources efficiently, and improve workflows in food and agriculture.

The IoF implements standard computable languages to create repositories of information, such as ontologies for describing food sector domain knowledge. Shishaev et al. describe ontology-based information for food security in the western Russian Arctic zone in Chapter 8 of this book. To effectively digitise food, everything from food processing plants to farms and grocery stores must be part of the same ecosystem. For example, a smart surveillance system to monitor food environments and fight food crime by tracking raw materials and commodities with sensors could be developed. However, there are concerns about the implications of mixing internet devices with the food supply, especially regarding malicious access and cyberattacks. Therefore, before deployment, IoF solutions should be validated using risk assessment and threat modelling techniques (Ramis Ferrer et al. 2017). However, a discussion of this issue is outside the scope of this chapter, which considers the ways the IoF can bridge food security gaps. Cybersecurity is critical and will be addressed in the design and development phases.

In the future, food security and safety in the EHN region will place more emphasis on the soil, air, and water as the most important inputs for the animals, fish, and plants that are grown, raised, and harvested in the region. A low-cost optical-chemical technique that uses smartphone cameras to detect bacteria on food or in water samples is already under development (Pearson et al. 2018). Organic farming in the EHN hold much promise for high quality organic foods, and the IoF offers a good opportunity to protect such foods from fraud with the use of digital technology. Natural and organic foods are becoming trendy in Europe and in emerging economies. As incomes increase, consumers become more willing to pay a little more for quality organic products that help improve their quality of life. Therefore, digital access to information on the origin of foods and the conditions under which foods were grown and harvested will be attractive to consumers.

Another key area for the future is the relationship between diet and human genetics, which is helping shape the personal nutrition and medicine. For instance, for a consumer with a polymorphic MTHFR gene, a lasagne could be prepared with more folate and more B-12 (Cechini 2017). The amount of calcium oxalate could be reduced for a consumer who is likely to get kidney stones. Consumers who refuse to eat food grown with phosphates can obtain information about the nutritional content of their food and even have a machine that makes personalised meals for them.

The value opportunities of Industry 4.0 can be described with the data – information – knowledge –wisdom (DIKW) model. Each stage builds on the next one and adds more value. Figure 12.1 shows how actions and decisions move from data (D) to information (I) to knowledge (K) to wisdom (W).

We argue that food industries will benefit from using analytics driven by Big Data to make critical decisions about pricing, product promotion, product development, and demand forecasting. Other benefits include improved product innovation, more effective sales, enhanced margins and profitability, extended customer reach, increased marketing return on investment (ROI), and greater customer satisfaction and loyalty. The following steps have been identified in the decision-making process using Big Data (Edwards 2017):

a Use data analytics tools to understand customer preferences to stock or serve the right products at the right time,
b carefully analyse collected data to uncover and address trends that may help or hurt the business,
c look for and evaluate promising new data analytics technologies and methods to keep pace with competitors and customer demands, and
d allow managers to access data and make quick changes based on the insights they receive.

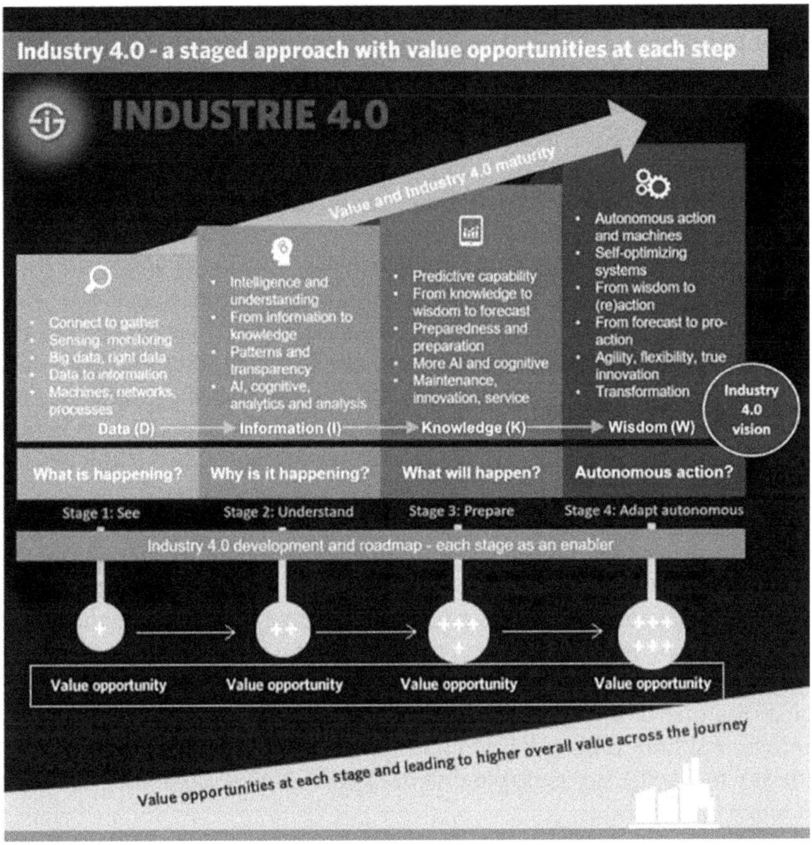

Figure 12.1 Value opportunities at each stage (adapted from i-SCOOP 2018; https://www.i-scoop.eu/industry-4-0/).

Climate change and the choice of food products in EHN

Diet change is one part of a successful climate change mitigation policy, and policies to improve the food system are also aimed at mitigating climate change (Hällstrom et al. 2017). In the future, it will be important to choose healthy diets that are also environmentally healthy. This is addressed in the previous chapter (Chapter 11 of this book) in Nilsson's reflection on Swedish food strategies from a Sami and an Arctic perspective.

Current trends in food choices point toward increased environmental impacts, and consumers' food choices affect the climate (Carlsson-Kanyama and Lindén 2001; Carlsson-Kanyama 2004). Previous studies (e.g., Carlsson-Kanyama 1998; Carlsson-Kanyama et al. 2003; Engström et al. 2007) have shown that food and diet choices can influence the energy requirements for the provision of human nutrition and the associated greenhouse gas (GHG) emissions. The GHG emissions of meals with similar caloric content may differ by a factor of two to nine (Carlsson-Kanyama 1998; Engström et al. 2007). An analysis of the energy inputs required to produce a large number of food items showed that different meals with similar nutritional value required GHG emissions that differed by up to a factor of four (Carlsson-Kanyama et al. 2003). All these studies have concluded that certain foods, such as animal products and vegetables produced in resource-intensive ways, require more resources and cause more pollution than other foods.

According to a Swedish study comparing farm-to-table emissions for 22 common food items, fresh vegetables, cereals, and legumes have the lowest emissions (Carlsson-Kanyama and Gonzalez 2009). The study found that meats and fruits transported by air have the highest total GHG emissions; eggs, certain fish, and frozen vegetables are in the midrange. Animal products range from 1.5 to 30 kg of GHG emissions/kg of food, and herring and eggs are on the lower end of the animal products (Carlsson-Kanyama and Gonzalez 2009). The study concludes that shifting toward a more plant-based diet could substantially contribute to mitigating GHG emissions. Another study of US household food purchases examined the association between the food-related GHG, household food spending patterns, and sociodemographic characteristics (Boehm et al. 2018). Households that generated the highest levels of GHG emissions spent a larger share of their food budget on protein-rich foods (Boehm et al. 2018). In Finland, food choices constitute one-sixth of the average carbon footprint of a Finn. On annual average, a standard meat-based diet in Finland produces 1.5 tonnes of carbon dioxide (CO_2), a vegetarian diet 0.9 tonnes, and a vegan diet 0.5 tonnes (Salonen et al. 2018).

Food choices are largely dictated by tradition and by what is locally produced or available in the EHN region. Community-led initiatives can improve food security and sovereignty (Herrmann et al.; Chapter 14 of this book). It is vital to discourage food imports, especially of foods that are nutritious and locally available in the EHN. In recent years, industrial food and agriculture have been driven by fossil fuels and have depended on chemicals to extend the shelf lives of food items. The EHN will need to encourage investments in locally available food that can promote a resilience and health in the climate and in local food and water.

In modern agriculture, insecticides are used extensively to manage insect populations and protect crops. Despite heavy regulation, misuse occurs, meaning that insecticides sometimes make their way into the surrounding environment via the air, water, or soil, ultimately entering the food chain. Once present, insecticides can have adverse health effects on humans in the area – or, due to the globalisation of industrial farming, over a wide area. The common insecticide fipronil was found to have contaminated millions of eggs that were subsequently distributed to 15 EU countries, Switzerland, and Hong Kong (WHO 2018). In July 2017, it was announced that fipronil had been mixed with another formulation and applied to chickens to protect them from ticks, lice, and fleas. In some cases, both the treated poultry and their eggs contained over 200 times the European Union's maximum residue levels (MRL) of fipronil and its metabolite, fipronil sulfone. Due to this contamination, a quick and reliable method for determining the presence and concentration of the insecticide in eggs, egg products, and possibly chicken meat as well is needed.

Food packaging labels contain nutritional information in the list of ingredients. Consumers are keenly interested in innovative 'clean labels', which are natural and free of chemical preservatives. Food choices often affect how food and nutrition promote consumer health.

Call for co-creation of and engagement in the IoF by EHN food businesses

The food sector will benefit from co-creation and sharing best practices regarding how digital methods can enhance consumers' trust in food and provide warnings when required in case of adulteration and frauds. Trust comes into play at a number of points in the food system: product information, product relationships (i.e., other products from the same manufacturer), geographical location of manufacturer and the detection of fraud or manipulation. Food business operators will gain consumers' trust when they provide official verifications and validations, endorsements, crowd annotation, and digital signatures. Digitalisation is expected to change consumption and purchasing behaviour (Su 2017). Information on the origins of food and on the conditions in which it is grown and harvested will benefit consumers. Exports to new markets will also benefit from networking and cooperation, activities that are also aided by digitalisation.

As manufacturing becomes increasingly globalised, customer-oriented manufacturing could improve service quality and competitiveness, particularly for SMEs. It is necessary to preserve the cultural and Indigenous knowledge associated with traditional and local foods in the EHN while promoting them through digitalisation. Ideally, natural food products could be cultivated or extracted even more efficiently in laboratories. Distributed manufacturing would support this kind of production.[4]

The ongoing discussion around the Internet and food aims to construct a framework for discussing how to facilitate openness and innovation in the food industry with the goal of feeding the planet in a healthy and sustainable way. The data structure can secure a common understanding and approach. Shared co-creation will emphasise data structures for food, asking if there is an existing one that should be implemented more broadly and how different data structures interact with other pieces of the digital puzzle. Specifically, the following questions will need to be answered: Is there an existing metadata description covering all aspects of food or are there many metadata descriptions? Are there any areas not covered by the gathered metadata? Do the existing descriptions follow a global standard (existing or *de facto*)? Could the descriptive standards be unified into one metadata standard suitable for digital distribution? Which fields or field types should be included in the metadata standards? (Fälström and Jörgensen 2015).

Digital solutions to gaps in the food system of EHN countries are on the rise. A few examples are worth mentioning. Wolt is a Finnish technology company that offers food delivery services in Finland, Sweden, Estonia, Denmark, Lithuania, Latvia, Norway, Georgia, Hungary, Czech Republic, Poland, Croatia, Greece, Israel, and Serbia. Wolt helps customers discover and order great food from nearby restaurants. It is promoted by the national Finnish airline (https://www.finnairshop.com/en/wolt).

Smartcart in shopping centres helps customers quickly and accurately identify products from the barcodes. It seeks to bring the benefits of self-checkout to the shopping cart itself. Smartcarts will soon be updated with top-of-the-line barcode scanning created by Swiss IT pioneers Scandit, adding a new dimension to an already slick shopping experience (Smartcart 2019).

A peer-to-peer (P2P) network that partially covers the Barents region is the retail and distribution model (REKO) system, which implements a new model of sustainable marketing channels based on social networking services (SNS) interactions. The system unites small food producers, farmers, and consumers and 'offer[s] consumers a way of ordering products directly from the producer, without the need for middlemen' (Raheem et al. 2019).

Another new technology is hyperspectral frame camera systems used for precision agriculture and for land use, vegetation, and forestry surveys. The device and the software are under continuous development, and the goal is to develop the technology so that it can function in narrow band indices and provide highly accurate remote sensing from a spatial and radiometric point of view (Mäkeläinen et al. 2013).

Biocode Finland provides services to help food companies, farmers, and consumers make ecological food choices based on carbon footprint. The code provides information about how products are made and how efficiently they are produced (Biocode, 2018).

Cloud manufacturing (CManufacturing) is regarded as the next generation of manufacturing models (Xu 2012). It enables ubiquitous, convenient, and on-demand network access to a shared pool of configurable manufacturing resources (e.g., manufacturing software tools, manufacturing equipment, and manufacturing capabilities) that can be rapidly provisioned and released with minimal management effort or service provider interactions (Xu 2012). CUsers gain access to the CComputing concept, manufacturing infrastructure, platform, and software application. In this system, all the objects and features of production function as a service.

The project 'Digital architecture as a roadmap for food business operators in Lapland' draws a roadmap for small and medium food businesses in Finnish Lapland to embrace digital solutions. It is funded by the European

Agricultural Fund for Rural Development; the main goal of the project is to spread the knowledge and abilities acquired through digitalisation for small- and medium-sized businesses. It also helps entrepreneurs face the challenges of digitalisation. Digitalisation can help to promote the traditional and cultural features of the Arctic thus providing businesses in the food sector with a competitive edge.

Adding value to local and traditional foods is an important way to preserve the culture and tradition associated with these foods. This will help empower local communities by creating jobs, thus strengthening food sovereignty in the EHN. The IoF could increase the percentage of sustainable production and thus bridge existing food security gaps. It will also help improve the social and economic conditions of communities in the region. Digital technology will enhance the unique brand of local foods through cross-border cooperation that takes advantage of shared practices amongst business enterprises in the region.

Concluding remarks and future outlook

The EHN is in the forefront of innovation in the application of 5G networks and digital technology. In the very near future, it will be possible to make food safer through efficient data management and the use of accurate, rapid bacteria-detecting chips. The adoption of the IoF will help close food security gaps and is therefore crucial to human health and the health of the planet.

Digitalisation is one piece of resilient strategies to environmental change. It can promote the efficient use of resources, reduce waste, and improve symbiotic relationships in the food value chain. In the future, obtained digital indicators can be used to design flow charts describing the overall food value chain; this will help make the value chain smarter, more efficient, and more cost effective. Such metrics will be mapped to key performance indicators like those adopted in other industries to monitor and make decisions about productivity and sustainability.

To successfully leverage the export potential of its traditional foods, the EHN must strengthen local networks that encourage cooperation in the digital sector. Cross-border collaboration will need to address international codes of conduct for data storage and sharing.

Notes

1 Here, the European High North is defined as the northernmost parts of Finland (Lapland), Sweden (Norrbotten), and Norway (Finnmark, Troms, and Nordland).

2 The Faroe Islands, Greenland, and Iceland.

3 The project focuses on product development in Iceland, Greenland, and the Faroe Islands, building on the work of the Nordic project Arctic Bio-Economy. It encourages SMEs and entrepreneurs who develop food products to use local bio-resources.

4 Distributed manufacturing, also known as distributed production, cloud producing, or local manufacturing, is a form of decentralised manufacturing in which a network of geographically dispersed manufacturing facilities coordinate their activities using information technology.

References

Beaumier, M.C. and J.D. Ford. 2010. Food insecurity among Inuit women exacerbated by socioeconomic stresses and climate change. *Canadian Journal of Public Health/Revue Canadienne de Sante'e Publique*, 101(3): 196–201.

Biocode. 2018. Ecological food experience. Retrieved 12 September 2019 from https://www.biocode.fi/?q=en.

Boehm, R., P.E. Wilde, M. Ploeg, C. Costello, and S.B. Cash. 2018. A comprehensive life cycle assessment of greenhouse gas emissions from U.S. household food choices. *Food Policy* 79: 67–76. doi: 10.1016/j.foodpol.2018.05.004.

Carlsson-Kanyama, A. 1998. Climate change and dietary choices – How can emissions of greenhouse gases from food consumption be reduced? *Food Policy* 23: 277–293.

Carlsson-Kanyama, A. 2004. Diet, energy and greenhouse gas emissions. In C.J. Cleveland (ed.), *Encyclopaedia of energy Amsterdam*, 809–816. The Netherlands: Elsevier.

Carlsson-Kanyama, A., M.P. Ekström, and H. Shanahan. 2003. Food and life cycle energy inputs: Consequences of diet and ways to increase efficiency. *Ecological Economics* 44: 293–307.

Carlsson-Kanyama, A. and A.D. González 2009. Potential contributions of food consumption patterns to climate change. *The American Journal of Clinical Nutrition* 89(5): 1704S–1709S. doi: 10.3945/ajcn.2009.26736AA.

Carlsson-Kanyama, A. and A-L. Lindén. 2001. Trends in food production and consumption: Swedish experiences from environmental and cultural impacts. *International Journal of Sustainable Development* 4: 392–406.

Cecchini, C. 2017. Welcome in the Internet of Food era. Retrieved 12 December 2018 from https://medium.com/feat/welcome-in-the-internet-of-food-era-18cd3c05a273.

Chan, H.M., K. Fediuk, S. Hamilton, L. Rostas, A. Caughey, H. Kuhnlein, G. Egeland, and E. Loring. 2006. Food security in Nunavut, Canada: Barriers and recommendations. *International Journal of Circumpolar Health* 65(5): 416–431.

Cormier, S. and D. Raheem. 2018. Food security in the Barents region. In K. Hossain and D. Cambou (eds.), *Society, environment and human security in the Arctic Barents Region*, 93–104. (Routledge explorations in environmental studies). Abingdon; New York.

Derqui, B., T. Fayos, and V. Fernandez. 2016. Towards a more sustainable food supply chain: Opening up invisible waste in food service. *Sustainability* 8(7): 693.

Dey, B.L., S. Alwi, F. Yamoah, S. Agyepong, H. Kizgin, and M. Sarma. 2019. Towards a framework for understanding ethnic consumers' acculturation strategies in a multicultural environment: A food consumption perspective. *International Marketing Review* 36(5): 771–804.

Dooley, D.M., E.J. Griffiths, G.S. Gosal, P.L. Buttigieg, R. Hoehndorf, M.C. Lange, … and W.W. Hsiao. 2018. FoodOn: A harmonised food ontology to increase global food traceability, quality control and data integration. *Science of Food* 2(1): 23.

Duhaime, G. and A. Godmaire. 2000. The conditions of sustainable food security: An integrated conceptual framework. *Sustainable food security in the Arctic: State of Knowledge,* 15–46. Groupe d'études inuit et circumpolaires (GÉTIC), Université Laval. Edmonton: CCI Press, Canada.

Edwards, J. 2017. *Developing a taste for big data.* HPE October 2017. Big data in the food industry: Lessons for leaders. Hewlett Packard Enterprise e-Newsletter. Retrieved from https://www.hpe.com/us/en/insights/articles/developing-a-taste-for-big-data-1710.html. Accessed 10 September, 2019.

Engström R, A. Wadeskog, and G. Finnveden. 2007. Environmental assessment of Swedish agriculture. *Ecological Economics* 60: 550–563.

Fälström, P. and J. Jorgensen. 2015. Internet of Food. When food gets networked, life changes. Retrieved 12 September 2018 from http://internet-of-food.org.

FAO. 2008. *Climate change and food security: A framework document.* Food and Agriculture Organisation of the United Nations (FAO), Rome.

FAO 2013. Indigenous Peoples' food systems and well-being: Interventions and policies for health communities. Centre for Indigenous Peoples' Nutrition and Environment. Food and Agriculture Organisation of the United Nations (FAO), Rome, Italy. Retrieved from http://www.fao.org/docrep/018/i3144e/i3144e.pdf.

FAO, 2017. *The future of food and agriculture: Trends and challenges.* Food and Agriculture Organisation, Rome. Retrieved from http://www.fao.org/3/a-i6881e.pdf.

Ford, J.D. 2006. Sensitivity of hunters to hazards associated with climate change: Iglulingmiut perspectives. In J. Oakes and R. Riewe (eds.), *Climate change: Linking traditional and scientific knowledge*, 202–235. Winnipeg: Aboriginal Issues Press.

Ford, J.D. 2009. Dangerous climate change and the importance of adaptation for the Arctic's Inuit population. *Environmental Research Letters* 4(2): 024006.

Ford, J.D., G. McDowell, and J. Jones. 2014. The state of climate change adaptation in the Arctic. *Environmental Research Letters* 9(10): 104005.

Ford, J.D. and T. Pearce. 2010. What we know, do not know, and need to know about climate change vulnerability in the western Canadian Arctic: A systematic literature review. *Environmental Research Letters* 5(1): 014008.

Godfray, H. J. C., I. R. Crute, L. Haddad, D. Lawrence, J.F. Muir, N. Nisbett, J. Pretty, S. Robinson, C. Toulmin, and R. Whiteley. 2010. The future of the global food system. *Philosophical Transactions of the Royal Society B* 365: 2769–2777.

Hallström, E., Q. Gee, P. Scarborough, and D.A. Cleveland. 2017. A healthier US diet could reduce greenhouse gas emissions from both the food and health care systems. *Climatic Change* 142(1–2): 199–212. doi: 10.1007/s10584-017-1912-5.

Hautamäki, A., J. Leppänen, R. Mokka, and A. Neuvonen. 2017. *From pause to play: Work and income in the next era.* Helsinki, Finland: The Finnish Innovation Fund Sitra. Retrieved 12 January 2019 from https://www.sitra.fi/en/publications/from-pause-to-play/#sources.

Iarovyi, S., W.M. Mohammed, A. Lobov, B. Ramis Ferrer, and J.L. Martinez Lastra. 2016. Cyber-physical systems for open-knowledge-driven manufacturing execution systems. *Proceedings of the IEEE* 104(5): 1142–1154.

ICRH 2018. The International Centre for Reindeer Husbandry. http://reindeerherding.org/herders/sami-norway/. Accessed 23 May 2019.

i-SCOOP. 2018. Industry 4.0: The fourth industrial revolution – Guide to Industry 4.0. BVBA, Belgium. Retrieved 10 December 2018 from https://www.i-scoop.eu/industry-4-0/.

Koistinen, K., V. Uusitalo, and A. Huostila. 2017. Systematic approach to sustainability of novel internet-based system for food logistics. *Logistics and Sustainable Transport* 8(1): 50–61.

Kuokkanen, A., V. Uusitalo, and K. Koistinen. 2018. A framework of disruptive sustainable innovation: An example of the Finnish food system. *Technology Analysis and Strategic Management* 31(7): 749–764.

Lambden, J., Receveur, O. and Kuhnlein, H.V. 2007. Traditional food attributes must be included in studies of food security in the Canadian Arctic. *International Journal of Circumpolar Health* 66(4): 308–319.

Lange, M. 2017. A language for legumes: Can the Internet of Food help us know what we eat? *Digital Trends*. Retrieved 12 January 2019 from https://www.digitaltrends.com/home/the-internet-of-food-dr-matthew-lange/.

Loboda, T.V. 2014. Adaptation strategies to climate change in the Arctic: A global patchwork of reactive community-scale initiatives. *Environmental Research Letters* 9(11): 111006.

Loring, P.A. and S.C. Gerlach. 2009. Food, culture, and human health in Alaska: An integrative health approach to food security. *Environmental Scientific Policy* 12(4): 466–478.

Mäkeläinen, A., H. Saari, I. Hippi, J. Sarkeala, and J. Soukkamäki. 2013. 2D hyperspectral frame imager camera data in photogrammetric mosaicking. *ISPRS – International Archives of the Photogrammetry, Remote Sensing and Spatial Information Sciences* 2: 263–267.

Müller-Wille, L. et al. 2008. Community viability and well-being in northernmost Europe: Social change and cultural encounters, sustainable development and food security in Finland's North. *International Journal of Business and Globalisation* 2: 4.

Natcher, D. 2018. Rethinking Canada's northern food systems: A basis for achieving zero hunger. In *Sustainable development goals: Shaping the future of the Arctic*, 16–18. The Circle, WWF Arctic Program. Retrieved from https://arcticwwf.org/newsroom/the-circle/sustainable-development-goals/rethinking-canadas-northern-food-systems/. Accessed 1 August 2018.

Nuttall, M. 1992. *Arctic homeland: Kinship, community and development in northwest Greenland*. Toronto: University of Toronto Press.

Nuttall, M., F. Berkes, B. Forbes, G. Kofinas, T. Vlassova, and G. Wenzel. 2005. Hunting, herding, fishing and gathering: Indigenous Peoples and renewable resource use in the Arctic. In *Arctic climate impact assessment*, 649–690. Cambridge: Cambridge University Press.

Pearson, B., A. Mills, M. Tucker, S. Gao, L. McLandsborough, and L. He. 2018. Rationalising and advancing the 3-MPBA SERS sandwich assay for rapid detection of bacteria in environmental and food matrices. *Food Microbiology* 72: 89–97.

Poppe, K.J., J. Wolfert, C.N. Verdouw, and A. Renwick. 2015. A European perspective on the economics of Big Data. *Farm Policy Journal* 12(1): 11–19.

Prause, G. and I. Boevsky. 2015. Smart rural development. *Agricultural Economics and Management* 60(4): 63–69.

Prowse, T.D, C. Furgal, R. Chouinard, H. Melling, D. Milburn, and S.L. Smith. 2009. Implications of climate change for economic development in northern Canada: Energy, resource, and transportation sectors. *Ambio* 38: 272–281.

Raheem, D., M. Shishaev, and V. Dikovitsky. 2019. Food system digitalisation as a means to promote food and nutrition security in the Barents Region. *Agriculture* 9: 168. doi: 10.3390/agriculture9080168.

Ramis Ferrer, B., O.S. Afolaranmi, and J.L. Martinez Lastra. 2017. Principles and risk assessment of managing distributed ontologies hosted by embedded devices for controlling industrial systems. IECON 2017 – 43rd Annual Conference on IEEE Industrial Electronics Society.

Ramis Ferrer, B., L. Gonzalez, S. Iarovyi, A. Lobov, J.L. Martinez Lastra, V. Vyatkin, and W. Dai. 2014. Knowledge-based web service integration for industrial automation. 12th IEEE International Conference on Industrial Informatics (INDIN*)*, 733–739.

Ramis Ferrer, B. and J.L. Martinez Lastra. 2017. Private local automation clouds built by CPS: Potential and challenges for distributed reasoning. *Advanced Engineering Informatics* 32: 113–125.

Salonen, A.O., J. Siirilä, and M. Valtonen. 2018. Sustainable living in Finland: Combating climate change in everyday life. *Sustainability* 10(1): 104.

Sentryo. 2017. The four industrial revolutions. Retrieved 26 September 2018 from https://www.sentryo.net/the-4-industrial-revolutions/.

Smartcart. 2019. Innovative personal shopping assistant. Retrieved from https://www.smartcart.fi/.

Sowinski, L. 2012. The global food supply chain. Retrieved from https://www.foodlogistics.com/technology/article/10657347/the-global-food-supply-chain.

Su, B. 2017. The evolution of consumer behaviour in the digital age: Three major shifts in marketing paradigms in the 21st century and where we are going next. Analytics for Humans. Retrieved 9 December 2018 from https://medium.com/analytics-for-humans/the-evolution-of-consumer-behavior-in-the-digital-age-917a93c15888.

Sundmaeker, H., C. Verdouw, S. Wolfert, and L. Pérez Freire. 2016. Internet of food and farm 2020. In O. Vermesan and P. Friess (eds.), *Digitising the industry Internet of Things connecting physical, digital and virtual worlds*, 129–151. Delft, Netherlands: River Publishers.

UNHCHR. 2010. *The right to adequate food.* Fact Sheet No. 34. Retrieved from http://www.ohchr.org/Documents/Publications/FactSheet34en.pdf. Accessed 18 July 2019.

Valsdottir, T. 2016, 13 October. Artisan/local food in the West Nordic countries. Finnish Championship in Food Craftmanship. Oral presentation, Novia University of Applied Sciences in Ekenäs, Raseborg, Finland.

Wang, N., N. Zhang, and M. Wang. 2006. Wireless sensors in agriculture and food industry – Recent development and future perspective. *Computers and Electronics in Agriculture* 50(1): 1–14. doi: 10.1016/j.compag.2005.09.003.

Wara, A.A. and S. Dugga. 2014. Enhancing user experience using Mobile QR-code application. *International Journal of Computer and Information Technology* 3(6): 1310–1315.

WEF. 2018. World Economic Forum Report. Innovation with a purpose: The role of technology innovation in accelerating food systems transformation. Retrieved from http://www3.weforum.org/docs/WEF_Innovation_with_a_Purpose_VF-reduced.pdf. Accessed 21 July 2019.

WHO. 2018. InFoSan activity report 2016/2017. 2018. Geneva: World Health Organisation and Food and Agriculture Organisation of the United Nations.

Xu, X., 2012. From cloud computing to cloud manufacturing. *Robotics and computer-integrated manufacturing* 28(1): 75–86.

Zheng, P., Z. Sang, R.Y. Zhong, Y. Liu, C. Liu, K. Mubarok, S. Yu, and X. Xu. 2018. Smart manufacturing systems for Industry 4.0: Conceptual framework, scenarios, and future perspectives. *Frontiers of Mechanical Engineering* 13(2): 137–150.

13 Food security and fertiliser supply

The role of Arctic deposits[1]

Hinnerk Gnutzmann and Piotr Śpiewanowski

Introduction

In 2008, during the most recent food crisis, food commodity prices rose by an average of 45% within a few months. This drastically reduced the real incomes of the world's poor, who spend approximately half of their income on food, and led to food insecurity and social instability in the developing world (Timmer 2010). While food prices have fallen since then, food prices continue to be volatile, and finding policy responses to agricultural price spikes remains an important challenge for food security research (see Grote 2014 for a survey).

This chapter addresses the cost-push causes of high food prices and the potential of mineral resources in the Arctic to mitigate these negative impacts. Since the second half of the twentieth century, agricultural yields have increased greatly due to the Green Revolution; drastically increased use of fertilisers is a central tenet of this revolution. It is estimated that up to 50% of the world's agricultural yield can be attributed to fertilisers (Stewart et al. 2005). Thus, securing the availability of affordable fertiliser has become a key piece of the food security puzzle (Gnutzmann and Śpiewanowski 2015).

Fertiliser consists of three nutrients: potassium, phosphorus, and nitrogen; they must be used in conjunction for optimal results. The first two nutrients are found in minerals. Thus, access to reserves is an important concern for the supply of these fertilisers. Producing the third major nutrient depends largely on energy, so the issues are different. According to Liebig's law in biology, it is not possible to substitute one nutrient for another: plant growth is limited by the most deficient nutrient. Thus, to ensure food security, supply of all three nutrients must be assured.

For mineral fertilisers, the risks to supply security are less related to the depletion of deposits than to supply interruptions, which may occur due to the concentration of deposits in small, specific regions. The bulk of world potassium production comes from a few mines in just three countries; this is discussed in more detail below. A single country – Morocco – holds over 70% of global phosphorus reserves. Thus, other reserves of these minerals outside the primary production locations can significantly boost resilience of supply. Arctic phosphorus deposits may play an important role in this.

DOI: 10.4324/9781003057758-13

While this chapter focuses on the role of fertiliser, particularly phosphates, in assuring food security, it is important to acknowledge the broader economic and environmental issues associated with heavy fertiliser use. First, beyond its downstream effects on food production, the fertiliser industry itself has considerable economic importance. It is estimated that global fertiliser sales are valued at more than 190 billion USD per annum (Soil Week 2016). Moreover, mining for fertiliser nutrients can have serious local environmental impacts (UNEP 2001). In addition, agricultural use of fertilisers can lead to environmental degradation; this is discussed briefly below.

This chapter first reviews the connection between food security and fertiliser supply. It then examines mineral fertilisers and describes the global reserves of phosphate rock (phosphates), and potash (potassium). These minerals are geographically highly concentrated. Next the chapter reviews the available phosphate reserves in the Arctic region in detail. Finally, it discusses the potential contribution of Arctic phosphates to food security, with a particular focus on EU food security.

Fertiliser supply and food security

Fertiliser is essential to sustaining crop yields. Stewart et al. (2005) conducted a meta-study of previous long-term studies of crop yield responses to fertiliser application. They found that, conservatively estimated, at least 30%–50% of crop yield is attributable to fertiliser application. Therefore, continued large-scale application of fertiliser is required to sustain a growing world population. Encouragingly, there is also evidence that the effectiveness of fertiliser has increased over time, with some studies finding a 35% increase in the United States from 1980 to 2000. Stewart et al. argue that this improvement can be attributed to several factors, including better farm and nutrient management as well as changes in crop genetics.

The intensive use of fertiliser is reflected in its cost share. The U.S. Department of Agriculture collects data on commodity costs and returns that demonstrate this point. For major crops, fertiliser accounts for 20% (for cotton) to more than 40% (for corn) of operating costs.

A time series analysis suggests that fertiliser price changes are passed through to agricultural crop prices to a very high degree. This means that an increase in fertiliser costs leads directly to a cost-push increase in food commodity prices. To test this hypothesis, we estimate a time series model in Gnutzmann and Śpiewanowski (2015). The goal of our model is to isolate the cost contributions of fertiliser and energy to food prices, using annual price indices provided by the World Bank's Global Economic Monitor. Following the literature, standard controls for time trends and manufacturing cost inflation are used in the model. The structure of the model allows us to estimate the cost shares of individual production factors from the available price data. We found that fertiliser accounts for 42% of cost in the World Bank's general food price index.

These findings are consistent with a competitive market for agricultural production. In a competitive market, prices reflect marginal cost. Thus, any changes in marginal cost – as proxied by operating cost – should lead to a one-for-one increase in commodity prices. For crops, this implies that the cost push of fertiliser should be large, and the empirical results indeed confirm this.

Fertiliser supply may be constrained in the short run by disruptions or in the long run by the depletion of mineral resources. It is important to distinguish these two causes because the needed policy responses differ significantly. Therefore, the next section will discuss the role of long-term supply and deposits; this section will focus on the short term. Here, again, a nutrient-specific approach is necessary. Nitrogen prices closely track natural gas prices (Gnutzmann and Śpiewanowski 2019). Thus, although there is significant price variation in the market, supply-side policies to stabilise nitrogen prices are unlikely to impact the factors underlying nitrogen prices as nitrogen production constitutes only 3% of global natural gas demand.

On the mineral side, supply shocks are more likely to be due to imperfect competition or the formation of cartels. Taylor and Moss (2013) provide strong evidence for the importance of imperfect competition in fertiliser markets. Just a few potash mines produce the bulk of world output. Furthermore, on the potash market, these few producers form a legal cartel that sets production quotas that hold global prices well above the competitive equilibrium (Gnutzmann, Kowalewski and Śpiewanowski. Forthcoming). Phosphate rock also has a long history of cartel episodes.

Since fertiliser markets are globally integrated, individual regions may be large net importers of nutrients. This dependence on foreign imports requires close attention in a study of regional fertiliser supply security. Indeed, 88% of phosphate rock used in the EU is imported (mostly from Morocco, Russia, Syria, and Algeria). There are no substitutes for phosphorus, and it is rarely recycled (European Commission 2017). The EU has acknowledged the importance of the stability of phosphate rock supplies for the European economy; the EU included phosphate, along with nineteen other minerals, on a 2014 list of critical raw materials. The EU relies much less on imported potash. Thanks to mining operations in Germany and to a smaller degree in Spain and the United Kingdom, the EU imports only 23% of potash used (European Commission 2018).

In conclusion, there is robust empirical evidence that fertiliser price shocks pass through to food prices in the short term. Thus, fertilisers are an important piece of the food security puzzle. Of the three nutrients in fertiliser, phosphate rock is the most important to European food security.

Global mineral fertiliser reserves

Globally natural reserves are not under an imminent threat of depletion. Nitrogen is the most abundant gas in the atmosphere; using currently available technology, natural gas is required for nitrogen fixation, but the use of alternative energy sources – including renewable sources such as hydro or solar energy – is already technically feasible with current technology. The data in the U.S. Geological Survey (Appendix, Table 13A.1) allow us to identify the key stylised facts of fertiliser reserves. At current rates of consumption, existing proven reserves of potassium should last for more than 70 years. Current consumption of phosphate rock could be sustained for more than 200 years at current rates. Moreover, in spite of ongoing exploitation, known reserves for these minerals have not declined in the past two decades, because new deposits were discovered at the same rate or even faster than existing resources were depleted. Therefore, compared to other natural resources, such as oil, the depletion of mineral fertiliser resources is not an imminent concern. This is reassuring news for food security, even with the more realistic assumption that continued population growth will increase future consumption rates.

The sustainability of mineral fertiliser reserves has been in doubt quite recently. Global interest in phosphorus peaked in the late 2000s when a fivefold increase in phosphate rock prices led to concerns about depletion: Perhaps the world had reached 'peak phosphorus' and needed to cope with dwindling availability of fertiliser in the future. The Global Phosphorus Research Initiative, led by Swedish and Australian scientists, predicted in a widely distributed paper that phosphorus production would reach its peak around 2030 and that global reserves would be depleted in 50–100 years (Cordell et al. 2009). However, soon after that publication, the U.S. Geological Survey updated its phosphate rock reserves estimates, from 15 billion tonnes in 2009 to 70 billion tonnes in 2018. This significantly postponed the threat of phosphate rock depletion.

The risk of resource depletion could be further decreased by an increase in recycling and resource use efficiency. However, despite recent advancements in precision agriculture (soil testing and remote sensing), phosphorus consumption in the EU has increased by around 30% since 2009 (IFADATA 2019); thus, more needs to be done in this area. At the same time, phosphate recycling rates remained at the same, very low, level. Excess use of fertilisers, especially phosphates, leads to eutrophication, a process that is especially dangerous in the vulnerable Arctic ecosystem.

The more immediate threats, however, stem from supply concentration. In total, 84% of potash reserves are held by the five countries with largest reserves. The phosphate rock supply is even more concentrated: Morocco alone accounts for more than 70% of known reserves; the country with the second highest reserves – China – accounts for less than 5%.

This concentration presents multiple challenges for resource security. First, the large producer countries are typically not large consumer countries; thus, fertiliser supply relies on international trade. International crises or conflicts that disrupt transportation networks could thus have drastic

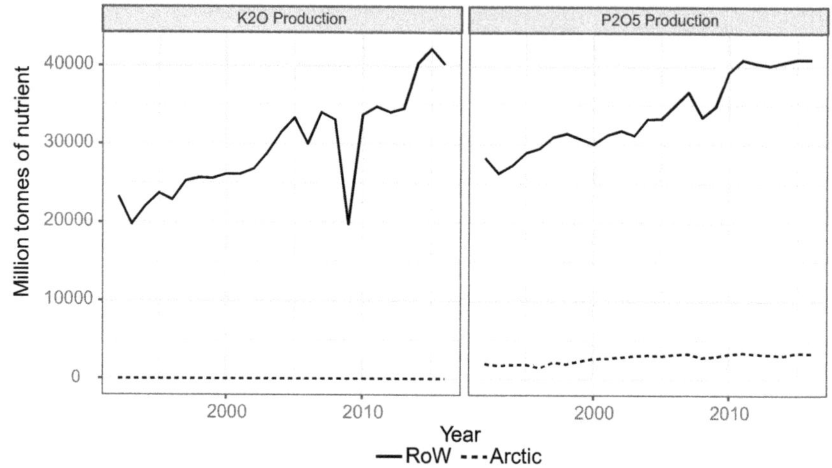

Figure 13.1 Phosphate rock and potash production, globally and in the Arctic, since 1991.

Source: IFA DATA; RoW means the Rest of the World.

effects on fertiliser availability in population centres. Second, this concentration can foster the formation of commodity cartels. Such cartels raise commodity prices in expectation of organic price increases. However, they can also make commodity prices more volatile as different cartels form and break down in turn.

While Arctic reserves of phosphate rock are a small share of global reserves and production (see Figure 13.1), they are outside the zones of concentration. For this reason, they could provide a valuable buffer in case of supply interruptions. Hence, the next section will discuss the potential for phosphate rock production in the Arctic region in more detail.

Arctic fertiliser resources

Since the late 1920s, the Arctic has been an important source of phosphate rock. Production in the then-Soviet part of the Kola Peninsula that started in 1929 has continued until today. Currently, the Arctic produces 13.5 million tonnes of phosphate rock – more than 5% of annual world production. Arctic phosphate rock production is worth around $100/t at current prices equalling nearly $1.5 billion per year.

Most potash deposits in the Arctic are of magmatic origin; the vast majority of deposits in other regions are sedimentary. The origin of potash can be determined by the amount of cadmium, a toxic metal, in the rock. Phytoplankton absorbs cadmium in surface waters, so sedimentary rocks formed from water debris naturally contain higher amounts of cadmium. The magmatic deposits in the Arctic have the lowest cadmium of any phosphate rock in the world (see Oosterhuis et al. 2000 for a review of various studies).

Magmatic reserves make up only 4% of the total phosphate rock reserves around the world, but around 50% of those deposits are located in the Arctic (Van Kauwenbergh 2010). A large share of world phosphate rock production would not be allowed to enter the European market under the new rule, as we discuss below.

The majority of phosphate rock deposits in the Arctic, as shown in Figure 13.2, are shared between Russia and Finland in Fennoscandia. Other Arctic regions have been less generously endowed. In Sweden, phosphorus occurs in apatite iron ore, but the process to recover phosphorus from iron ore mining waste is under development. Some phosphate rock deposits have been found in Norway, Greenland, Canada, and Alaska, though exploration has been limited at best. However, production of phosphate rock in the Arctic region may increase in the coming years thanks to an increase in exploration in search of rare earth elements (REE). REE-rich deposits in Greenland, Alaska, and Finland also contain substantial amounts of phosphates, which could be mined as a by-product.

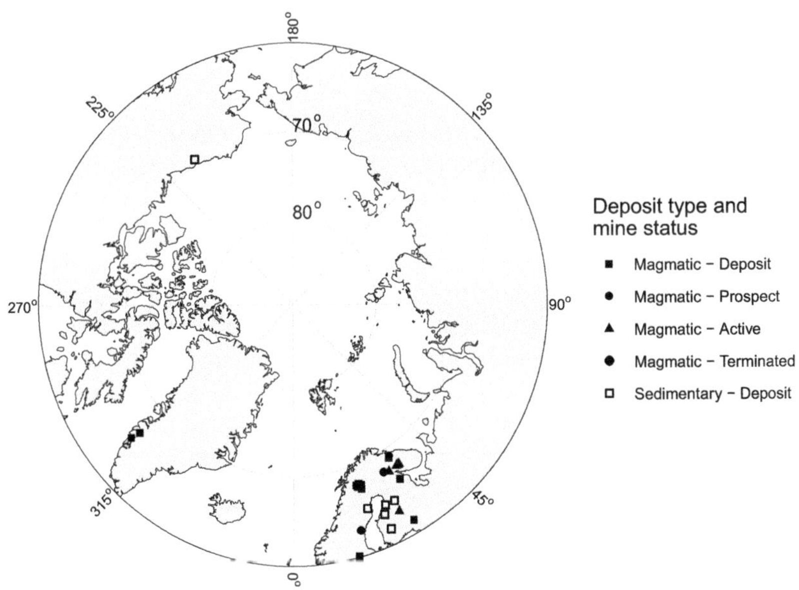

Figure 13.2 Phosphate rock mines and deposits in the Arctic (north of 60°N). Source: Orris and Chernoff (2002) and Vihtakari (2018).

Russia

Russia has an annual production of 12.5 million metric tons (MMT) of phosphate rock, making it the world's fourth largest phosphate rock producer (almost 5% of global output) (USGS 2018) and the world's largest producer of phosphate rock of magmatic origin. Over 75% of Russian phosphate rock production is exported to other countries (IFADATA 2019). The entire Russian production is located in the Arctic, and reserves are estimated at 700 MMT.

The largest explored phosphate rock deposit is located at Khibny Alkaline Igneous Complex and was discovered in the late nineteenth century. The deposit is located north of the town of Kirovsk, about 160 km south of the Arctic port of Murmansk. Production in Khibny began in 1929, and currently there are five mines and two beneficiation plants (Phosagro 2019). The mining complex is owned and mined by Apatit, a subsidiary of the Russian fertiliser producer Phosagro, and this operation has a capacity of 10 MMT per year. The resource base contains significant reserves of rare earth elements (more than 41% of all Russia's reserves).

In the late 2000s, high phosphate rock prices inspired the development of new phosphate rock mining projects in the Kola Peninsula. CJSC North-West Phosphorus, a subsidiary of the Russian fertiliser company Acron, opened a new, open-pit mine at Oleniy Ruchey, adjacent to the Vostochny mine of the Khibny Complex. State licenses to develop the ore deposits were obtained in 2006. In 2012, the first stage of the mine was completed, and production started. Currently, the mine produces approximately 1.8 MMT of apatite concentrate per annum (North-Western Phosphorus Company 2019). Another project in the area, Sordinsky mine, which is at the Vyatsko-Kamskoe deposit, was being developed by Uralchem. The company acquired a licence to mine phosphate rock from Russia in 2007 and planned a mine. However, plans to develop a mine of 0.5 million t/y ceased in 2010 after a slump in phosphate rock prices.

The only other operational phosphate rock mines in Russia are the igneous deposits at Kovdor Complex, located close to the Finnish border and some 150 km WSW of the Khibny Complex. The complex is primarily a source of iron ore (magnetite) and has been mined using open-pit methods since 1962. The mine is operated by Kovdorskiy GOK, a subsidiary of Eurochem. It has an adjacent beneficiation plant and produces 2.5 MMT of apatite concentrate per year, along with substantial amounts of iron ore and baddeleyite as co-products (Eurochem 2019).

Until 2006, Russia also mined phosphate rock deposits at Kingisepp, which is approximately 110 km south-west of St Petersburg at 59°10' N, 29° E. Unlike the other two deposits described here, the Kingiseppp deposits are sedimentary and are essentially a continuation of those along the Estonian coast. The mine is owned by Eurochem and produced only 0.3 MMT phosphate rock per year before it was closed.

The Russian Arctic also hosts unexplored phosphate rock deposits at Elisenvaara, Tikshozero, and in the Gremyakha-Vyrmes Alkaline Complex. These unexplored resources are estimated to contain 500 million tonnes of phosphate rock. However, the low phosphorus content of the ore makes it unlikely that mining operations will start in the very near future.

Finland

Finland is the second largest phosphate rock producer in the Arctic Region and the only country in Western Europe that produces phosphate rock. The country's entire production of approximately 0.95 million t/y comes from an open-pit mine at Siilinjärvi in Eastern Finland Province, 25 km north of Kuopio. Phosphate rock production in Finland covers approximately double the local demand for that fertiliser (IFADATA 2019). The mine has been operated by Yara International since 2007, when the company acquired Kemira Growhow, the previous owner of the mine. The rich magmatic deposits contain over 1500 million tonnes of phosphate rocks (Ahokas 2015). The mine also has some potential for REE recovery. Apatite concentrates from Siilinjärvi indicate REE contents between 3000 and 4000 ppm (O'Brien et al. 2015). The ore deposit was found in 1950, it was decided to establish a mine and phosphoric acid production facilities in 1967, and full-scale production started in 1979 (Yara Suomi 2019). Since then, over 300 million tons have been mined, and 19.5 million tons of apatite concentrate have been produced (Yara 2013). All the rock mined there is transformed into phosphoric acid on site.

Another economically significant phosphate rock deposit is located near Sokli, 15 km east of Russian border and 60 km east of Russia's Kovdor phosphate rock deposits. The Sokli ore reserves of 'soft-rock' phosphate-rich materials consist of about 114 MMT with 15% phosphorus pentoxide (P2O5) content and approximately 75 Mt of weathered bedrock ore with about 5.6% P2O5 (O'Brien and Hyvönen 2015).

The Sokli deposit was discovered in 1967. At the end of 1970s, and again in the late 1980s, deposit owners conducted extensive research and engaged in development for production, but these attempts were terminated before production started. The latest attempt to start production was undertaken in the late 2000s. Phosphate rock production was supposed to start in 2015, but instead the project was cancelled (Siirama 2009)

Finland also has igneous phosphate rock deposits in southern Ostrobothnia (Kärkkäinen and Appelqvist 1999; Ahokas 2015). Furthermore, more than twenty occurrences of uraniferous phosphorite and phosphatic rocks have been identified in Finland (Pohjolainen 2015). However, the phosphorus content of those deposits is lower than Finland's average, so these reserves will probably not be commercially utilised in the near future (Ahokas 2015).

Sweden

Sweden currently produces no phosphate rock. However, it has substantial P2O5 resources in the apatite iron ores located mostly in the Norrbotten region of Northern Sweden. Sweden's iron ore reserves are one of the world's largest and amount to 3.5 billion tonnes or 2% of the global total (USGS 2018). Apatite iron ores contain an average of 1%–5% P, and multiple attempts have been made to recover apatite as a by-product. Between 1985 and 1988, about 0.4 MMT apatite concentrates were produced via flotation from Kiirunavaara ore (Pålsson et al. 2014). Sweden may soon restore its phosphate rock production. The state-owned mining company LKAB, the largest iron ore producer in Sweden, is piloting the process of upgrading mine waste from iron ore production to allow phosphorus recovery at one of its iron ore mines. With this method, LKAB can produce phosphorus without heavy metals that equals more than five times Sweden's annual demand – approximately 0.5 MMT/year of phosphate rock (LKAB 2018).

Norway

Despite Yara's long-standing production of compound fertilisers in Norway, no phosphate rock is produced in Norway. The country's overall potential is quite limited as the iron ore and phosphate deposits in the northern part of the country are rather narrow and low grade.

Recently, the Kodal deposit, which is 100 km south of Oslo and contains ore reserves calculated at 70 MMt with 4.9 P2O5 content (Orris and Chernoff 2002), has been explored as a potential prospect for a combined operation extracting phosphate and REEs. However, this project has not reached the construction stage.

Arctic Canada, Alaska, Greenland, and Iceland

No phosphate production has ever taken place, nor has any exploration concession been licenced in these regions. Individual phosphate deposits are probably too small to be economically viable. However, in the future, small-scale phosphate production may be possible as a by-product of REE mining. The most promising prospect for this is the Sarfatoq deposit in Greenland. It is estimated at 1,000 million tonnes with an average of 3.5% P2O5 content (Orris and Chernoff 2002).

Conclusion: what can Arctic phosphate rock deposits contribute?

Today, the Arctic is a small but reliable producer of phosphate rock. There is some potential for increased production, especially in the Arctic EU countries. However, Arctic reserves of phosphate make up a vanishingly small share of global reserves. Moreover, the difficult climatic conditions, the complex geology, and the need to protect the Arctic environment all justify caution before exploiting Arctic phosphorus.

In the Arctic, Indigenous People's perspectives must also be taken into consideration in the expansion of any extractive industry. In Finland, Russia, and the European Arctic, phosphate rock reserves are located in the traditional homelands of the Sami people. These Indigenous residents have significant needs for land that compete with such extraction; their food security relies heavily on the production of traditional foods, as described in other chapters of this book (Bogdanova et al., Chapter 4 and Nilsson, Chapter 11).

Arctic phosphorus deposits may have some unique advantages from a food security perspective. First, since they lie outside the existing zones of concentration, mining in the Arctic could increase competition in the phosphate industry, leading to lower fertiliser and food prices. Second, today, the EU depends almost entirely on imported phosphate fertiliser. Given current knowledge of world reserves, increasing Arctic production is probably the only way to reduce the EU's dependence on phosphate imports.

The oil market demonstrates the value of an elastic, competitive segment in market supply, even if the production cost of that segment is relatively high. Fracking is expensive, and depending on oil prices, wells are turned on or off. This effectively puts a ceiling on the market, which improves the security of oil supply. A similar solution may be needed for fertiliser, and the Arctic could provide such a ceiling for phosphorus.

Arctic phosphate is also more environmentally friendly than phosphate from other regions during consumption. Sedimentary phosphate deposits naturally contain cadmium, a potentially harmful substance, that can accumulate in the soil. In contrast, the cadmium content of the magmatic phosphate deposits in the Arctic and Brazil is negligible. Health concerns have led the European Commission to propose cadmium content limits on fertilisers. After a lengthy debate weighing health concerns and the opinions of various interest groups, the EU has proposed a limit of 60 mg/kg P2O5, though limits as low as 20 mg were discussed. The preliminary agreement still has to pass a number of legislative stages, including a vote by the EU Parliament and approval by the EU Council of Ministers (European Parliament 2018).

This seemingly innocuous piece of legislation has significant implications for fertiliser production. Cadmium is not evenly distributed across phosphate deposits, and there is currently no economically viable technology for extracting cadmium from phosphates. Given the current import structure, only around 8% of the EU's phosphate fertilisers would exceed the 60 mg limit (Toplensky and Foy 2017). However, if the limit were cut to 20 mg (as originally proposed), only phosphates from the magmatic deposits from Russia, Finland, and Brazil and some of the sedimentary deposits in Florida, Israel, and Jordan would be allowed in the European market (Oosterhuis et al. 2000; Roberts 2014). In such a case, Europe's food security would rely even more heavily on Arctic phosphate resources.

Note

1 We thank the seminar participants in Stockholm and Umea for their insightful comments. The authors also gratefully acknowledge financial support from the DFG 'Beethoven' grant FINEXCA (Gnutzmann) and the Narodowe Centrum Nauki (NCN) 'Beethoven' grant (UMO 2014/15/G/HS4/03629) (Śpiewanowski).
Hinnerk Gnutzmann: hinnerk@gnutzmann.info.
Piotr Śpiewanowski: Piotr.spiewanowski@inepan.waw.pl, ORCID: 0000-0001-9812-4674.

References

Ahokas, L. 2015. Finland's phosphorus resources are more important than ever. *Geological Survey of Finland* February. Retrieved from http://verkkolehti.geofoorumi.fi/en/2015/10/finlands-phosphorus-resources-are-more-important-than-ever/.

Chernoff, C.B., G.J. Orris. 2002. Data set of world phosphate mines, deposits, and occurrences—Part A. Geologic Data; Part B. Location and Mineral Economic Data: U.S. Geological Survey Open-File Report 02-156, https://pubs.usgs.gov/of/2002/0156/.

Cordell, D., J.O. Drangert, and S. White. 2009. The story of phosphorus: Global food security and food for thought. *Global Environmental Change* 19(2): 292–305.

Eurochem. 2019. Retrieved 29 January 2019 from http://www.eurochemgroup.com/en/project/kovdorskiy-gok-expansion/.

European Commission. 2017. Communication from the commission to the European parliament, the council, the European economic and social committee and the committee of the regions on the 2017 list of Critical Raw Materials from the EU, COM (2017) 490. Retrieved from https://eur-lex.europa.eu/legal-content/EN/TXT/?uri=CELEX:52017DC0490.

European Commission. 2018. Raw materials information system – Potash, 323–334. Retrieved from http://rmis.jrc.ec.europa.eu/uploads/rmprofiles/Potash.pdf.

European Parliament. 2018. Fertilisers/cadmium: Parliament and council negotiators reach provisional deal. Retrieved 29 January 2019 from http://www.europarl.europa.eu/news/nl/press-room/20181119IPR19407/fertilisers-cadmium-parliament-and-council-negotiators-reach-provisional-deal.

Gnutzmann, H. and P. Śpiewanowski. 2015. Did the fertiliser cartel cause the food crisis? Retrieved from https://papers.ssrn.com/sol3/papers.cfm?abstract_id=3106757 https://papers.ssrn.com/sol3/papers.cfm?abstract_id=2534753.

Gnutzmann, H. and P. Śpiewanowski. 2019. Pass-through and international linkages. Retrieved from http://www.fleximeets.com/eaere2019/getpaper.php?fid=1513

Gnutzmann, H., O. Kowalewski, and P. Śpiewanowski. Forthcoming. Market structure and resilience: Evidence from potash mine disasters. *American Journal of Agricultural Economics*. doi: 0.1093/ajae/aaz041.

Grote, U. 2014. Can we improve global food security? A socio-economic and political perspective. *Food Security* 6(2): 187–200.

International Fertiliser Industry Association (IFADATA). 2019. Retrieved 1 October 2019 from http://ifadata.fertiliser.org/ucSearch.aspx.

Kärkkäinen, N. and H. Appelqvist. 1999. Genesis of a low-grade apatite-ilmenite-magnetite deposit in the Kauhajärvi gabbro, western Finland. *Mineralium Deposita* 34(8): 754–769.

LKAB. 2018. LKAB develops new technology for producing strategic minerals from mine waste. Retrieved 29 January 2019 from https://www.lkab.com/en/news-room/press-releases/lkab-develops-new-technology-for-producing-strategic-minerals-from-mine-waste/.

North-Western Phosphorus Company. 2019. Retrieved 29 January 2019 from https://www.szfk.ru/en/project/steps.

O'Brien, H., E. Heilimo, and P. Heino. 2015. The Archean Siilinjärvi carbonatite complex. In W.D. Maier, R. Lahtinen, and H. O'Brien (eds.), *Mineral deposits of Finland*, 327–343. Elsevier.

O'Brien, H and E. Hyvönen. 2015. The Sokli carbonatite complex. In W.D. Maier, R. Lahtinen, and H. O'Brien (eds.). *Mineral deposits of Finland*, 305–325. Elsevier.

Oosterhuis, F.H., F.M. Brouwer, and H.J. Wijnants. 2000. A possible EU wide charge on cadmium in phosphate fertilisers: Economic and environmental implications. *Final Report for the European Commission*. Retrieved from http://ec.europa.eu/environment/enveco/taxation/pdf/cadium.pdf.

Orris, G.J. and C.B. Chernoff. 2002. *Data set of world phosphate mines, deposits, and occurrences: Part B. Location and mineral economic data*. U.S. Department of the Interior, US Geological Survey. Retrieved 29 January 2019 from https://pubs.usgs.gov/of/2002/0156/pdf/OF02-156B.pdf.

Pålsson, B.I., O. Martinsson, C. Wanhainen, and A. Fredriksson. 2014. Unlocking rare earth elements from European apatite-iron ores. In *Proceedings of the First European Rare Earth Resources Conference (ERES2014)*, 211–220.

Phosagro. 2019. Retrieved 29 January 2019 from https://www.phosagro.com/about/holding/item636.php.

Pohjolainen, E. 2015. Uranium deposits of Finland. In *Mineral deposits of Finland*. doi: 10.1016/B978-0-12–410438-9.00025-X.

Roberts, T.L. 2014. Cadmium and phosphorus fertilisers: The issues and the science. *Procedia Engineering* 83: 52–59.

Siirama, L. 2009. The Sokli Mine project. Retrieved from http://fem.lappi.fi/c/document_library/get_file?folderId=48927andname=DLFE-2796.pdf.

Soil Week. 2016. The fertiliser industry: Plant food in a bag, firms with a common cause. Retrieved 10 February 2019 from https://gsf.globalsoilweek.org/wp-content/uploads/2016/04/THE-FERTILISER-INDUSTRY.pdf.

Stewart, W.M., D.W. Dibb, A.E. Johnston, and T.J. Smyth. 2005. The contribution of commercial fertiliser nutrients to food production. *Agronomy Journal* 97(1): 1–6.

Taylor, C.R. and D.L. Moss. 2013. The fertiliser oligopoly: The case for global antitrust enforcement. AAI Working Paper No. 13–05. The American Antitrust Institute.

Timmer, C.P. 2010. Reflections on food crises past. *Food Policy* 35(1): 1–11.

Toplensky, R. and H. Foy. 2017. Fertiliser mines row leaves EU between a rock and a hard place. *Financial Times*, 12 July 2017. Retrieved 29 January 2019 from https://www.ft.com/content/0ec328aa-6634-11e7-8526-7b38dcaef614.

United Nations Environment Programme. 2001. Environmental aspects of phosphate and potash mining. Retrieved 10 January 2019 from https://wedocs.unep.org/bitstream/handle/20.500.11822/8071/-Environmental%20Aspects%20of%20Phosphate%20and%20Potash%20Mining-20011385.pdf?sequence=2andisAllowed=y.

USGS. 2018. Mineral commodity summaries 2018: U.S. Geological Survey, 122–123. doi: 10.3133/70194932.

Van Kauwenbergh, S.J. 2010. *World phosphate rock reserves and resources*. Muscle Shoals: IFDC.

Vihtakari, M. 2018. PlotSvalbard: PlotSvalbard – Plot research data from Svalbard on maps. R package version 0.7.1. Retrieved from https://github.com/MikkoVihtakari/PlotSvalbard.

Yara. 2013. Siilinjärvi phosphate mine – Fourth decade ongoing. Retrieved from http://fem.lappi.fi/c/document_library/get_file?folderId=1405164andname=DLFE-20769.pdf.

Yara Suomi. 2019. Retrieved 29 January 2019 from https://www.yara.fi/tietoa-yarasta/yara-suomi/toimipaikat/siilinjarvi/.

Appendix

Table 13A.1 Global reserves of mineral fertilisers

(a) Potash	*Reserves (K2O equiv.)*	*Share of global reserves*
Canada	1,000,000	25.64%
Belarus	750,000	19.23%
Russia	500,000	12.82%
China	360,000	9.23%
Israel	270,000	6.92%
Top five share of reserves	73.85%	
Total world reserves	3,900,000	
Remaining years	93	
(b) Phosphate	*Reserves (Mil. Tons)*	*Share of global reserves*
Morocco	50,000,000	71.43%
China	3,300,000	4.71%
Algeria	2,200,000	3.14%
Syria	1,800,000	2.57%
Brazil	1,700,000	2.43%
Top five share of reserves	84.29%	
Total world reserves	70,000,000	
Remaining years	266	

Source: USGS 2018 mineral reports for potash and phosphate rock.

Table 13A.2 Phosphate rock mines and deposits in the Arctic

	Location	*Country*	*Type*	*Production*	*Lat.*	*Long.*	*P2O5 resources*
1	Siilinjarvi	Finland	Magmatic	Active	63.08	27.67	18.6
2	Sokli complex	Finland	Magmatic	Prospect	67.75	29.25	27.15
3	Kainuu area/ Paltamo/ Nuottijärvi	Finland	Sedimentary	Deposit	64.42	27.83	0.1

(Continued)

	Location	*Country*	*Type*	*Production*	*Lat.*	*Long.*	*P2O5 resources*
4	Peräpohja area/ Mustamaa	Finland	Sedimentary	Deposit	63.50	24.00	0.0625
5	Pohjanmaa area/Vihanti/ Lampinsaari	Finland	Sedimentary	Deposit	64.42	25.15	0.075
6	Qaqarssuk carbonatite complex	Greenland	Magmatic	Deposit	65.38	−51.67	0.198
7	Sarfartoq carbonatite complex	Greenland	Magmatic	Deposit	66.50	−51.25	35
8	Elisenvaara	Russia	Magmatic	Deposit	61.42	29.77	
9	Tikshozero	Russia	Magmatic	Deposit	66.18	32.03	40
10	Gremyakha-Vyrmes alkaline complex	Russia	Magmatic	Deposit	68.92	32.50	105
11	Khibiny	Russia	Magmatic	Active	67.67	33.22	600
12	Khibiny complex/ Apatitovy Tsirk	Russia	Magmatic	Active	67.75	33.73	75
13	Khibiny complex/ Kukisvumchorr	Russia	Magmatic	Active	67.75	33.70	69.87
14	Khibiny complex/ Rasvumchorr	Russia	Magmatic	Active	67.63	33.80	75.06
15	Khibiny complex/ Vostochny	Russia	Magmatic	Active	67.67	34.17	75.05
16	Khibiny complex/ Yukspor	Russia	Magmatic	Active	67.63	33.75	64.94
17	Kovdor complex (Kovdorskoe)	Russia	Magmatic	Active	67.57	30.40	53
18	Sebl'yavr carbonatite complex	Russia	Magmatic	Prospect	68.97	33.08	37.25
19	Skellefte district/ Boliden	Sweden	Sedimentary	Deposit	64.87	20.38	
20	Grängesberg district	Sweden	Magmatic	Deposit	60.08	14.98	
21	Ekströmsberg	Sweden	Magmatic	Deposit	67.78	19.52	
22	Kirunavaara	Sweden	Magmatic	Deposit	67.83	20.28	
23	Malmberget/ Gallivare	Sweden	Magmatic	Deposit	67.20	20.67	
24	Pattok	Sweden	Magmatic	Deposit	67.57	19.70	
25	Svappavaara area	Sweden	Magmatic	Deposit	67.63	21.00	
26	Tuolluvaara	Sweden	Magmatic	Deposit	67.85	20.33	
27	Alnö (Alno Island)	Sweden	Magmatic	Terminated	62.45	17.18	0.5
28	Shublik formation/ Sadlerochit river	United States	Sedimentary	Deposit	69.45	−145.23	

Source: Orris and Chernoff (2002).

14 Community-led initiatives as innovative responses

Shaping the future of food security and food sovereignty in Canada

Thora Martina Herrmann, Philip A. Loring, Tricia Fleming, Shirley Thompson, Annie Lamalice, Marion Macé, Véronique Coxam, Geraldine Laurendeau, and Sylvie Blangy

Introduction

The hunting, harvesting, sharing, processing, storing, and consumption of traditional food in Indigenous communities in Canada's Arctic and Provincial North, have social and cultural benefits, which are key determinants of health (Council of Canadian Academies 2014, 76, 81). Moreover, land use and land-based cultural practices involving traditional foods (e.g., spending time on the land harvesting, preparing, and sharing food with relatives) also improve the nutritional quality of Indigenous People's diets (Kuhnlein and Receveur 2007; Sheehy et al. 2015). These practices, are still key to Indigenous wellbeing and identity, even if a large portion of contemporary northern diets is composed of store-bought food (Cuerrier et al. 2012; Kant et al. 2013; Boulanger-Lapointe et al. 2019).

In recent decades, settler colonialism across the Canadian Arctic and Provincial North including forced sedentarisation and the construction of permanent infrastructures (e.g., airports for bringing food to the North) have profoundly affected aboriginal peoples' ability to meet their nutritional needs and desires in a way that responds to their cultural needs and holistic health, which acknowledges the relationships between people, the food they consume, their culture and lifestyles and nature (Loring and Gerlach 2015; Ford et al. 2016; Delormier et al. 2017; Lamalice et al. submitted; Herrmann et al. 2020).

This shift in diet has reduced the quality of many communities' diets. Many people are consuming more store-bought foods with relatively low nutritional value and fewer nutrient-dense country foods (Huet et al. 2012; Rosol et al. 2016; Sheehy et al. 2013, 1). Schroeder and Smaldone (2015, 274) define food insecurity as 'uncertain ability or inability to procure food, inability to procure enough food, being unable to live a healthy life, and feeling unsatisfied' and identify 'three layers of food insecurity (individual, community,

DOI: 10.4324/9781003057758-14

society)'. In the Canadian Arctic and Provincial North, food insecurity is a chronic problem, particularly for Indigenous communities (Loring and Gerlach 2015). The 2007–2008 Inuit Health Survey of Nunavut, the Inuvialuit Settlement Region, and Nunatsiavut reported that 68.8%, 43.3%, and 45.7% of Inuit households in these communities (respectively) have a high level of food insecurity (Rosol et al. 2011, 488). The dietary transition away from local, nutrient-dense, traditional foods has major implications for the health of remote northern and Arctic communities, such as increases in the rates of obesity and cardiovascular diseases (Willows et al. 2012; Halseth 2015).

Moreover, the northern regions of Canada have become highly dependent on supplies of provisions from the corporate, globally integrated food system (Lamalice, 2019). This has eroded their food sovereignty, which is 'the right of peoples to healthy and culturally appropriate food produced through ecologically sound and sustainable methods, and their right to define their own food and agriculture systems' as defined in the Nyéléni declaration (Vía Campesina 2007). Traditional food systems support food sovereignty (Turner and Turner 2008; Thompson et al. 2011). For Indigenous communities in the North, food sovereignty is central to food security, as the Inuit of Alaska demonstrate: Food sovereignty is represented by the handle of the traditional Inuit *qilaut'* drum, and food security is the drum itself (Inuit Circumpolar Council–Alaska 2015, 12).

To tackle food insecurity and (re)establish food sovereignty in Inuit and First Nations communities, to adapt to the food crisis and to make communities more resilient, an array of community-based food initiatives are arising all over the Canadian North. These initiatives promote access to land-based activities and to local food and knowledge of local food. They also support local food production and seek to reduce inequalities in access to a healthy diet (e.g., Papatsie et al. 2013; Delormier et al. 2017; Kenny et al. 2018; Chen and Natcher 2019). As Delormier et al. (2017, 2) note, '[...] resilient Indigenous Peoples are reclaiming their food security, nutrition, and health by revitalising their food systems in ways that privilege Indigenous worldviews, livelihoods, and governance'.

As Berkes and Ross (2013, 17) express it, 'community resilience is not only about theory; it is equally about practice: How can adaptive capacity, self-organisation and agency be supported and fostered through processes such as community development and community-based planning?' For Ford et al. (2016), such community-based adaptation initiatives require university-affiliated non-Indigenous researchers and practitioners to work 'in close collaboration with communities, in which decision-making is shared and underpinned by frequent dialogue' (Ford et al. 2016, 130). This chapter describes community-based adaptation initiatives centred on food security and food sovereignty. Two questions are of relevance: (1) What factors effectively enable community-based food initiatives to improve food security and food sovereignty? and (2) How do the communities perceive these programs?

To address these questions, this chapter will first present three specific examples from across Canada of community-based and in some cases community-led projects. In these projects, all co-authors join forces with their community partners to address food insecurity and enhance food sovereignty and thereby community resilience. The three projects considered here are: (i) the greenhouse and community gardening projects in Kuujjuaq and Kangiqsujuaq in Nunavik (province of Quebec), (ii) the clam garden restoration in the Gulf Islands region (province of British Columbia), and (iii) the country food program and Meechim farm in Garden Hill First Nation (province of Manitoba). This chapter examines how these community initiatives enhance access to affordable, nutritious food in Northern communities while also improving food security, food sovereignty, and wellbeing; maintaining Indigenous identity and culture; and preventing diet-related chronic diseases. Next, the chapter analyses the communities' perceptions of these programs. It concludes with a discussion of the synergies and trade-offs of such projects and then provides some recommendations for future food security and food sovereignty projects in Northern communities.

Indigenous communities' concerns about food security and food sovereignty

Inuit community food issues in Nunavik

Nunavik is the northernmost part of the province of Quebec. It is in Eastern Subarctic Canada, which has no land routes for vehicles. Nunavik is 443,685 km^2 and was established under the James Bay and Northern Quebec Agreement of 1975. The 13,188 inhabitants, of whom 89.5% are Inuit, live in 14 coastal northern villages (2016 census, Statistics Canada 2017a). Over the last 50 years, the proportion of country foods (based on subsistence land-use activities such as hunting, fishing, and gathering) consumed by Inuit in Nunavik has dropped (Blanchet and Rochette 2008). Almost exclusive consumption of imported foods poses financial and nutritional problems for families; these problems are so extreme that, according to the 2012 Aboriginal Peoples Survey, 'in Nunavut and in Nunavik, at least 55% of Inuit adults (aged 25 and over) lived in a household that experienced food insecurity' (Arriagada 2017, 1). This Westernisation of nutrition and the health problems associated with it (e.g., type 2 diabetes and obesity) particularly affect children; 49.7% of children in Nunavik face food insecurity, which increases the prevalence of anaemia and slows growth (Pirkle et al. 2014, 233).

Poverty is a major impediment to food security in Nunavik, where '37.5% of Nunavik Inuit households live in poverty' (Kativik Regional Administration 2014, 8). For families who cannot take part in subsistence activities because they lack the necessary expensive equipment such as all-terrain vehicles, snowmobiles, and boats, store-bought food is the main alternative

(Meakin and Kurvits 2009). However, grocery store prices are also a limiting factor due to high transportation and storage costs (Damman et al. 2008, Lamalice 2019). Indeed, consumer prices in Nunavik are much higher than in southern Quebec: 'the cost of food is 52% higher than elsewhere in Québec'(Kativik Regional Administration 2014, 8). In sum, about 44% of the income of Inuit households in Nunavik is devoted to food, compared to only 12% in the rest of the province of Quebec (Kativik Regional Administration 2014, 8).

A closely related, but often overlooked, challenge to food insecurity that has been an issue for several decades in Nunavik is food sovereignty. The growing dependence on southern food supply networks erodes the Nunavimmiut's (the Inuit of Nunavik) control of their lands and their food system, destroying the food sovereignty that characterised the traditional Inuit food system (Lamalice 2019). The low frequency of food shipments (once a month) and the irregularity of ship arrivals due to climate hazards further weaken this already weak control. Federal food aid programs designed by southern government officials, such as Nutrition North Canada, also affect food sovereignty by outsourcing decision-making power (Lamalice 2019).

The Niqiliriniq Project (which means 'caring for food' in Inuktitut) was begun in 2015 in the northern villages of Kuujjuaq, the administrative capital of Nunavik, and in Kangiqsujuaq, a village 500 km north of Kuujjuaq. In Kuujjuaq, 72% of the 1,353 inhabitants identify as Inuit (2016 Census, Statistics Canada 2017b). In Kangiqsujuaq, 705 of the 750 inhabitants, or 94% of the population, self-identify as Inuit (2016 Census, Statistics Canada 2017c) (Figure 14.1).

First Nations community food challenges in northern Manitoba

First Nation communities in northern Manitoba have traditionally relied on a diet of meat, fish, and birds, along with berries and plants (Thompson et al. 2012). The merits of this diet of wild food are evident from archaeological findings of ancient Indigenous skeletons with excellent dental health and no evidence of arthritis despite advanced ages (Price 1939). Kuhnlein (2006) reports on the major pharmacologic and therapeutic benefits of wild foods, which are low in unhealthy fats, sodium, carbohydrates, and sugar but high in good-quality complete proteins and other nutrients (Batal et al. 2017). Other physiological benefits stem from the aerobic and muscle-building activities of harvesting, gathering, and preparing wild food. These foods also protect against chronic diseases such as diabetes, cancer, cardiovascular diseases, and obesity, as well as many other negative health conditions (Kuhnlein and Receveur 1996; Thompson et al. 2012; Willows et al. 2012).

Harvesting involves a complex social system requiring ceremonies, stewardship and protocols to maintain the balance between the earth and the people (Thompson, Harper and Whiteway, 2019). In the First Nation worldview, land and all life are considered a gift from the Creator; animals and plants

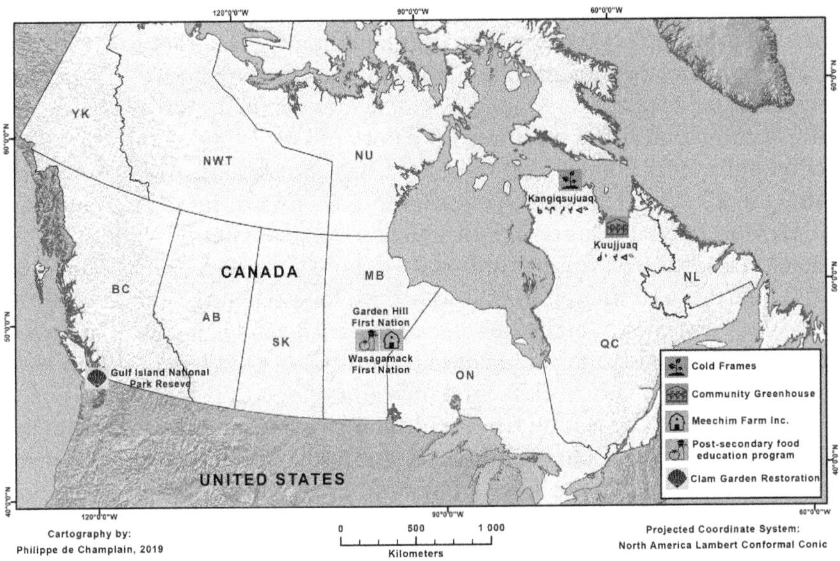

Figure 14.1 The locations of the community-led food initiatives studied here and their partner communities.

agree to sacrifice themselves for food, clothing, and housing in exchange for human stewardship of the land (Thompson, Harper and Whiteway, 2019). Spiritual practices are an important part of all cultural practices, including harvesting. These practices include 'reciprocity' or making an offering before the harvest; taking only what is needed; and offering a feast from the first harvest – for the spirits and all elements of the earth to celebrate (Thompson, Harper and Whiteway, 2019).

Before Canada's colonisation by European settlers, Indigenous hunter-gatherers were nomadic. In small family units Indigenous people lived and ate off the land with in local foodsheds and in ancestral territories. Living in these smaller groups was a sustainable strategy to prevent overharvesting of any single ecosystem. All Indigenous People in Canada employed such careful practices to sustain the animal populations they relied on.

Many Indigenous People in Northern Manitoba still rely heavily on hunting and fishing. Hunting, fishing, trapping, and gathering are integral to Indigenous culture and livelihood (Kuhnlein 2006). According to the Northern Food Prices Project (NFPSC 2003), even in the late 1950s, aside from flour, sugar, and some other similar items, Northern communities relied on their own foods. But then Indigenous People experienced a radical dietary shift and began to depend more on store-bought foods then traditional foods (Batal et al. 2017). Reducing traditional food practices eventually created a vicious cycle of increased dependence on market foods and adoption of southern food practices (Thompson et al. 2012). To make matters worse, in 1976, a hydro development megaproject intentionally flooded the O-Pipon-Na-Piwin Cree Nation (OPCN) and displaced this community. The traditional territories of many northern Manitoba Cree communities were flooded, negatively impacting many northern Manitoba Indigenous communities, displacing the OPCN community. The traditional territory of OPCN and many other Cree communities was flooded and their food shed was negatively impacted, resulting in a higher dependence on market foods (Waldram 1988; Thompson 2015). Market food is lower in iron, calcium, folacin, vitamins, and fibre but higher in fat and sugar than traditional foods (Willows et al. 2012; Batal et al. 2017). Northern communities that were once food self-sufficient and healthy now struggle with very high rates of food insecurity and ill health (Thompson et al. 2012). The monopoly and commodification of food by non-Indigenous corporations has made healthy food unavailable or cost prohibitive.

The nutritional transition from traditional food to store-bought food has increased the rate of chronic diseases and unhealthy lifestyles. Problems of food access in remote fly-in communities in Manitoba include a limited selection of healthy foods, high food prices, the uncertainty of travel on winter ice roads, high poverty rates, and the declining consumption of country foods (Thompson et al. 2012). Food security is highest in remote communities – 88% in Garden Hill and 86% in St. Theresa – and 100% in the flood-impacted communities of OPCN (Thompson et al. 2012). A lack of safe drinking water also impacts food safety and food security in Northern Manitoba, including in the Garden Hill, Wasagamack, and OPCN First Nation communities. Buckets of water and slop pails increase the risk of food contamination, with safe water being a requirement for hygienic food preparation (Thompson, Bonnycastle and Hill, 2020). The lack of piped water and of piped sewage, contamination of cisterns, and inadequate kitchens all present obstacles to food security (Davy 2016). Without hygienic conditions for food preparation due to a lack of plumbing and of functioning faucets, sinks, and toilets, community members are more liable to become ill. Indigenous food sovereignty requires the human right to safe drinking water and sewage services as well as greater control over land and community control of food markets.

The reinvigoration of local food production is key to food access (NFPSC 2003). Healthy foods, such as fruits and vegetables. Cost four times as much as unhealthy foods. Because of their long shelf life, junk foods are cheaper than healthy ones in Northern Manitoba. Soft drinks and potato chips do not expire quickly, so they can be brought in on winter roads and stored in isolated communities that rely on airplanes for access, as is the case for most Northern Manitoba communities. Nutrition North Canada subsidised 635,680 kg of food (worth CA$915,275) between 1 April 2014 and 31 March 2015 for the four communities in Island Lake (Nutrition North Canada 2016) but refuses to subsidise local production, hunting, trapping, fishing, or conservation activities. Nutrition North Canada does not subsidise chicken feed, seed potatoes, fishnets, hunting traps or gas for fishing. Farmers receive government subsidies for fuel, but fishers and hunters do not.

Policies also forbid the serving of wild foods in public venues. For example, the Garden Hill schools could not include wild foods in their school lunch programs and feasts, despite their limited budget; the local Environmental Health Inspector allowed only commercial meats and fish. In a community with high rates of poverty and unemployment, even recovering the costs of hunting and trapping is difficult, as residents may not sell wild meat or local fish (Thompson et al. 2012). Hunting and trapping can be expensive as most trap lines are located far from the community; with no possibilities to recover the costs, these activities are not economically possible for disadvantaged people in Garden Hill and other Indigenous communities.

First Nations community food challenges in the Gulf Islands

Seafood is a core component of health and culture for coastal First Nations. It nourishes the body and supports ties within communities and between the community and the land (Donatuto 2008). Although salmon often receives the most attention in research on food systems in this part of the world, clams are also extremely important to Indigenous Peoples on the Northwest Coast of North America nutritionally, culturally, and economically, and have been for as many as 11,000 years (Kuhnlein and Humphries n.d.; Deur et al. 2015; Toniello et al. 2019). Clams contain many healthy nutrients: proteins, omega-3 fatty acids, vitamin B-12, and various essential minerals (Hamed et al. 2015; Moll and Davis 2017). Traditionally, clams were a dietary staple, used during times of decreased access to or availability of other country foods. They were prepared in several ways for immediate consumption or preserved for later use (Deur et al. 2015; Kuhnlein and Humphries n.d.). Culturally, clams are revered Coast Salish communities. They are often considered an important relative as their families and social structures are similar to those of humans (Deur et al. 2015; Kuhnlein and Humphries n.d.). Clams often feature prominently in cultural stories. Furthermore, they are an essential focal component of various ceremonies and annual gatherings for Indigenous communities in this region (Donatuto 2008; Kuhnlein and Humphries n.d.). Economically, clams were also historically used for trade. They helped coastal Indigenous communities acquire resources such as berries, root vegetables, herring eggs, and eulachon oil. Following European contact, some Indigenous communities used clam gardens as an inroad into the cash economy, selling clams to help feed communities and maintain traditions (Deur et al. 2015). Today, commercial clamming is an important source of income in many Coast Salish communities (Evans et al. 2005).

There is extensive evidence that communities along the Northwest coast used clam gardens, a beach modification that involves the construction of rock walls at the lowest tide line, to improve clam habitats and support food security for at least 3,500 years (Duer et al. 2015; Smith et al. 2019).

Historically, clam gardens are thought to have been governed by a tenure system and social rules designed to maximise community benefits and health (Lepofsky and Caldwell 2013; Lepofsky et al. 2015; Moss and Wellman 2017). Their construction and maintenance created optimal habitats for clams and also strengthened social and cultural ties, food security, and economic stability (Groesbeck et al. 2014; Deur et al. 2015; Jackley et al. 2016). The use of clam gardens diminished after European contact, likely because of various factors, including population decline from epidemics and dislocation and settler government policies including residential schools, changing food options, and loss of access to traditional clamming beaches. Today, some clam gardens are still maintained and harvested while others have been unused for generations (Augustine and Dearden 2014; Deur et al. 2015).

Today, multiple First Nations are developing projects to restore and recover these ancient clam gardens. These projects support numerous goals related to food and environmental health; in addition to providing nutritious and culturally appropriate food (Groesbeck et al. 2014), they are reconnecting people to their traditional lands and practices, providing a platform for strengthening environmental education and elder-youth relationships, and creating a venue for these First Nations to assert their sovereignty and tenure in these coastal places (Clam Garden Network 2015).

Community-led initiatives for tackling food insecurity and enhancing food sovereignty: what is working?

Inuit gardening and greenhouse initiatives: the Niqiliriniq project in Nunavik

The greenhouse project in Kuujjuaq

Public consultations held in Kuujjuaq in 2010 as part of a previous research project (Avard 2015) confirmed the interest of the population in a gardening program and secured funding to launch the project. It was supported by the Northern village of Kuujjuaq; the Quebec Horticultural Council; the Quebec Ministry of Agriculture, Fisheries, and Food (MAPAQ); the Makivik Corporation; and the Kativik Regional Government (KRG). Eight greenhouse horticulture microprojects were set up. Local actors, including the Ungava Supervised Apartments, the school, and local residents, were interested in gardening. In the second year, these eight microprojects were grouped together under the Kuujjuaq Greenhouse Project. Significant progress was made that year, particularly in the compost project, which was joined by Kuujjuaq's grocery stores and restaurants, KRG, and the Ungava Supervised Apartments. The project's success led to the construction of a second greenhouse in 2012. The two buildings, both 133 m^2, now stand beside each other on the edge of the village. The greenhouses are divided into 46 individual garden beds, each about 4 m^2, which are distributed by lottery to interested individuals or families for each annual growing season, which usually runs from mid-May to September. A crop monitoring protocol put in place by our team during the 2016 growing season showed that the greenhouses' production capacity is around 1.15 tons of vegetables, fruits, and aromatic herbs (Lamalice et al. 2018; Lamalice 2019).

Constructing cold frames in Kangiqsujuaq

With the increasing success of the Kuujjuaq community greenhouse project and with a favourable provincial climate for funding this kind of project, several northern villages in Nunavik expressed their interest in implementing their own gardening initiatives. Kangiqsujuaq was one of them, and our Niqiliriniq team organised a first visit to this community in October 2015 to meet local actors and define the project to build together (Lamalice 2019). This co-construction process took two years. Project members visited the community three more times to organise consultations and workshops on the topics of food, Northern agriculture (Lamalice 2019). During these exchanges, it was decided to start with a small, low-tech project to test the interest of the local population in gardening. The community wanted the school to be the main anchor of the project so that the youth would take ownership of this new way of producing food locally. The village also wanted to include the cultivation of traditional plants to increase their presence in the village and to promote the transmission of knowledge related to this traditional food source.

In August and September 2017, Arsaniq school classes participated in the construction of four cold frames (unheated wooden frames with a glass top used to protect seedlings and plants from the wind) (Lamalice, 2019). In summer 2018, after multiple visits and workshops, the four cold frames were installed in different locations in the village: at the town hall building (Northern Village), the clinic, the Elder house, and the Landholding corporation. This made it possible to study the effect of location on plant growth. The cold frames contained different soil mixes to test the quality of soil and the possibility of using the local substrate (i.e., a compost–peat–black soil mix and a black soil–seaweed–compost mix). An employee at the associated institution was in charge of managing each cold frame. Based on the cold-frame managers' interests and the climate conditions, different types of edible plants were grown in each frame: leafy and root vegetables (e.g., lettuce, carrots, and spinach), aromatic herbs (e.g., chives and parsley), and local berries and plants (*Empetrum nigrum*, *Vaccinium vitis-idaea*, *Rhodiola rosea*, and *Oxyria digyna*) (Figure 14.2a–d). The cold frame containing local plants was installed at the Elder's house. Our Inuit partners call the cold frames *Pirursiivik*, 'a place to grow stuff' in Inuktitut or 'big boxes'. Due to the very short growing season, several plants were started in the Kuujjuaq Greenhouse before being transplanted to the cold frames in Kangiqsujuaq. This collaboration provided significant support for the first year of the Kangiqsujuaq project.

Figure 14.2 Cold frames in Kangiqsujuaq. (a) Constructing cold frames; (b) Planting cold frames; (c) Cold frame with root vegetables and aromatic herbs in front of the municipality; (d) Cold frames with traditional plants in front of the Elders' house. (Photo credits: Marion Macé 2018.)

Building local capacity: community gardening, nutrition, and cooking workshops

With the goal of building interest in and technical knowledge of food gardening, multiple hands-on workshops were offered during winter 2018 in Kangiqsujuaq and Kuujjuaq (Lamalice 2019). The researchers from the Niqiliriniq project and managers of different local institutions (e.g., the town hall, the school, and the Elder house) organised a wide variety of workshops on different aspects of nutrition, including growing, harvesting, and cooking healthy, culturally relevant food. Participants in yearlong indoor food gardening workshops grew sprouts, microgreens, aromatic herbs, and local plants (e.g., *Qunguliq* [*Oxyria digyna*], mountain sorrel in English). These

Figure 14.3 Gardening workshops and creating interactive maps of gardening activities. (a) Monitoring Qunguliq (*Oxyria digyna*) at the Makivik Corporation in Kuujjuaq; (b) Kangiqsujuaq municipal staff members learn how to grow aromatic herbs, spices, and vegetables at the Nassivik adult education centre in Kangiqsujuaq; (c) Screenshot of *Niqiit Piruurtut* – an interactive online map of local gardening and greenhouse initiatives in Kangiqsujuaq and Kuujjuaq (Macé 2018). (Photo credits: Marion Macé.)

plants can be grown at home and can represent a significant source of nutrients and fresh food. Cooking and nutrition workshops were also offered to school students and the public in order to build awareness of balanced diets and micronutrients (Figure 14.3a–c).

The multiple workshops offered during the winter preceding the first growing season in Kangiqsujuaq played an important role in teaching local residents about the possibilities of growing food at home. Starting with simple, short-growth plants such as sprouts, microgreens, and aromatic herbs allowed participants of different ages to increase their interest in and technical knowledge of indoor food gardening. At the end of the workshops, several participants expressed interest in getting personally involved with the cold frame food-growing project.

Regular updates about the project were widely shared with communities via social media, community radio, and direct communication. Technical documentation and knowledge sharing tools (e.g., a step-by-step compost guide and a step-by-step greenhouse guide) were produced to support the management and development of the initiative but also to encourage full local ownership of the project. These actions helped bring a wide range of local people on board; some of them became long-term leaders ('local champions') of the food gardening project in their respective communities.

Interactive online maps of food habits: telling my food story

Inuit culture is deeply rooted in oral tradition, in stories that relate every day, historical, mythical, or personal experiences. All these stories have a spatial dimension. Interactive, dynamic story mapping (Caquard and Cartwright 2014; Caquard and Joliveau 2016) is an emerging approach to better understand the relationships among individuals, communities, and the environment. This project used online mapping applications, particularly the StoryMap JS platform developed by Northwestern University Knight Lab,[1] to map the stories of Inuit youth in Kuujjuaq and Kangiqsujuaq related to food sources and locations of food supplies. This cartographic platform is closely related to the ESRI Story Maps software. In their comparative analysis of the main applications for online narrative mapping, Caquard and Dimitrovas (2017, 10) call the ESRI Story Maps software well suited for

> [...] the representation of travel stories or documentary stories in which places and journeys are clearly identified. [...] ESRI Story Maps offer the ability [...] to design maps collaboratively, to represent different stories on the same map [...] and to distinguish graphically different links between places.[2]

The youth received training on StoryMap JS (e.g., how to organise information in a map, how to import matrix and vector data, and how to link different media such as photos or videos with a point), they produced – individually or in groups – personal interactive narrative food maps online (for a detailed description, see Macé 2018). These interactive narrative maps are thus the tangible link between the oral narratives and the current role of the land as a food provider for the Nunavimmiut (Figure 14.4a). Different types of narrative spaces that characterise these 'feeding landscapes' can be identified on the story maps:

- Places are mentioned frequently: 'I often go to the co-op to buy pork chops' (a young Inuk from Quaqtaq 2018).

- Specific locations are mentioned, including local supermarkets such as the Newviq'vi/Tullik or Northern Store in Kuujjuaq; for example: The 'community freezer [a communal refrigerated storage space; most of the time Hunters sells their harvested traditional food to the hunter support which donate it to the community] is where I get my wild food, like Arctic char' (StoryMap of a young Inuk from Quaqtaq 2018).
- Narrators provide precise, detailed geographical descriptions of some places: 'Roberts Lake (Tasirjuakuluk) is where we fish for lake trout and landlock char (Nutilliq)' (StoryMap of a young Inuk from Quaqtaq 2018).
- Other places are described in general terms: 'On the land. Ptarmigan. You chase it for six hours in a skidoo' (StoryMap of students at Kuujjuaq High School 2018).

The digital story maps show specific places (e.g., hunting routes) and items (e.g., food species) but also integrate local Inuit knowledge as well as the emotions and cultural relationships that Nunavimmiut associate with particular places and that mark the land as a provider of food. Hence, these maps are a spatiotemporal analytical tool that can clarify the different forms of lived and perceived spaces that structure narratives about foodscapes (on this topic, see also Lamalice et al. submitted). The narrative interactive maps helped engage youth in discussions about their representation of contemporary Inuit food systems (Macé 2018, 55).

Niqiit piruurtut: creating interactive, dynamic online maps of local food initiatives in Kangiqsujuaq and Kuujjuaq

Free and open source geospatial software allows new forms of dynamic and interactive cartography to be created online and promotes citizen science (Kar et al. 2016). Inuit communities are increasingly using new GeoWeb mapping tools to monitor their communities (e.g., Inuit Sea Ice Knowledge and Use http://sikuatlas.ca/index.html; Inuit Trails Atlas http://paninuittrails.org). In the Niqiliriniq project, the open-source platform uMap was used to create an interactive online map showing the progress of local gardening and greenhouse initiatives in Kangiqsujuaq and Kuujjuaq (for a detailed description, see Macé 2018). Multimedia data (photos, videos, and oral testimonies) shared by local gardeners were incorporated into the map and associated with the locations where they took place (Macé 2018, 56). Since the mapping platform is online, local villagers could easily update it to reflect the progress of each initiative and could use it as a community-based monitoring tool (Macé 2018, 55). The interactive features of the Niqiit Piruurtut map combined with multimedia content helped residents and decision makers in Kuujjuaq and Kangiqsujuaq understand different aspects of the local gardening initiatives (Fig. 3d) (see also Laurini 2004; Joliveau 2011). This interactive map can also be used as a reference tool in both municipalities for land use planning as well as for decision-making in regional governmental institutions such as the Kativik Regional Government or the Makivik Corporation.

The StoryMap JS platform was also used to create an interactive atlas of all Northern community-based food gardening initiatives in the Canadian Arctic and Subarctic. Developed at the request of the Kuujjuamiut, this atlas is a knowledge-sharing platform. It provides key information and links to relevant contacts, websites, or social media pages, and publications associated with gardening projects. The map was embedded into the website developed by the Niqiliriniq project researchers dedicated to sharing information related to Northern agriculture (https://www.polarharvest.com/en/cartographie-interactive/) (Lamalice 2019).

Community-led food initiatives in northern Manitoba

Community-based food action is an important way to tackle food insecurity (Thompson et al. 2014). A 2012 survey of 14 First Nation northern Manitoba communities determined that communities with country foods programs led by First Nations were more food security (Thompson et al. 2012). Thompson et al. (2012) credit the country food program in the remote northern Manitoba community of Nelson House First Nation with significantly improving food security there, as well as increasing other livelihood assets, including natural and social assets (Thompson et al. 2012). Country food programs encourage residents to share traditional food resources (wild food and garden produce) rather than depending on the market system. OPCN and many other flood-impacted communities in northern Manitoba have also developed country food programs (Thompson and Pritty, 2020).

Through focus groups the priorities in northern Manitoba were found to be promoting traditional knowledge about hunting, trapping, and gathering food in school programs. Another priority was to provide funding for residents to obtain boats/motors, skidoos, or gas so they can engage in traditional activities. This connection to the land and stewardship of the land is considered a sacred trust and is important for cultural identity. Communities can create Indigenous food systems by building community gardens, implementing country food programs, providing education about the land, and establishing farms, stores, and restaurants that provide healthy and affordable food as well as jobs for local people.

In OPCN 90% of 87 families surveyed agreed that there was a need for a community store; 10% were satisfied with the Northern Store, the corporation that has a monopoly on supermarkets (Thompson and Pritty, 2020). To meet this need, a pop-up community-run market was trialled in 2016; this market offered good, healthy food at affordable prices. Pop-up markets only opened on days that community members received social assistance and family assistance pay and were very popular. This trial demonstrated that local stores present a real alternative to the Northern Store.

OPCN, and other remote and rural First Nation communities, would benefit from something similar to the Outback stores, which were funded by the Australian government but owned by remote Indigenous communities. These food stores provided access to nutritious food and developed local economies (Davy 2016). In contrast, the Canadian government supports a corporate, non-Indigenous monopoly, the Northern Store, which competes with non-subsidised local food production and local stores (Thompson et al. 2012).

Post-secondary education to teach Indigenous food systems has been identified as a powerful way to encourage young people to tackle food issues in their communities. This community-led applied education program would also improve food security by engaging youth in fishing, hunting, gardening and managing community grocery stores. Meechim Farm, Inc., a community enterprise in Northern Manitoba, employs and trains 15 young people every summer. This farm began hiring young people in 2015 and (at this writing) is still the only farm in in the many First Nations on the East-side of Lake Manitoba to do so (Thompson et al. 2014). The farm's goal is to improve food security, train young people, and build sustainable livelihoods. This community-owned social enterprise also runs a pop-up store that sells healthy imported food as well as local food. The farm is run by a manager and offers employment training and mentorship programs for youth and unemployed people. Young interns are hired and trained to cultivate, produce, harvest, process local food and sell local food.

Garden Hill and Wasagamack First Nations have identified a need for post-secondary education in Indigenous food systems. Better training and advanced education are needed to effectively reduce food insecurity and build local food systems (e.g., food stores, farms, country food programs, and traditional camps). Communities need agroecology education as well as knowledge of traditional medicine and skills in fishing, hunting, gathering and preparing food. The Mino Bimaadiziwin partnership is currently developing a curriculum for providing community-led to offer post-secondary education for communities to grow, harvest and sell local food using a train-the-trainer model. Colonial systems have impoverished northern Manitoba communities by creating reserves as economic dead zones and then underfunding education and training programs. At present, to be sustainable, local food educational programs must dip into several meagre pots of government funding, in the different sectors of education, employment training, and community development. The applied education program will create and increase the capacity of local businesses (e.g., grocery stores, restaurants, and commercial fisheries) (Figure 14.4a–d). By harvesting food locally from their vast traditional territory, First Nations steward and protect their land.

According to the people of Island Lake, the land is perfect as it is. They do not want any new developments that would interfere with traditional land uses, such as hunting, gathering, fishing, and traditional ceremonies

Figure 14.4 Local food initiatives in northern Manitoba. (a) O-Pipon-Na-Piwin Cree Nation (OPCN) country food coordinator, John Bonner; (b) Meechim, Inc. Healthy Food Market in Garden Hill with local and imported food, run by the community on welfare and paydays; (c) Youth centre educational gardens are fertilised with fish guts; (d) Meechim Farm poultry and acres of land. (Photo credits: Shirley Thompson.)

(Thompson et al. 2019). Many people in Red Sucker Lake FN, Garden Hill FN, and Wasagamack FN created map biographies to demonstrate how fully they use their ancestral land for traditional activities. For example, 49 active Indigenous harvesters in Wasagamack FN created map biographies that explore the locations where wild food is found in traditional land use mapping (a summary map is provided in Figure 14.5). According to this traditional land-use study, the community's foodshed is approximately 13,378 km^2 (Figure 14.6).

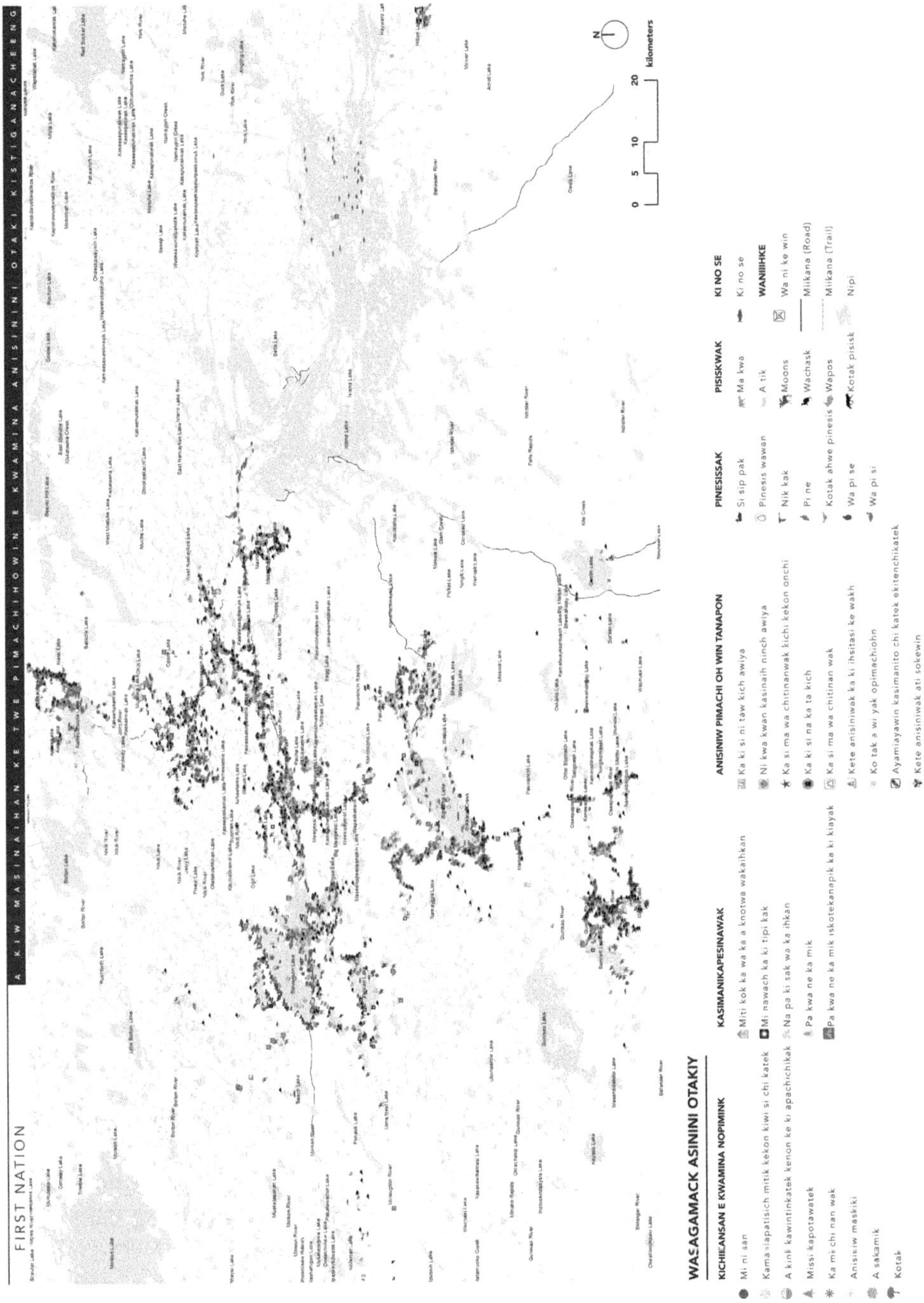

Figure 14.5 Map of food harvesting and other traditional land uses for Wasagamack First Nation.

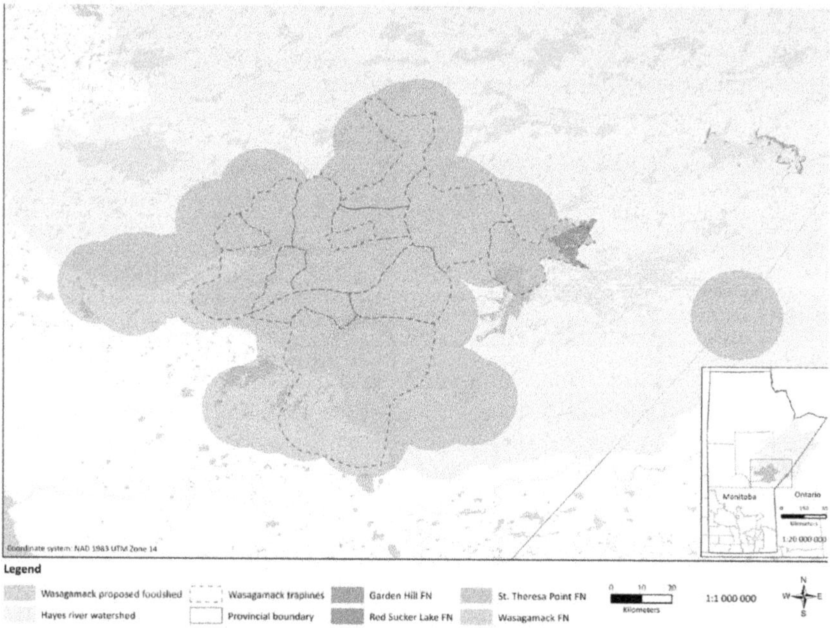

Figure 14.6 Map of the foodshed of Wasagamack First Nation.

First Nations new mariculture initiatives in coastal British Columbia: clam garden restoration in the Gulf Islands region

One program by First Nations in Canada that could offer a template for building community food security is the clam garden restoration project in the Gulf Islands region of British Columbia. Clam gardens are a beach modification that coastal Indigenous Peoples have created and maintained for over 1,500 years to increase shellfish production and harvest (Clam Gardens n.d.; Harper et al. 1995; Augustine and Dearden 2014; Deur et al. 2015). They are found in the Pacific Northwest region of North America, from Washington to Alaska. They are usually constructed in semi-protected inlets by building a rock wall at the low tide line across a bay or cove. The rock walls encourage sand deposits on the landward side, creating a gradually sloping beach terrace that is ideal for clams (Figure 14.7a, b) (Clam Gardens n.d.; Harper et al. 1995; Augustine and Dearden 2014; Groesbeck et al. 2014; Deur et al. 2015; Lepofsky et al. 2015; Jackley et al. 2016; Moss and Wellman 2017). To maintain clam gardens, large clams are harvested; the sediment is aerating with a digging stick and by rolling rocks; new or uncovered rocks and large clam shells are removed; gravel and crushed clam shells are added; and the garden is protected from competitors and predators (Augustine and Dearden 2014; Groesbeck et al. 2014; Deur et al. 2015).

Figure 14.7 Clam gardens (a) Harvesting clams; (b) Clam garden. (Photo credits: Philip Loring.)

Currently, only a handful of communities are actively working to reclaim and restore clam beaches. One such project offers valuable insights into how to link food security, ecosystem restoration, and community health and wellbeing. This project is located in the Gulf Islands National Park Reserve (GINPR) and is a collaborative initiative of Parks Canada and multiple local First Nations, including the Hul'qumi'num Treaty Group (HTG) and the W̱SÁNEĆ First Nations. These groups are restoring multiple historical clam gardens as a part of a multifaceted project. The goal of the project is to determine whether traditional marine stewardship practices can support better habitats for shellfish in the GINPR while simultaneously strengthening Indigenous food security and cultural practices (Bouevitch 2016). Specifically, the project aims, among other goals, to: (1) create an opportunity for local First Nations people to reconnect with traditional activities and sites of cultural significance, (2) use traditional knowledge and Western science to evaluate the efficacy of traditional beach management, and (3) increase the access of local First Nations to traditional foods. The project was initiated in 2013 after years of consultation among Parks Canada and multiple First Nations. It is an example of community-directed ecological stewardship that seeks to support food security, coastal biodiversity, heritage reclamation, self-governance, and the health of communities and ecosystems (Lepofsky et al. 2015).

Considering the significance of clams to coastal Indigenous peoples, clam gardens are important. They increase clam productivity and the ease of clam harvests while simultaneously providing human health benefits on individual, social, and community levels (Groesbeck et al. 2014; Deur et al. 2015). On the individual level, the construction, maintenance, and harvesting of the gardens supports physically active lifestyles and connections with the land. Activities include walking on the beaches, digging clams, moving rocks, building the rock wall, and aerating the soil (Deur et al. 2015). Socially, the gardens become a forum for intergenerational sharing of cultural practices, language development, and socialisation. Elders and youth work together in the gardens, building relationships in the process (Augustine and Dearden 2014). On the community level, clam gardens may prove to be an important source of food, strengthening food security (Groesbeck et al. 2014). Community control over clam garden activities also supports sovereignty and self-determination, which have been identified as critical factors in Indigenous health. They also support reconciliation (Tsosie 2007; CSDH 2007; TRC n.d; Fleming 2019).

Clam gardens could also add resilience to local and regional food systems. Generally, country food systems rely on a portfolio of primary and secondary food resources (Loring and Gerlach 2009). This portfolio allows people to respond to changing environmental circumstances and also be flexible with the changing seasons (Kofinas et al. 2011). Clam gardens increase clam production and possibly allow clams to be produced in areas that they would not naturally. Therefore, they could be key contributors to the resilience of First Nations' food systems in the future.

Insights from the Clam Garden Project could be relevant to other projects for improving community food systems in the North. Three aspects in particular stand out. First, the project explicitly seeks to link human and environmental health. Clam gardens promote positive health outcomes for people (as described above). But the gardens and beach stewardship also improve beach health. For example, the rock walls promote biodiverse life in the intertidal zone. Second, the project seeks not just to produce food but to use the gardens as a platform to connect people. At the gardens, youth interact with Elders and also with each other in a new, positive context. Third, the clam gardens have symbolic value for reconciliation and tribal sovereignty. They are a venue in which First Nations people can reassert their right to territory and their capacity to be effective stewards of the land. These linked biophysical, ecological, psychosocial, and political dimensions arguably make clam gardens a durable, resilient, and effective food security intervention that addresses not just diets but all of the many ways that food can be central to our lives and societies (for a detailed description on Clam Garden Project impacts see also Fleming 2019).

Challenges and constraints of community-based food initiatives

Technological challenges

Nunavik's communities, like other Arctic communities, rely heavily on diesel-powered generators for energy (Cherniak et al. 2015). Due to this dependence on fossil fuels, remote communities are more vulnerable to global market variations or shortages in diesel and to emissions of greenhouse gases and pollutants. Hydro Quebec (2013, 47) estimates that the electricity demand of the 14 Nunavik communities will increase by 2.9% per year from 2013 to 2023 due to anticipated population growth. Despite the availability of wind and hydroelectricity resources near these communities, renewable energy installations are still absent from Nunavik. One main challenge to implementing greenhouse technology in the Arctic is ensuring that greenhouse projects do not threaten energy and water security. The Kuujjuaq greenhouse has no heating system and operates only during spring and summer, from May to September. Technical solutions and greenhouse energy optimisations are currently being investigated to enable the Kuujjuaq Greenhouse to extend the growing season and remain energy efficient, such as the installation of a thermal storage system (Lamalice et al. 2018, 335; Blangy et al. 2018, 318).

Other renewable-energy food growing installations have recently opened in Kuujjuaq. The Kuujjuaq Hydroponic Container runs since December 2018. This is the first year-round commercial gardening initiative in this northern village. It also functions as a waste-to-energy plant and is equipped with a with a waste oil heating system (Makivik Society 2018).

Another key limitation to local food production in the Canadian Arctic is the absence of nutrient-rich growth substrate. This was an issue for the cold frame initiative in Kangiqsujuaq. However, many locally available components can be used to produce nutritious, 'class-A' soil for agricultural use via composting. Local resources with significant potential as organic waste in Nunavik include kelp seaweed and other macroalgae (Sharp et al. 2008); wood waste from construction projects; post-consumer waste (e.g., households, grocery stores); sand, clay and silt from riverbeds. It would be interesting to test the potential of composting technology and different soil mixes in the future growing season.

Challenges to participatory interactive web-based cartography such as the Niqiit Piruurtut map include a lack of access to computers and limited access to a reliable Internet connection in the North (Macé 2018, 59).

Environmental risks

Technological challenges are one barrier to implementing programs such as the ones described in this chapter. Environmental hazards are another. For example, First Nations mariculture initiatives, such as the clam garden restoration, can increase communities' risk of exposure to ocean acidification, harmful algal blooms, other environmental contamination such as paralytic shellfish poisoning (Fleming 2019).

Lack of knowledge and training; merging different types of expertise

One barrier to the food growing project was a lack of prior knowledge about farming or gardening in a greenhouse setting. Vocational schools in Nunavik offer no training in horticulture or agriculture. Here, the post-secondary education program in food developed by Wasagamack FN and Garden Hill FN can serve as an example for the development of agricultural skills.

Most greenhouses in the Arctic today use a container-like architecture or are arenas transformed into greenhouses. Local traditional knowledge then becomes important; ideally, greenhouses should merge traditional ways of storing food such as semi-subterranean dwellings (e.g., Qarmaq, Yupik house, Qaggi) with modern greenhouse designs to develop culturally appropriate greenhouses (Laurendeau 2018, 18).

Organisational challenges

Skinner et al. (2014, 8) studied the greenhouse at Fort Albany First Nation and described the challenges around ownership. Several community members did not think of the greenhouse as a communal structure that belonged to everyone. Similarly, Chen and Natcher (2019, 150) highlight the challenges of managing local greenhouses in relation to the ethnic and socioeconomic background of the local participants. They pose a question:

> [...] are participants in community gardens and greenhouses those most in need or do they represent a more transient population (e.g., schoolteachers or government employees) who are either accustomed to having regular access to fresh foods or are drawn to the company of others with similar socio-economic standing? If the latter, participation in community gardens and greenhouses could be viewed negatively by permanent community members, or even considered elitist and used only by those who can afford membership (e.g., membership and plot fees) and have flexible schedules to volunteer their time. This situation could then result in those who are most vulnerable to food insecurity being excluded from any potential benefits.
>
> (Chen and Natcher 2019, 150)

Participation varies at the Kuujjuaq greenhouses; 46 individual lots are allocated by lottery to individuals or families interested in gardening for the annual growing season.

Tables 14.1a–c overviews the different priorities and expectations of local stakeholders and the outcomes of the food growing initiatives considered in this chapter.

Table 14.1 Community members' expectations and outcomes for the three different initiatives studied (a) The Niqiliriniq project; (b) Community-led food initiatives in northern Manitoba; (c) Clam garden restoration project

Niqiliriniq project	
Short-term expectations	Realised outcomes
• Start with a small, low-tech project => *'We need to test the community's interest towards gardening' (resident of Kangiqsujuaq)* • Learn to garden => *'It's complicated, someone needs to teach us' (resident of Kuujjuaq)*	• The project started with cold frames which are light, inexpensive, and easy to repair structures • 29 gardening workshops between 2016 and 2018 • Cold frames were built with grade 3 to 5 high school students and many gardening activities were held with students of all ages as well as day-care children
• Involve the youth => *'To teach children how to grow own food plants' (resident of Kuujjuaq)* • Include local plants => *'Would be nice to have berries in the village' (resident of Kangiqsujuaq)* • Have a pleasant pastime => *'It would give people something to do' (resident of Kuujjuaq)*	• Crowberry, lingonberry and mountain sorrel were planted in the cold frames • Most of the participants from the 2018 growing season enjoyed it and asked for more cold frames for next spring
Long-term expectations	Partially realised outcomes
• Local capacity building and empowerment • Embellish the village with gardening and landscaping project • Opportunity to have a small greenhouse at home => *"Individual projects at home would be great, even inside the house" (resident of Kangiqsujuaq)* • Secure the supply of fresh fruits, vegetables and herbs year-round => *'Fruits are missed very much during the winter' (resident of Kangiqsujuaq)* • Improve health and well-being • Save some money => *'Veggies and fruits are too expensive' (resident of Kuujjuaq)*	• Participation of a local leader in a regional meeting on greenhouse agriculture • Landscaping proposition was designed by a Niqiliriniq team member and will be presented to the community in 2019

(Continued)

Community-led food initiatives in northern Manitoba	
Short-term expectations	Realised outcomes
• Provide youth opportunities to garden and farm to produce food for the community • Provide some educational opportunities in elementary and secondary school regarding country foods and gardening	• Realised a community pop-up store of healthy reasonably priced food at Garden Hill, Wasagamack, and OPCN that sells once or twice per month • Developed a permaculture design with community to expand farm over 20 years in Garden Hill • Country food programs provide food for elders at OPCN • Developed a farm that employs 15 people and farms 999 poultry
Long-term expectations.	Partially realised outcomes
• Develop community store for selling affordable foods as well as nets and materials to get country foods • Developing a post-secondary Indigenous food system which includes hunting, fishing, gathering, cooking, food-business, food safety, gardening, animal raising, and starts businesses in other areas (e.g., community store, local restaurant). • Starting a local restaurant or canteen selling some wild foods (fish, Labrador tea, etc.)	• Improved food security slightly through country foods, farming, and pop-up stores • Jobs for youth in food production at Meechim Farm • Some income from food production • Growing sustainable food businesses • Mino Bimaadiziwin committed to start post-secondary food program
Clam garden restoration in the Gulf Islands region[3]	
Short-term expectations	Realised outcomes
• Restore two clam garden sites according to the traditional practices of Hul'q'umi'num and WSÁNEĆ peoples. • Create opportunities for elders and youth to connect through beach restoration activities • Determine whether clam gardens can be used as an effective resource management tool in the region	• Multiple years of beach restoration activities in summer and winter months, including elders, youth, clam experts, and research scientists • Science and culture camps are held each year in association with beach restoration activities
Long-term expectations	Partially realised outcomes
• Improve ecosystem health in intertidal zone • Build community food security • Facilitate opportunities for Coast Salish peoples to engage with their ancestral territories and practice their traditional harvesting rights	• Increased awareness of beach stewardship and cultivation by First Nations through knowledge mobilisation that includes news articles and short documentary videos, e.g., https://youtu.be/j2wPVx4sCN0

Conclusion

This chapter adds to the growing body of literature providing empirical support for local strategies to enhance food security and sovereignty in remote communities across the Canadian North.

The study cases presented and analysed in this chapter – the Kuujjuaq greenhouse, Kangiqsujuaq cold frames, the Meechim Farm, and the Clam garden project – show that community agriculture initiatives and initiatives to restore traditional mariculture systems offer valuable alternatives to the unsustainable import of nutrient-poor foods to markets in Canada's North. These local food procurement strategies provide local access to fresh, nutritious, affordable foods. They effectively reduce food insecurity and build or rebuild local food systems where communities of First Nation and Inuit food growers and farmers are exercising increased control over their food systems and the socioecological aspects of food production. This reinforces food sovereignty and creates economic opportunities in local food production. Our results corroborate earlier findings (e.g., Allan 2014; Skinner et al. 2014) that the benefits of community gardens, farms, and greenhouses go beyond their capacity to produce food and to increase consumption of healthy food. They also provide community social services. The community food initiatives described in this chapter are examples of fruitful collaborations between First Nations/Inuit communities and university-affiliated researchers or government agencies (i.e., Parks Canada) that meet the food needs identified by local Northern residents.

However, as shown by our case studies, community growing initiatives in the North still face several challenges. These include costs, technological challenges, organisational and management challenges, sustainability, and ownership.

Clearly, local food production alone cannot solve all the food-related issues Northerners are facing today. Food insecurity in the North requires effective policy responses that tackle the root causes of food insecurity (e.g., poverty, the abnormally high cost of imported foods, the colonial legacy, environmental factors, the export of food decisions to the South). Nonetheless, community food initiatives like those presented in this chapter are part of the solution. They are an important piece of the puzzle on the path toward true food security and sovereignty in the North.

Several recommendations for community projects targeting food security and sovereignty in remote communities emerge from our study:

1 A range of strategies allows better access to all types of food (country food, market food, locally produced food) and empowers local communities to gain control over their food systems.
2 Empowering governance and policies are needed to help communities navigate economic and natural resource development.
3 Arctic and First Nation food innovation and collaboration is needed; examples of collaboration include the UArctic program and the Mino Bimaadiziwin partnership.

4 Existing local knowledge should be used in food initiatives. For example, traditional food storage methods can be merged with modern greenhouse designs to develop a culturally appropriate greenhouse.
5 Community-led post-secondary programs on Indigenous food sovereignty should be created to develop local capacity throughout the food system.
6 The development of community-led education programs on Indigenous food sovereignty for young people provides them with new skills and experiences.
7 Community-level champions should be identified and supported. These are community members who are interested in leading local actions around food systems and who can clearly communicate about these actions and engage other community members.
8 More long-term funding should be allocated to community food programs.
9 Programs that offer training and support for raising animals should be developed and implemented in Northern communities that are starting to farm to provide assistance and supports for animal husbandry and food production.

Ultimately, the case studies examined in this chapter provide examples for other community-led initiatives.

Acknowledgements

T.H., A.L., M.M., V.C., G.L., and S.B. would like to thank the LabEx DRIIHM (Dispositif de Recherche Interdisciplinaire sur les Interactions Hommes-Milieux), French program 'Investissements d'Avenir' (ANR-11-LABX-0010) – and more specifically the OHMi (Observatoire Homme Milieu international) Nunavik; the Conseil Franco-Québecois de Coopération Universitaire (SEQINEQ project number 2017-FQ-202580, Programme de développement de partenariats stratégiques en matière d'enseignement et de recherche); the Social Sciences and Humanities Research Council of Canada; the Réseau DIALOG, The Royal Canadian Geographical Society (James Bourque Northern Doctoral Scholarship) and the Northern Scientific Training Program for the financial support kindly provided.

Notes

1 https://knightlab.northwestern.edu/about/.
2 Translation by the authors; original French passage (Caquard and Dimitrovas 2017, 10):

> [...] la représentation de récits de voyages ou de récits documentaires dans lesquels les lieux et trajets sont clairement identifiés. [...] ESRI Story Maps offrent la possibilité d'importer des fichiers shapefile, [...], de concevoir les cartes de manière collaborative, de représenter différentes histoires sur une même carte [...] et de distinguer graphiquement différents types de liens entre les lieux. [...]

3 Note this is not intended as a comprehensive list of goals as represented by project partners, but instead a cross section of such goals as observed by Loring.

References

Allan, T. 2014. Costs and benefits of a northern greenhouse. In S. Seefeldt and D. Helerich (eds.), *Sustainable agriculture and food security in the circumpolar north. MP 2014–16.* Proceedings of the Eighth Circumpolar Agricultural Conference andand Inaugural University of the Arctic Food Summit, held 29 Sept. – 3 Oct. 2013 in Girdwood, Alaska. Fairbanks, Alaska: Agricultural andand Forestry Experiment Station. 58–63. Retrieved 19 March 2019 from https://www.uarctic.org/media/1002371/sustainable-agriculture-and-food-security-in-the-circumpolar-north-2.pdf.

Arriagada, P. 2017. Insights on Canadian society: Food insecurity among Inuit living in Inuit Nunangat. Statistics Canada. Catalogue no. 75-006-X, ISSN 2291-0840.12 p. Retrieved 24 May 2020 from: https://www150.statcan.gc.ca/n1/pub/75-006-x/2017001/article/14774-eng.htm.

Augustine, S. and P. Dearden. 2014. Changing paradigms in marine and coastal conservation: A case study of clam gardens in the Southern Gulf islands, Canada: A values-based approach to coastal conservation. *The Canadian Geographer | Le Géographe Canadien* 58(3): 305–314. doi:10.1111/cag.12084.

Avard, E. 2015. *Northern Greenhouses: An Alternative Local Food Provisioning Strategy for Nunavik*, PhD Thesis, Université Laval, Québec.

Batal, M., L. Johnson-Down, J.C. Moubarac, A. Ing, K. Fediuk, T. Sadik, C. Tikhonov, L. Chan, and N. Willows. 2017. Quantifying associations of the dietary share of ultra-processed foods with overall diet quality in First Nations peoples in the Canadian provinces of British Columbia, Alberta, Manitoba and Ontario. *Public Health Nutrition*: 1–11. doi:10.1017/S1368980017001677.

Berkes, F. and H. Ross. 2013. Community resilience: Toward an integrated approach. *Society andand Natural Resources* 26: 5 – 20.

Blanchet, C. and L. Rochette. 2008. *Nutrition and food consumption among the Inuit of Nunavik. Nunavik Inuit Health Survey 2004, Qanuippitaa? How are we?* Quebec: Institut national de santé publique du Québec (INSPQ) andand Nunavik Regional Board of Health and Social Services (NRBHSS).

Blangy, S., M. Bernier, N. Bhiry, J.P. Dedieu, C. Aenishaenslin, S. Bastian, L. Chanteloup, V. Coxam, A. Decaulne, J. Gérin-Lajoie, S. Gibout, D. Haillot, E. Hébert-Houle, T.M. Herrmann, F. Joliet, A. Lamalice, E. Lévesque, A. Ravel, and D. Rousse. 2018. OHMi-Nunavik: A multi-thematic and cross-cultural research program studying the cumulative effects of climate and socio-economic changes on Inuit communities. *Écoscience* 25(4): 311–324.

Bouevitch, N. 2016. *Eco-cultural restoration as a step towards co- management: Lessons from the Gulf Islands National Park Reserve.* Master Thesis. Simon Fraser University, Vancouver, p. 133.

Boulanger-Lapointe, N., J. Gérin-Lajoie, L. Siegwart Collier, S. Desrosiers, C. Spiech, G.H.R. Henry, L. Hermanutz, E. Lévesque, and A. Cuerrier. 2019. Berry plants and berry picking in Inuit Nunangat: Traditions in a changing socio-ecological landscape. *Human Ecology*: 1–13. doi:10.1007/s10745-018-0044-5.

Caquard, S. and W. Cartwright. 2014. Narrative cartography: From mapping stories to the narrative of maps and mapping. *The Cartographic Journal* 51(2): 101–106. doi:10.1179/0008704114Z.000000000130.

Caquard, S. and S. Dimitrovas. 2017. StoryMaps and Co. The state of the art of online narrative cartography. *M@ppemonde* 121: 1–31.

Caquard, S. and T. Joliveau. 2016. Penser et activer les relations entre cartes et récits. *M@ppemonde* 118: 1–7.

Chen, A. and D. Natcher. 2019. Greening Canada's Arctic food system: Local food procurement strategies for combatting food insecurity. *Canadian Food Studies | La revue Canadienne des études sur l'alimentation* 6(1): 140–154.

Cherniak, D., V. Dufresne, and K. Lawrence. 2015. *Report on the state of alternative energy in the Arctic*. Canada: School of Public Policy and Administration, Carleton University.

Clam Gardens. n.d. The Clam Garden Network. Retrieved from https://clamgarden.com/clamgardens/.

Clam Garden Network. 2015. "Eco-cultural restoration and the Gulf Islands National Park Reserve." *The Clam Garden Network* (blog). January 30, 2015. https://clamgarden.com/research-2/restoration-in-gulf-islands/.

Council of Canadian Academies. 2014. *Aboriginal food security in Northern Canada: An assessment of the state of knowledge*. Ottawa, ON: The Expert Panel on the State of Knowledge of Food Security in Northern Canada, Council of Canadian Academies.

Commission on the Social Determinants of Health (CSDH). 2007. *Social determinants and Indigenous health: The international experience and its policy implications*. Report on specially prepared documents, presentations and discussion at the International Symposium on the Social Determinants of Indigenous Health Adelaide, 29-30 April 2007 for the Commission on Social Determinants of Health (CSDH). Retrieved on 24 May 2020 from https://www.who.int/social_determinants/resources/indigenous_health_adelaide_report_07.pdf.

Cuerrier, A., A. Downing, J. Johnstone, L. Hermanutz, L. Collier-Siegwart, elders and youth participants of Nain and Old Crow. 2012. Our plants, our land: Bridging aboriginal generations through cross-cultural plant workshops. *Polar Geography* 35(3–4): 195–210.

Damman, S., W.B. Eide, and H.V. Kuhnlein. 2008. Indigenous Peoples' nutrition transition in a right to food perspective. *Food Policy* 33(2): 135–155.

Davy, D. 2016. Australia's efforts to improve food security for Aboriginal and Torres Strait Islander Peoples. *Health and Human Rights Journal* 18(2): 209–218. Retrieved from https://cdn2.sph.harvard.edu/wp-content/uploads/sites/13/2016/12/Davy-final-1.pdf.

Delormier, T., K. Horn-Miller, A.M. McComber, and K. Marquis. 2017. Reclaiming food security in the Mohawk community of Kahnawà:ke through Haudenosaunee responsibilities. *Maternal and Child Nutrition* 13(S3): e12556.

Deur, D., A. Dick, K. Recalma-Clutesi, and N.J. Turner. 2015. Kwakwaka'wakw 'Clam Gardens': Motive and agency in traditional northwest coast mariculture. *Human Ecology* 43(2): 201–12. doi:10.1007/s10745-015-9743-3.

Donatuto, J. 2008. *When seafood feeds the spirit yet poisons the body: Developing health indicators for risk assessment in a Native American fishing community*. University of British Columbia. doi:10.14288/1.0066720.

Evans, B., Gardner, J., Thom, B., L. Joe. 2005. *Shxunutun's Tu Suleluxwtst, In the Footsteps of our Ancestors: Interim Strategic Land Plan for the Hul'qumi'num Core Traditional Territory*. Ladysmith. Hul'qumi'num Treaty Group.

Fleming, T.B. (2019). *Health, Risk, and Environmental Justice for Indigenous Shellfish Harvesters in British Columbia, Canada*. Master Thesis, University of Saskatchewan, Sakatoon, 73p.

Ford, J., J. Petrasek Macdonald, C. Huet, S. Statham, and A. MacRury. 2016. Food policy in the Canadian North: Is there a role for country food markets? *Social Science and Medicine* 152: 35–40.

Groesbeck, A.S., K. Rowell, D. Lepofsky, and A.K. Salomon. 2014. Ancient clam gardens increased shellfish production: Adaptive strategies from the past can inform food security today. *PLoS ONE* 9(3): e91235. doi:10.1371/journal.pone.0091235.

Halseth, R. 2015. *The nutritional health of the First Nations and Métis of the Northwest Territories: A review of current knowledge and gaps.* Prince George, BC: National Collaborating Centre for Aboriginal Health. Retrieved from: https://www.ccnsa-nccah.ca/Publications/Lists/Publications/Attachments/141/2015_04_21_RPT_DietNutritionNWT_EN_Web.pdf.

Hamed, I., F. Özogul, Y. Özogul, and J.M. Regenstein. 2015. Marine bioactive compounds and their health benefits: A review. *Comprehensive Reviews in Food Science and Food Safety* 14(4): 446–465. doi:10.1111/1541-4337.12136.

Harper, J.R., J. Haggerty, and M.C. Morris. 1995. Broughton Archipelago clam terrace survey: Coastal and Ocean Resources. Final Report. Sidney, BC.

Herrmann, T.M., Lamalice, A., Coxam, V. 2020. Tackling the question of micronutrients intake as one of the main levers in terms of Inuit food security. *Current Opinion in Clinical Nutrition and Metabolic Care* 23(1): 59-63, doi:10.1097/MCO.0000000000000613.

Huet, C., R. Rosol, and G.M. Egeland. 2012. The prevalence of food insecurity is high and the diet quality poor in Inuit communities. *The Journal of Nutrition* 142(3): 541–547.

Hydro-Québec. 2013. Plan d'approvisionnement 2014-2013 des réseaux autonomes, Annexes. R-3864-2013. Montreal, Canada: Hydro-Quebec. Retrieved 24 March 2019 from http://publicsde.regie-energie.qc.ca/projets/232/DocPrj/R-3864-2013-B-0010-Demande-Piece-2013_11_01.pdf.

Inuit Circumpolar Council Alaska. 2015. *Alaskan Inuit food security conceptual framework: How to assess the Arctic from an Inuit perspective.* Anchorage, AK: Inuit Circumpolar Council Alaska. Retrieved 29 January 2019 from: https://iccalaska.org/wp-icc/wp-content/uploads/2016/03/Food-Security-Summary-and-Recommendations-Report.pdf.

Jackley, J., L. Gardner, A.F. Djunaedi, and A.K. Salomon. 2016. Ancient clam gardens, traditional management portfolios, and the resilience of coupled human-ocean systems. *Ecology and Society* 21(4): art20. doi:10.5751/ES-08747-210420.

Kant, S., I. Vertinsky, B. Zheng, and P.M. Smith. 2013. Social, cultural, and land use determinants of the health and wellbeing of Aboriginal peoples of Canada: A path analysis. *Journal of Public Health Policy* 34(3): 462–476.

Kar, B., R. Sieber, M. Haklay, and R. Ghose. 2016. Public participation GIS and participatory GIS in the era of GeoWeb. *The Cartographic Journal* 53(4): 296–299. doi:10.1080/00087041.2016.1256963.

Kativik Regional Administration. 2014. *Parnasimautik – Report on the consultations carried out with 1260 Nunavik Inuit in 2013.* Retrieved 29 January 2019 from: http://parnasimautik.com/wp-content/uploads/2014/12/Parnasimautik-consultation-report-v2014_12_15-eng_vf.Pdf.

Kenny, T.A., S.D. Wesche, M. Fillion, J. MacLeane, H.M. Chan. 2018. Supporting Inuit food security: A synthesis of initiatives in the Inuvialuit Settlement Region, Northwest Territories. *Canadian Food Studies | La revue canadienne sur l'alimentation* 5(2): 73–110.

Kuhnlein, H. 2006. Indigenous Peoples' food systems for health: Finding interventions that work. *Public Health Nutrition* 9(8): 1013–1019. doi:10.1017/PHN2006987.

Kuhnlein, H.V. and M.M. Humphries. n.d. *Traditional animal foods of Indigenous Peoples of the northern North America: The contributions of wildlife diversity to the subsistence and nutrition of Indigenous cultures.* Retrieved from http://traditionalanimalfoods.org/marine-invertebrates/bivalves/page.aspx?id=6504.

Kuhnlein, H. and O. Receveur. 2007. Local cultural animal food contributes high levels of nutrients for Arctic Canadian Indigenous adults and children. *Journal of Nutrition* 137(4): 1110–1114.

Kuhnlein, H.V. and O. Receveur. 1996. Dietary change and traditional food systems of Indigenous Peoples. *Journal of Nutrition Annual Review* 16: 417–442.

Lamalice, A. 2019. Géographie du système alimentaire des Inuit du Nunavik: du territoire nourricier au supermarché. Doctoral dissertation, University of Montreal, Montreal, Canada and University of Montpellier, Montpellier, France.

Lamalice, A., D. Haillot, M.A. Lamontagne, T.M. Herrmann, S. Gibout, S. Blangy, J.L. Martin, V. Coxam, J. Arsenault, L. Munro, and F. Courchesne. 2018. Building food security in the Canadian Arctic through the development of sustainable community greenhouses and gardening. *Écoscience* 25(4): 325–341.

Lamalice, A., T.M. Herrmann, S. Rioux, A. Granger, S. Blangy, V. Coxam, and M. Macé. Submitted. Imagined foodways: Social and spatial representations of an Inuit food system in transition. *Polar Geography.*

Laurendeau, G. 2018. Étude des caractéristiques et typologie pour une serre adaptée au climat nordique. Unpublished document, Université de Montréal, Montréal.

Laurini, R. 2004. Systèmes d'information pour la participation des citoyens aux décisions relatives à l'aménagement du territoire. *Techniques, territoires et sociétés* 37: 209–219.

Lepofsky, D. and M. Caldwell. 2013. Indigenous marine resource management on the northwest coast of North America. *Ecological Processes* 2(1): 12. doi:10.1186/2192-1709-2-12.

Lepofsky, D., N.F. Smith, N. Cardinal, J. Harper, M. Morris, E.W. Gitla (Elroy White), R. Bouchard, et al. 2015. Ancient shellfish mariculture on the Northwest Coast of North America. *American Antiquity* 80(2): 236–59. doi:10.7183/0002-7316.80.2.236.

Loring, P.A., and Gerlach, S.C. 2009. Food, culture, and human health in Alaska: An integrative health approach to food security. *Environmental Science & Policy* 12(4): 466 – 478.

Loring, P.A. and C. Gerlach. 2015. Searching for progress on food security in the North American North: A research synthesis and meta-analysis of the peer reviewed literature. *Arctic* 68: 380–392.

Macé, M. 2018. *The role of participatory geomatics technologies in supporting community-based food gardening initiatives in northern Canadian Indigenous communities.* Unpublished Master's Thesis. Department of Geography, Planning, and Environment, Concordia University, Montréal.

Makivik Society. 2018. Kuujjuaq hydroponic container – Growing fresh produce in Nunavik. Retrieved 24 March 2019 from: https://www.makivik.org/kuujjuaq-hydroponic-container-growing-fresh-produce-in-nunavik/.

Meakin, S. and T. Kurvits. 2009. *Assessing the impacts of climate change on food security in the Canadian Arctic.* Arendal, Norway: GRID-Arendal.

Moll, R. and B. Davis. 2017. Iron, vitamin B-12, and folate. *Medicine* 45(4): 198–203. doi:10.1016/j.mpmed.2017.01.007.

Moss, M.L. and H.P. Wellman. 2017. The Magoun Clam Garden near Sitka, Alaska: Niche construction theory meets traditional ecological knowledge, but what about the risks of shellfish toxicity? *Alaska Journal of Anthropology* 15(1&2): 7–24.

Northern Food Prices Project Steering Committee (NFPSC), (2003). Northern Food Prices Project Report. Healthy Child Committee of Cabinet, Manitoba Government. Retrieved from: https://www.gov.mb.ca/inr/food_prices/2003report.html. Accessed May 24 2020.

Papatsie, L., L. Ellsworth, S. Meakin, and T. Kurvits. 2013. *The right to food security in a changing Arctic: The Nunavut Food Security Coalition and the Feeding My Family campaign.* Retrieved 29 January 2019 from https://www.nunavutfoodsecurity.ca/sites/default/files/files/Resources/External_Resources/DublinConferenceOnHungerNutritionAndClimateJustice_NunavutCaseStudy.pdf.

Pirkle, C.M., M. Lucas, R. Dallaire, P. Ayotte, J.L. Jacobson, S.W. Jacobson, E. Dewailly, and G. Muckle. 2014. Food insecurity and nutritional biomarkers in relation to stature in Inuit children from Nunavik. *Canadian Journal of Public Health* 105(4): e233–e238.

Price, W. 1939. *Nutrition and physical degeneration: A comparison of primitive and modern diets and their effects.* New York: Paul B. Hoeber, Inc., Medical Book Department of Harper andand Brothers.

Rosol, R., C. Huet, M. Wood, C. Lennie, G. Osborne, and G.M. Egeland. 2011. Prevalence of affirmative responses to questions of food insecurity: International Polar Year Inuit Health Survey, 2007–2008. *International Journal of Circumpolar Health* 70(5): 488–497.

Rosol, R., S. Powell-Hellyer, and H.M. Chan. 2016. Impacts of decline harvest of country food on nutrient intake among Inuit in Arctic Canada: Impact of climate change and possible adaptation plan. *International Journal of Circumpolar Health* 75: 31127. doi:10.3402/ijch.v75.31127.

Schroeder, K. and A. Smaldone. 2015. Food insecurity: A concept analysis. *Nursing Forum* 50(4): 274–284.

Sharp, G., M. Allard, A. Lewis, R. Semple, and G. Rochefort. 2008. The potential for seaweed resource development in subarctic Canada; Nunavik, Ungava Bay. In M.A. Borowitzka, A.T. Critchley, S. Kraan, A. Peters, K. Sjøtun, and M. Notoya (eds.), *Nineteenth International Seaweed Symposium. Developments in Applied Phycology, vol 2.* Dordrecht: Springer, pp. 41–48.

Sheehy, T., Roache, C., and S. Sharma. 2013. Eating habits of a population undergoing a rapid dietary transition: portion sizes of traditional and nontraditional foods and beverages consumed by Inuit adults in Nunavut, Canada. *Nutrition journal*, 12, 70. https://doi.org/10.1186/1475-2891-12-70.

Sheehy, T., F. Kolahdooz, S.E. Schaefer, D.N. Douglas, A. Corriveau, and S. Sharma. 2015. Traditional food patterns are associated with better diet quality and improved dietary adequacy in Aboriginal peoples in the Northwest Territories, Canada. *Journal of Human Nutrition and Dietetics* 28(3): 262–271.

Skinner, K., R.M. Hanning, J. Metatawabin, and L.J.S. Tsuji. 2014. Implementation of a community greenhouse in a remote, sub-Arctic First Nations community in Ontario, Canada: A descriptive case study. *Rural and Remote Health* 14: 2545.

Smith, N. F., D. Lepofsky, G. Toniello, K. Holmes, L. Wilson, C.M. Neudorf, and C. Roberts. 2019. 3500 years of shellfish mariculture on the Northwest Coast of North America. *PLOS ONE* 14 (2): e0211194. doi:10.1371/journal.pone.0211194.

Statistics Canada. 2017a. Région du Nunavik [Health region, December 2017], Quebec and Quebec [Province]. Census Profile. 2016 Census. Statistics Canada Catalogue no. 98-316-X2016001. Ottawa. Retrieved 30 January 2019 from https://www12.statcan.gc.ca/census-recensement/2016/dp-pd/prof/index.cfm?Lang=E.

Statistics Canada. 2017b. Kuujjuaq [Population centre], Quebec and Quebec [Province]. Census Profile. 2016 Census. Statistics Canada Catalogue no. 98-316-X2016001.

Ottawa. Retrieved 30 January from https://www12.statcan.gc.ca/census-recensement/2016/dp-pd/prof/index.cfm?Lang=E.

Statistics Canada. 2017c. Kangiqsujuaq, VN [Census subdivision], Quebec and Nord-du-Québec, CDR [Census division], Quebec. Census Profile. 2016 Census. Statistics Canada Catalogue no. 98-316-X2016001. Ottawa. Retrieved 30 January 2019 from https://www12.statcan.gc.ca/census-recensement/2016/dp-pd/prof/index.cfm?Lang=E.

Thompson, S. 2015. Flooding First Nations and environmental justice in Manitoba: Case studies of the impacts of 2011 flood and hydrodevelopment in Manitoba. *Manitoba Law Journal* 38(2): 220–259.

Thompson, S., A. Gulrukh, M. Ballard, B. Beardy, D. Islam, V. Lozeznik, and K. Wong. 2011. Is community economic development putting healthy food on the table? Food sovereignty in northern Manitoba's aboriginal communities. *Journal of Aboriginal Economic Development* 7(2): 14–39.

Thompson S., J. Wiebe, A. Gulrukh, and A. Ashram. 2012. Analysing food-related economic development in Indigenous communities in northern Manitoba for impacts on food sovereignty, food security and sustainable livelihoods. *Canadian Journal of Nonprofit and Social Economy Research* 3(2): 43–66.

Thompson, S., Rony, M., Temmer, J., & Wood, D. 2014. Pulling in the Indigenous fishery cooperative net: Fishing for sustainable livelihoods and food security in Garden Hill First Nation, Manitoba, Canada. *Journal of Agriculture, Food Systems, and Community Development*, 4(3), 177–192. http://dx.doi.org/10.5304/jafscd.2014.043.016.

Thompson, S., Thapa, K & Whiteway, N. 2019. Sacred Harvest, Sacred Place: Mapping harvesting sites in Wasagamack First Nation. *Journal of Agriculture, Food Systems, and Community Development*, 9 (1), 1–29. https://doi.org/10.5304/jafscd.2019.09B.017.

Thompson, S. & Pritty, P. 2020. Damming Food Sovereignty of Indigenous Peoples: A Case study of Food Security at O-Pipon-Na-Piwin Cree Nation. In Settee and Shuklah (Eds.), *Indigenous Food Systems: Concepts, Cases, and Conversations*. Toronto: Canadian Scholars Press.

Toniello, G., D. Lepofsky, G. Lertzman-Lepofsky, A. K. Salomon, and K. Rowell. 2019. 11,500 y of human–clam relationships provide long-term context for intertidal management in the Salish Sea, British Columbia. *Proceedings of the National Academy of Sciences* 116(44): 22106–221014. doi:10.1073/pnas.1905921116.

Tough, F. 1996. *'As their natural resources fail': Native Peoples and the economic history of northern Manitoba, 1870–1930*. Vancouver: UBC Press.

TRC. n.d. *Indian residential schools truth and reconciliation commission*. Retrieved 11 May 2018 from http://www.trc.ca/websites/trcinstitution/index.php?p=39.

Tsosie, R. 2007. Indigenous People and environmental justice: The impact of climate change. *University of Colorado Law Review* 78, p. 1625.

Turner, N.J. and K.L. Turner. 2008. Where our women used to get the food: Cumulative effects and loss of ethnobotanical knowledge and practice; case study from coastal British Columbia. *Botany* 86(2): 103–115.

Vía Campesina. 2007. Nyéléni declaration Sélingué, Mali: Forum for Food Sovereignty. Available from: https://nyeleni.org/spip.php? article290, Accessed 24 May 2020.

Willows, N.D., A.J. Hanley, and T.A. Delormier. 2012. A socioecological framework to understand weight-related issues in Aboriginal children in Canada. *Applied Physiology, Nutrition, and Metabolism* 37(1): 1–13.

15 Building traditional food knowledge

An approach to food security through North-South dialogue

Anne Ouma

Introduction

The relationship between governance and food security in the Arctic region is evolving to include more complex partnerships among members of civil society, non-governmental actors, academia, economic agents, and various levels of government (Petrov et al. 2016). Interdisciplinary collaboration is needed more than ever to study and comprehend the challenges of food security and solutions to it. These collaborations for instance must examine governance partnerships and provide recommendations (Petrov et al. 2016). Addressing some key components of the governance of natural resources could ensure sustainable and equitable food security in the current rapid socioeconomic and sociocultural transformation in the Arctic region (Kurtz 2013). Indigenous populations′ worldview define sustainable development and land stewardship by emphasising the importance of these activities to maintain ecological health and food security, as well as to secure access to resources from the land. Indigenous societies and communities argue that stewardship of the land partially defines their identity, which is viewed as significant to decisions pertaining to the sustainable governance of food systems and natural resources (Berkes 2009). Indigenous knowledge systems are based on reciprocity and mutual relationships between humans and nature within a holistic worldview that is anchored in specific spatial contexts (Cunningham 1993). They are combined knowledge systems that encompass technological, social, economic, philosophical, educational, legal, and governance systems. In Oguamanam's description, the ideology of nature is portrayed as an organic entity and humanity as part of an integral whole. This is a theory of life 'whereby unity in the diversity of life, forms a synthesis of materialism and deep spiritualism' (Oguamanam 2006, 53).

This chapter discusses some of the immediate and long-term challenges currently facing the Arctic. The Arctic (and other parts of the globe) currently faces diverse environmental, economic, and social challenges that impact food (in) security. This chapter's empirical contribution is informed by participatory, community-based research that builds on knowledge and understanding of the local dimensions of food. The chapter explores how we

DOI: 10.4324/9781003057758-15

might find a qualitative grounding for the ideas of biocultural diversity and food (in) security within the personal stories shared in the talking circle. It also examines how various actors perceive current governance structures and the direct or indirect relationship of land stewardship to food (in) security. The chapter further explores the foundation for Indigenous governance of the environment, which is based on spiritual principles and provides just, forward-looking governance of natural resources, particularly food resources. It discusses how the findings are applicable in the broader context of Arctic research with examples from other communities with in Northern and Southern countries. The stories from the talking circle examine processes of socio-ecological change that impact food security. They assess the research, education, and communicative strategies required to address possible change and active community wellbeing. This includes developing strategies for relationship building, which is the foundation for Indigenous governance of food and natural resources.

Methodology

The 2011 Coastal CURA People in Places Conference focused on the inclusion of Indigenous communities in integrated resource management. This focus created an ideal opportunity for sharing knowledge in a format based on Indigenous talking circle traditions. Talking circles are one of several Indigenous methodologies which have a non-hierarchical form of dialogue. Participants sit in a circle; this symbolises equal respect for and the equal importance of the ideas and stories of all participants. All participants are given the opportunity to speak without interruption (Simmons et al. 2012). An object obtained from nature (e.g., a stick, feather, or tobacco pouch) is passed around the circle to indicate whose turn it is to speak. A participant can also decide to keep silent and pass the object to the next person. Indigenous Peoples have traditionally used talking circles to solve problems. A talking circle is a very effective way to remove barriers and to allow people to express themselves with complete freedom. As Muin'iskw and Crowfeather (2016, 1) explain, 'The symbolism of the circle, with no beginning and with nobody in a position of prominence, serves to encourage people to speak freely and honestly about things that are on their minds'. Traditional talking circles have been used to create a culturally safe space for Indigenous women to talk about healthcare experiences (Kurtz 2013).

Because the Coastal CURA People in Places Conference focused on the inclusion of Indigenous communities, talking circles were chosen as a methodological portion of the conference proceedings. The aim of the talking circle was to provide a space for dialogue and for an understanding of the role of traditional environmental governance knowledge in natural resource management. This discussion acknowledged the idea of agency that underpins how Indigenous Peoples think and relate to nature. It aimed to start new or (in some cases) renewed relationships that could generate new possibilities and confidence for turning knowledge sharing into solidarity in action.

It was imagined that the forum could help build relationships and that participants might feel comfortable continuing conversations from the talking circles during the rest of the conference or even beyond the conference. The conference talking circles focused on sharing experiences with Indigenous traditional knowledge related to environmental governance. The participants in the talking circles were conference attendees who enrolled in the talking circle workshop. A total of 34 people sat in the circle for at least one of the two 1.5-hour afternoon sessions. This schedule was planned to allow conversations to fully develop. Chairs were set up in a circle. The 34 participants shared their homelands or research sites, and the geographical spread was impressive. They came from across Canada, the United States, Latin America and the Caribbean, and Africa. The talking circle revealed a remarkable range of struggles and positive experiences for both Indigenous researcher–practitioners and non-Indigenous allies. It provided a space where participants (including elders of First Nations in Canada; master's students from Kenya, Canada, and Sweden; practitioners from Ecuador; and university researchers) from around the world, with diverse backgrounds and experiences, could feel comfortable sharing their experiences in working with Indigenous or traditional knowledge of food resources. The workshop attracted considerable interest, partly because of the format itself. As one participant said, 'I come to circles because I know there won't be somebody talking with a PowerPoint'.

The synthesis of key messages was enhanced by the range of perspectives provided in participant's narratives. The talking circle was very fertile in a number of respects. Contributions were oral and not based on prepared written documents or slide presentations. There were three facilitators; one opened the discussion with a quiet moment and then passed the tobacco pouch to the next speaker. The tobacco pouch went around the circle several times. The talking circle was audio recorded so that contributions could be transcribed and included in published proceedings. The session coordinators provided participants with a formal opportunity to provide consent for these forms of documentation. The audio recording was supplemented by written notes to ensure that the names of speakers were recorded as a reference for the transcriber. Photos were taken, and, to provide a visual representation of the geographical/cultural spread of the participants, a Google Earth map was displayed, and participants' homelands and/or research locations were marked. As envisioned by the co-facilitators, the talking circle approach was necessarily hybrid in this international context as it bridged multiple non-Indigenous and Indigenous cultures. It contained ceremonial aspects rooted in Cree traditions. [For other accounts of talking circle processes, see Simmons et al. (2012).] The talking circle allowed for a certain flexibility and informal approach that enabled people from different backgrounds to feel comfortable.

This chapter uses a form of thematic narrative analysis to examine the content of the stories. A thematic narrative approach, according to Sandberg (2011, 46), focuses on themes in a story to give a sense of direction and purpose. The emphasis then is on the content of the message rather than language, form, or interactions. A thematic narrative analysis can aid comprehension of how stories are understood in the context of social, spatial, and environmental processes and spaces. It can help unearth contextual aspects of the stories given the diverse locations represented in the circle. In the transcribed stories, similarities and differences appear, and common themes emerge and evolve through the analysis. A contextual interpretation was used to develop these themes. Several common themes and concerns emerged from the stories shared in the talking circle. These themes were interwoven in a complex fabric that incorporated participants' multiple, diverse stories. Three key themes were used to code the transcripts: socioecological and governance changes that impact Indigenous knowledge of food resources; conflicts around food security and threats to traditional ecological knowledge (TEK); and strategies for partnership and relationship building (including resilience). Wiles et al. (2005) discuss a contextual analytical approach for interpreting stories. To better understand the themes and their implications, it was necessary to place the issues and understandings in context. This helped reveal some salient tacit meanings in the stories that could be used to address the key questions. Context was particularly important given the diverse geographical spread of the circle participants.

Consent to use the transcribed stories and narratives was obtained from the primary facilitator, who was also responsible for documentation.

Figure 15.1 Talking circle, People in Places: An International Conference, 26–29 June 2011, Halifax, NS.

Theoretical framework

In the Indigenous worldview, humans are part of rather than separate from the natural world. This worldview recognises links between cultural identities, language, and land resources, including food resources (Blythe and McKenna Brown 2004). Morphy (1995) discusses the inherent tie between Indigenous knowledge and the land, while Frawley (1999) recognises the connections and understanding of place and space found in Aboriginal knowledge of natural resources.

Other literature discusses the role of the land in place and space with an emphasis on 'maintaining the integrity of the land and food resource itself' (Battiste 2002, 13). This interface between nature and culture exists within the complex interactions between people and the environment over time (Rössler 2005). Maffi (2005) discusses people's knowledge and practices, highlighting the role of language as a vehicle for communicating and transmitting cultural values (which include knowledge of food security and practices), while also mediating interactions and mutual adaptations between humans and the environment. According to Battiste (2002, 2), about 20% of the world's population engages with TEK systems, which are described as

> knowledge which comprises the complex set of technologies developed and sustained by Indigenous civilisations; this knowledge which is often oral and symbolic, is transmitted through the structure of Indigenous languages. It is passed on to the next generation through modelling, practice and animation, rather than through the written word.

Silverman (2011) proposes that culture and links to place are both contested concepts. Distinct local communities are attached or linked to their local biological environments through a network of feedback webs that have been subjected to scholarly debate. This is in contrast to viewpoints that depict humans as external to and separate from nature; in this view, human interactions with nature are based on efforts to dominate it (Maffi 2005).

TEK is 'the sum total of local knowledge and skills on ecology, unique to places and societies, which people in a particular geographic area possess and which enables them to get the most out of their natural environment' (Grenier 1998, 2). The knowledge of food systems embedded in TEK can often be difficult to systemise as it is embedded in collectively owned socio-cultural values and belief systems, rituals and practices, governing institutions, and relationships (Tanzania Gateway 2010). Battiste and Henderson (2000) call TEK a complete knowledge system embedded within wider knowledge systems; all these systems are dynamic and have spatial, temporal, and place-specific dimensions as well as their own philosophical, epistemological, scientific, and logical validity. Some scholars further expand the definition of TEK to describe a holistic worldview that is embedded in communities. While contextually based within the local language,

TEK includes environmental practices that are based on reciprocity, obligations to community laws, communal resources, and other life forms. They are management institutions based on sharing and meaning (Oguamanam 2006). TEK can be said to invoke mutual relationships between humans and nature (within the complex kinship systems of relationships among people, animals, the earth, the cosmos, etc.). They are founded on a sociocultural milieu that sustains a belief in complex spiritual and social relationships among all life forms (Ermine 1995).

Gibson (2004) situates TEK in relation to increased mobility and governing economic dynamics. The argument is that rights to traditional knowledge and food resources are determined by historical, temporal, geographical, and spatial dimensions. Gibson argues that, within communities, individuals emerge who claim commercial rights to what have previously been seen as communal resources. Hart (2002) discusses the extreme inequalities that exist during the application of neoliberal policies of local governance and land policy. Hart also provides tangible recommendations for land governance policies and activism, advocating for the 'disarticulation of land from natural resource management and agriculture and the re-articulation of its significance in terms of its potential contribution to a social wage' (Hart 2002, 305).

Frawley (1999) also considers the relationship of TEK to place and space, discussing the interconnectedness and understanding of place, space, and cultural diversity as part and parcel of Aboriginal knowledge of land stewardship:

> For Aboriginal Australians, nature and culture are inextricably bound together in the Dreaming – the time when the world, including Aboriginal people and their law, was created. Belief systems associated with the Dreaming link specific places with Dreaming events and give every person, living and dead, a place within a physically and spiritually united world. The landscape is not therefore a composite of external objects, but is made up of culturally defined features of mythical significance.
>
> (1999, 272)

Indigenous communities are and have been engaged in an array of diverse efforts to reverse the erosion of TEK, by reclaiming a reverential relationship with the sacred and spiritual worlds in an attempt to restore balance. It is argued that this could provide a framework to address food security. Sacred place(s) and space(s) are inherently tied to specific landscapes where harvest and planting ceremonies are held; sacred places define spiritual connectedness (Morphy 1995). Frawley (1999) adds that this interconnectedness and an understanding of place and space are central features of knowledge of food resources and the mythical significance of those resources. UNESCO's discussion of approaches during the UN Decade on Biodiversity (2010–2020) refers to 'moral values, norms, and traditions that will be needed to reshape our relationship with the living world of which we are part of (UNESCOS Biodiversity Initiative 2011, 4).

Morphy's (1995) research on spiritual connectivity is exemplified by the harvest and planting season ceremonies in Misungwi Tanzania. These ceremonies include the use of staple fodder and animal products such as milk, ghee, sorghum, and millet. Participants adhere to a special dress code in a specific ecological space that is recognised as sacred. These spaces are believed to enable food resources to acquire spiritual power before and after the planting and harvesting seasons (Ouma 2013).

The erosion of TEK and specifically of knowledge of food security has been widely discussed. Hoppers (2002) argues that 'the erosion of people's knowledge in general in Africa associated with food and natural resources is said to be under greater threat than the erosion of the food and natural resources themselves' (2002, 7). Some postcolonial alterations in land policies and land rights in Tanzania and Kenya influenced the management of TEK. For instance the abolition of chiefdoms in Tanzania and therefore their role as administers and overall managers of the custodianship of TEK within communities was truncated. New alternative administrative and managerial structures were put in place by the colonial project; (Bukurura 1994). The resulting alterations affected the control and management systems of TEK (Mshana 2002). The original chiefdom systems played a central role in managing TEK of food resources, land tenure, farming systems and practices, and related governing structures. In Tanzania, the Dagashida Institution of the Sukuma Societies, which has existed for hundreds of years, works with decision-making processes, with linkages to TEK governance and its ability to resolve conflicts (Brandström 1990). This could have contributions that relate to food security in communities. Previous research in Mwanza Tanzania (Ouma 2013) provides an example of the contemporary use of local TEK to predict the success of crop harvests which could determine food (in) security for the local communities (Figure 15.2).

Whittaker (2015) examines the role of TEK in sustainable food resourcing in rural Yuan Yang China and discusses how the use of TEK in food sourcing provides communities with stable and diverse diets and enables various ecological niches to connect into a resilient whole.

Food Tank (2018) Indigenous Network (https://foodtank.com) discusses ways to improve the resilience of food systems. It also highlights the key role Indigenous Peoples play in sustainable farming and food security around the world due to their traditional knowledge and understanding of land governance, ecological systems, and local biodiversity. The resilience of Indigenous farming and its contributions to food security can be seen in examples of traditional farming practices from around the globe. Examples include the use of agroforestry by some Amazonian Indigenous communities and in Latin America, the Caribbean, and Ghana; crop rotation practices in Malawi; sustainable honey harvesting by Ogiek Indigenous communities, amidst biodiversity erosion through deforestation and challenges to ancestral lands; the Center for Integral Small Farmer Development (CEDICAM) in Mexico, which integrates sustainable agricultural techniques to enhance local food security; Soils, Food, and Healthy Communities (SFHC) in Malawi, which uses local Indigenous knowledge and agro ecological methods to improve food security and nutrition in Malawi; and Navdanya, an NGO founded by scientist Vandana Shiva that is involved in the rejuvenation of Indigenous knowledge and culture through biodiversity-based organic farming methods and the development of locally adaptable cropping practices that provide sustainable food resources.

A study on the socioeconomic, environmental, and governing structure pressures on food resources in diverse ecological settings in Thailand demonstrates that appropriately used and adapted TEK could play a significant role in addressing the food security of small rural farmers (Ekarin et al. 2016).

This chapter discusses presentations from talking circle discussions by First Nation members, community-based management and networking organisations, members of civil society, community leaders, small-scale natural food resource industries, students, and academic scholars that were compiled after the second International Coastal CURA Conference. This conference, titled People in Places: Engaging Together in Integrated Resource Management, took place in June 2011 in Halifax, Canada. Coastal CURA was a six-year project (2006–2012) and was structured as a community–university research alliance. Partners included First Nations communities, fishery organisations, university researchers, students, and government agencies. It was funded by the Social Sciences and Humanities Research Council of Canada. The project aimed to build knowledge and capacity across the Canadian Maritimes and to support community involvement in managing food resources from coasts and oceans. It is linked to actors supporting the ecological, social, and economic wellbeing of place-based communities by exploring how communities, food resource organisations, and governments could work together to manage coastal and ocean food resources. The project reviewed past experiences and engaged in innovative participatory research of the management approaches of community-led and government organisations across the Canadian Maritimes. The CURA project applied three key strategies: research, capacity building, and knowledge mobilisation. Its defined goals were: to improve the effectiveness of governance of natural/coastal resources; to enhance community capacity to participate in coastal management; to develop a Maritimes network allowing fishers to engage in community-level coastal governance; and to research innovation and the generation of local food knowledge.

The key questions of this chapter are: (1) What key shared issues and challenges in Indigenous environmental governance impact food (in) security? (2) How can North–South dialogue about Indigenous environmental governance initiatives inform policy on food (in) security?

Results

Negotiating a governance system: Links among places, natural resources, culture, and food in Indigenous contexts

Stories from the talking circle particularly emphasised the worldview that different pieces of the biocultural diversity web are linked. Some links among place, food, culture, and natural resources in Indigenous contexts are mentioned in the following quote:

> Aboriginal people, I'm learning now, don't live by numbers. They live by concepts like balance, ecosystem. ʔehtsə̀o was the happiest when he went to a certain place and he found fish. He went to another place and he found moose. Another place and he found caribou, beaver. When he saw that, he was happy because he was part of that, balancing those things. That's what made him happy.
>
> (First Nations talking circle participant Canada)

Some communities are attempting to negotiate a balanced approach to environmental governance in order to secure sustainable food resources. This resonates with earlier research on the interface between human, other beings, place, and space in relation to Indigenous knowledge. For example, Blythe and McKenna Brown (2004) and Morphy (2005) discuss the inherent relationships and links between all life forms and the Indigenous view of humans as one actor in a whole ecosystem. In this view, balance is a key concept. A participant shares a collective East African way of managing food resources:

> We do not sell fish. We share. In every coastal community in Kenya, [in] the small-scale fishing communities you visit, you'll find that we share with the whole. We are all one. In our country we call it the spirit of *ubuntu*. We have to care about other people. But why have we lost the way of our forefathers? Why does the tide change?
>
> (Indigenous talking circle participant from Kenya)

The African term *ubuntu* is the belief in a universal bond of sharing that connects all humanity. These Kenyan communities view the fish they catch as a shared resource for the community. However, this view is under threat due to changes in governance and livelihoods (Hoppers 2002). Another quote also emphasises the Indigenous view of the interconnectedness of everything in the universe, including food resources:

> I'm making the choice today to walk a Dene path today, a true path, the path of a true human being. I want to be out on the land. I want to be thinking like ʔehtsə̀o. Harvesting. Teaching. Passing my language on to my grandchildren. ʔehtsə̀o's worldview is centred on Néwhehtsı̨nę̨, the

> Creator. But he never talked about Néwhehtsı̨nę̨. His belief is learning about Dırınéné, the land, the water, food resources everything in the Universe. The more you learn about this, the closer you are to Néwhehtsı̨nę̨. Everything has to be related and must be treated with respect. If things are not related, then there's a problem. And things have to be alive to build relationships. To respect something, it has to be alive, it has to have its own being. In our language there's no such thing as death.
>
> (First Nations talking circle participant from Canada)

> We saw Bear River history in [the] Mi'kmaw and English languages. F. has a wonderful way of relating to this earth, himself, and the creator. He talked about the seven signs in his tepee. I recall six of these: respect, bravery, honour, humility, sharing, and love. He explained how being at the camp teaches our children how to be true human beings. The land teaches us to be human. We listened to some of their music as well and shared ours (I carry an iPod with most of my grandfather's songs).
>
> (First Nations talking circle participant Canada)

A similar idea about guiding signs (as mentioned in the aforementioned citation) is also used in Tanzanian harvest ceremonies:

> Four ways are used: south, north, east, and west; spears and fly whisk, *mkia wa nyumbu* [the tail of a wild animal] … black clothes, special songs; there are women who *piga vigelegele* [sing ceremonial songs].
>
> (Talking circle participant from Tanzania)

Some readings depict humans as external to and separate from nature; in this view, human interactions with nature are based on efforts to dominate it (Anderson 2010). Indigenous environmental governance views humans as a part of the natural world rather than separate from it. This view is common to Indigenous societies around the world (see Figure 15.4) that define biocultural diversity through intense links to the land, culture, and food resources. The ethos of this worldview entails treating all living things, including food resources, with respect given the interrelationships that sustain everything in the universe. The role of language in transmitting TEK (see Maffi 2005) is illustrated in the quotations above and will be discussed in more detail in a later section of this chapter on conflict and resolution strategies. The central role of respect is also mentioned in the context of the governance of forest resources, food, and water:

> For example, we spent three years working on a new Wildlife Act for the Northwest Territories. I was with that group. We had four lawyers and all the co-management boards working together on the new Act. All that time I was thinking, 'We can't put these laws into ʔehtsə̀o's world.' One day somebody said, 'Let's put our Dene concept of respect in the

> Act.' Right off the bat, we were told that this wouldn't be possible. And it's true. How can a person be charged for not respecting wildlife or the forest or the water or food resources from the land? The community themselves have to maintain their own principles and laws. That's what self-government is all about. I asked myself, 'Why is it so hard to do this?'
>
> (First Nations talking circle participant)

In some cases, governance of food and land resources by formal governing bodies in the form of agreements, treaties, and joint decisions is problematic:

> What is frustrating is that our government is not listening to us regarding the livelihoods around fishing. Nobody listens.
>
> (Talking circle participant from Kenya)

> But it's very delicate. You can kill Aboriginal knowledge very easily by using the wrong methodology in the government work with agriculture.
>
> (Talking circle participant from Mexico)

> Without doubt, the government doesn't really want to be part of what I prefer to call joint management. They use 'co-management' and actually recognise only advisory processes. My view of joint management is that it should be about jointly making decisions. I don't know if there's any setup anywhere in this country where First Nations are true partners with government. I could never understand why, because they're newcomers to our land, our territories, and our resources. The five First Nations within this region have not been part of land use planning. The land use planning in this region is not totally finished, but you might as well say it's finished, because the government's come and told us they've done all the work. I've always been a rebel and I've always had an issue with the decision-making process, because it has never involved us. I was never any kind of decision maker because the system always made sure that people like myself were out of the process.
>
> (Talking circle participant from Canada)

The hegemonic, unequal dominance between formal and informal (Indigenous) land governance structures is clear in agreements and treaties on land governance. This provides avenues for conflict within communities that could relate to Indigenous land stewardship using local food knowledge and food security. De Sousa Santos (2007) and Gibson-Graham (2006) suggest counter-hegemonic processes, advocating for actors governing land and natural resources to re-engage with activism and tangible recommendations for land governance policy. Formal agreements and treaties should be realigned with a TEK base so they are in tune with communities' ways of governing and managing food resources. These stories and scholars describe the historical and contemporary governance structures and organisational approaches to land stewardship that have defined and continue to define a contested space within communities. Formal and informal governance of land, marine, and related food resources is often contested. This chapter considers the relationships among land stewardship, governance, and decisions about food systems based on talking circle discussions. The discussion on conflicts will be developed later in the chapter.

Food security, conflict, and threats to TEK

The stories shared by talking circle participants point out some conflicts of interest between Indigenous communities and non-Indigenous fishermen with official fishing licenses. This poses threats to TEK. Hoppers (2002) and Oguamanam (2006) discuss the often contrasted and divided paradigm between formal and informal land stewardship. Indigenous communities' worldview positions them as custodians of the land (see Figure 15.3 &15.4), while formal governing structures honour formal agreements:

> The fishermen and the natives were not cooperating at all. Boats were being burned. Wharves were being burned. In my area, there were a lot of fishermen coming with boats, with their lobster boats inside the harbour blocking the harbours. Sitting inside a boat for two or three days is probably not what those fishermen wished to do. They were going to the taverns a lot and they were getting drunk, and they had rifles aboard. It was getting to be a very trickish situation.
>
> (Talking circle participant from Canada)

Here, the Indigenous worldview of humans as custodians of fishery resources conflicts with the actions of formally licenced fishers. The lack of cooperation could be due to the absence of collaborative processes and frameworks that would enable both groups to use and manage marine food resources. Formal postcolonial natural resource policies and rights regarding the control and management of TEK (Mshana 2002) created changes and then conflicts over communities' access to food resources, especially as licences were required to access communal food resources. This conflict appears in quotes about other countries as well:

> I am from the Western Cape of South Africa, and today with a democracy we feel it is worse, as access to our food and livelihoods is cut off by our own government. That is our experience. That is why I have so much interest in what we call in our language 'Indigenous knowledge.' The sharing of cultural practices and beliefs and some lessons my grandparents, aunties, and uncles told me they learned when growing up. This transfer has been completely disconnected.
>
> (Youth talking circle participant from South Africa)

> My big concern is that, in the context of South Africa – given the colonial and the apartheid past and the layers of law that have led to marginalisation of coastal communities – along with that has been an erosion of peoples' customary food practices and also their knowledge systems ... In many instances, fishers have become even more marginalised. Although there are signs of change, and there is new hope, I think. ... It's probably a rage I have about the lack of respect amongst many in our scientific community for the value of local food knowledge. I want to hear from other people about how one gains the respect of scientists. I'm not using that word in a blanket sense; I'm referring to certain members of the scientific community who are in very powerful positions, who are informing management recommendations or informing government about policies and about management measures about the value of hearing about the knowledge from local fishing communities. I'm also interested to learn how it's possible to incorporate this local knowledge into mainstream scientific processes.
>
> (Talking circle participant from South Africa)

These stories describe the hierarchical relationship discussed by Dahlström et al. (2006), which dictates and is closely related to the power of the preferential right of interpretation. This includes what is considered appropriate knowledge, as well as access to customary food practices and their related knowledge systems about the land and governance of natural resources. Talking circle participants' views that knowledge about community-based stewardship was not treated with respect, placing TEK in a hierarchical relationship with formal governance. This view is in line with Beyer (2009), who discusses how societies evolve into cultures that devalue their own Indigenous knowledge.

Participants described some of the threats to TEK:

> The *Namgis* have a traditional territory of about 2,800 square miles that includes the largest watershed on Vancouver Island ... [It's] famous for its sockeye salmon, the Nimpkish River, and for the height and the breadth of the trees. It's been horribly abused, as have been most of the watersheds on Vancouver Island. Since 1950, there have been sixty million cubic meters of wood taken out by Western Forest Products.
>
> (Talking circle participant from Canada)

> One issue that we're working on is fish farming. In the territory of the *Namgis* and the surrounding neighbours, there are thirty Atlantic salmon open net cage fish farms with a million fish each. Two and a half kilograms of faeces out of each fish in its lifetime. That stuff wafts, fills up clam beaches. The mooring buoys for the fish farms keep people away from places where they've fished for thousands of years. The scientists on both sides build up their cases.
>
> (Talking circle participant from Canada)

These quotes describe the threat to TEK posed by activities related to fish farming and forestry. These could also be viewed as threats to communities' food security, specifically to their access to marine food resources.

Conflict mediation, consensus building, and resilience

These stories describe conflicts between Indigenous and non-Indigenous communities and between communities and governing structures. However, emerging strategies for mediating conflict over food resources were also described. One example is the adoption of the talking circle model to resolve tensions and conflicts over marine food resources:

> With the ever-increasing tensions and violence between Indigenous and non-Indigenous fisher people over the lobster fishing during that period, F. was starting to think that he should resign as Chief. … Later, outside people wanted to know what happened. They all looked surprised when they were told that the group had not talked about fish – they had talked about everything else except fish. Amazingly, they seem to have resolved the issue, without talking about it. And so, without realising it, F. received the gift of the talking circle, our native way of having meetings.
>
> (Talking circle participant from Canada)

> I mean, there was a conflict going on that was fishery related. Tensions were rising in our area. A few days later there was a secret meeting with the Chief from Acadia First Nations and another Chief and five fishermen's representatives.
>
> (First Nations talking circle participant)

> Regarding natural resource management, Aboriginal natural resource management in Mexico varies a lot from region to region – the land holdings that we have, the form of tenure that we have in Mexico. We have two forms of land tenure and that is *ehildo* – is common holding – and also *communidad* – not in a geographic meaning but *communidad* as a form of land tenure. In these common holdings, working food agriculture is like community-based management. They decide on what they want to plant and about the [inaudible] harvest as well. Regarding all their resources – for example, rain forest – some of the common holding …require permits for commercial logging. Some others, they're small size. Nearly 5,000 hectares. They don't have permission, and they only exploit the rain forest and gathering of non-commercial forest food products.
>
> (Talking circle participant from Mexico)

> However, in 1994, when we secured our Interim Measures Agreement with the government of British Columbia for co-management of land and resource use and operations in Clayoquot Sound during treaty

> negotiations, that was joint management. We have responsibilities in ecotourism, forestry, energy, and other sectors. We're busy every day. Some days we cry almost at the lack of respect and the barriers placed in the way of community-based land stewardship. Every day, whether we're challenged or not, luckily, there's also a lot of laughter. It's addictive, this work, because it's the real work. It's bringing power back to place.
>
> (First Nations talking circle participant)

The narratives of the participants propose learning, communicative strategies, and policy recommendations that could inform active community wellbeing through engagement in food resources and governance practices (Berkes 2009). The development of strategies to build relationships is the foundation of Indigenous governance and land stewardship (Frawley 1999). This could create space for developing models of food security that could complement other official models and address wellness in communities.

These examples of emerging forms of collaboration among communities around marine, land, and food resources demonstrate the struggle which has sometimes characterised relationships between non-Indigenous and Indigenous Peoples, as discussed in Fatnowna's (2002) research on Aboriginal communities in Australia.

Talking circle participants' stories also resonate with the work of Berkes et al. (2017), who discuss the interrelatedness of different parts of the environment and propose that the key to sustainable development is in balancing socio-ecological systems. The idea of a balanced ecosystem was referenced frequently during the talking circle. It was mentioned in reference to access to fish, caribou, and moose as food resources (e.g., Figure 15.3). In this discussion, joint agreements coupled with resilience approaches that place reconnecting with the land and the revitalisation of land-based activities at the forefront were proposed.

Indigenous food knowledge in an international dialogue

Stories from the talking circle unveil a form of global dialogue between Indigenous and non-Indigenous Peoples. Some scholars argue that TEK is inadequate for guiding sustainable harvesting (Mollel 1994). This is partly because they consider TEK dated and unable to adapt quickly to rapid global ecological change. The dynamic and non-static nature of TEK has, however, been widely discussed (Fatnowna et al2002; Gibson 2002; Ekarin et al. 2016). Berkes (2009) argues that the processes of acquiring and transmitting Māori traditional knowledge about food security are essentially similar to those used in Canada. Berkes' research argues that debating the merits of 'science' versus 'traditional knowledge' is a waste of time and effort. Instead this debate should be reframed as a dialogue and partnership between science and traditional knowledge. This chapter argues that approaches to addressing food security through governance and communal management are fundamentally similar around the globe.

The following quotations speak to the global nature of these challenges and opportunities:

> I learned that there are many Indigenous People in other places who think the same as we do in the Sahtu Region of the Northwest Territories. When you experience this, you realise that ours is truly a global community. We have to talk to a lot of people, other First Nations across Canada and the world, to look for solutions to our own challenges regarding access to land and natural resources.
>
> (First Nations talking circle participant from Canada)

> Interpretation of our common world system is a complex task. It's easier for me to use both systems. Now we're saying, 'Let's put it on paper.' And we can make one more step of taking stories and making them into policies, using our languages, our concepts. That would be a huge step forward for Indigenous communities. That's one of my goals.
>
> (Talking circle participant from Ecuador)

The stories in the talking circle highlight the importance of Indigenous and non-Indigenous Arctic communities' engagement with communitics in other parts of the world. The stories advocate for and encourage an exchange of knowledge and approaches to Indigenous governance of food systems. Perspectives are emerging that emphasise the national and global nature of the challenges for Indigenous communities around food and natural resource governance.

My previous research (Ouma 2013, 21) discusses the historical and contemporary globalisation processes that influence governance structures and organisational approaches to land stewardship.

Discussion, perceptions, and recommendations

This chapter presents arguments supporting the need to reverse the ongoing erosion of biocultural diversity and knowledge associated with its governance. One way to do this is by re-engaging and realigning local and regional agreements with Indigenous knowledge of the environment and governance. The chapter discusses local Indigenous knowledge of and practices related to food. This food knowledge is embedded in knowledge of the environment and of biocultural diversity and in local governance structures (Figures 15.2 and 15.3). The chapter suggests that engaging efforts that embrace Indigenous cultural practices related to food systems, food security, and governance could help engage and inform mainstream official policies around the challenges of food (in) security in the Arctic and beyond.

Lessons learned in one region could be used to strengthen Indigenous environmental governance of food security elsewhere. Indigenous and non-Indigenous Peoples in various geographical locations have similar approaches and face similar challenges to environmental governance that

Figure 15.2 A seasonal farming prognosis station used to provide information on which crops to plant and an estimate on the season's prognosis for harvesting for rice and sorghum. Munguwapili (village), Mwanza Region, Tanzania. Photo by the author.

impact food security. The stories shared in the talking circle advocate more North–South dialogue, something that could build and expand relationships and networks to create a truly global community to face global challenges. This approach could help include and uphold Indigenous knowledge of food resources in policies related to food security.

The importance of the relationship between a healthy environment and physically active ways of accessing food illustrate the strengths of Indigenous Peoples, who use traditional knowledge and governance systems to act as stewards of the environment (Figure 15.3 & 15.4). This stewardship ensures sustained access to food, which in turn guarantees communities' food security (Figure 15.3).

As mentioned during the talking circle discussions, the issues discussed there could support and inform similar forums in the future, which could then build on this shared, collectively created knowledge. Participants also expressed this hope when consenting to have their oral contributions translated into textual proceedings. Their messages might help strengthen the case for traditional knowledge in general and traditional food knowledge specifically as a valid, meaningful basis for the wise governance of natural resources and food systems within and beyond Indigenous communities and their traditional territories around the globe. This advocacy could contribute to policy in the current discourse around food (in) security. Together, the understanding of the interrelatedness of all components of the environment and the belief in the power and spirituality of nature have enabled

Figure 15.3 This is an example of community governance in the Canadian Arctic. The community of Délı̨nę, which has recently achieved self-governing status, has also developed Canada's first caribou conservation plan. This has implications for food security. Some years ago, the community of Délı̨nę was recognised as the steward of the world's first Indigenous-sponsored international biosphere reserve. This is one of the examples of Indigenous environmental governance that were discussed at the 2011 People in Places Indigenous talking circle.

communities worldwide to live sustainably in their local environments for millennia. Naturally, this includes secure access to food.

To address the challenges of access to food resources and dilemmas around the governance of land resources that are now evident globally due to the loss of biodiversity, strategies based on TEK – which exists in some form all over the world – should be recognised and applied. This could help humans create a much better relationship with our environment and therefore a more sustainable way to manage food resources and food security (Figure 15.4).

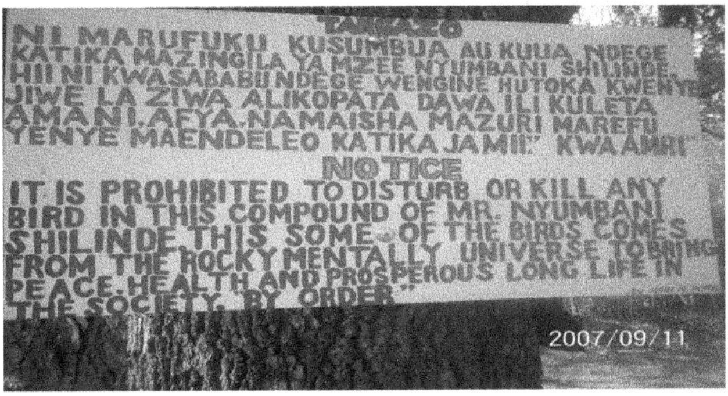

Figure 15.4 An example of local traditional governance of bird life. Situated in the Bondo Ward Missungwi District of Mwanza Region, Tanzania in a village food clinic and hospital run by Shamans. This example illustrates how messages from the talking circle on Indigenous governance and sustainable stewardship of natural resources are relevant to a global discourse. Photo by the author, 2007.

References

Anderson, J. 2010. *Understanding geography; places and traces.* London: Routledge.

Battiste, M. 2002. *Indigenous knowledge and pedagogy in First Nations education. A literature review with recommendations.* National Working Group on Education and the Minister of Indian Affairs. (INAC): Ontario, Ottawa: Indian and Northern Affairs Apamuwek Institute Publications.

Battiste, M. and Henderson, J.Y. 2000. *Protecting Indigenous knowledge and heritage: A global challenge.* Saskatoon: Purich Publications.

Berkes, F. 2009. Indigenous ways of knowing and the study of environmental change. *Journal of the Royal Society of New Zealand* 39(4): 151–156. doi:10.1080/03014220909510568.

Berkes, F. 2017. Environmental governance for the Anthropocene? Social-ecological systems, resilience, and collaborative learning. *Sustainability* 9(7): 12–32. doi:10.3390/su9071232.

Beyer, S. 2009. *Singing to the plants: A guide to Mestizo shamanism in the upper Amazon.* Albuquerque: University of New Mexico Press.

Blythe, J., and McKenna Brown, R. (eds.) 2004. *Maintaining the links, language, identity and the land.* Proceedings of the Seventh Conference of the Foundation of Endangered Languages, Broome, Australia, Foundation for Endangered Languages, Bath, UK.

Brändström, P. 1990. *Boundless universe: The cultural expansion among the Sukuma Nyamwezi.* Ph.D. diss., Uppsala University.

Bukurura, S.H. 1994. *Indigenous communication systems: Lessons and experiences from among the Sukuma and Nyamwezi of West-Central Tanzania.* Paper presented at the National Conference on Indigenous Knowledge (IK) and Sustainable Development–Sokoine University of Agriculture (Morogoro) Tanzania.

Cunningham, A.B. 1993. *African medicinal plants: Setting priorities at the interface between conservation and primary health care.* People and Plants working paper, No. 1. UNESCO, Paris.
Dahlström, L. and Mannberg, J. (eds.) 2006. *Critical educational visions and practices in neo-liberal times.* Umeå and IICBA Addis Ababa: Global South Network Publisher, 127–132.
De Sousa Santos, B. 2007. *Cognitive justice in a global world: Prudent knowledges for a decent life.* Lanham, MD: Lexington Books.
Ekarin P., Kulvadee, K. and Opart, P. 2016. Traditional ecological knowledge in Thailand: Mechanisms and contributions to food security. *Kasetsart Journal of Social Sciences* 37(2): 82–88.
Ermine, W. 1995. Aboriginal epistemology. In M. Battiste and J. Barman (eds.), *First Nations education in Canada. The circle unfolds.* Vancouver: University of British Colombia Press, 101–112.
Fatnowna, S. and Pickett, H. 2002. Establishing protocols for an Indigenous-directed process. Perspectives from Australia and the region. In H. Odora (ed.), *Indigenous knowledge and the integration of knowledge systems towards a philosophy of articulation.* Claremont: New African Books, 67–95.
Food Tank. 2018. Retrieved 23 February 2019, from https://foodtank.com/.
Frawley, K. 1999. A 'green' vision: The evolution of Australian environmentalism. In K. Anderson and F. Gale (eds.), *Cultural Geographies.* Melbourne: Longman, 265–293.
Gibson, J. 2002. Traditional knowledge and the international context for protection. *SCRIPT-ed Open Licence, SOL* 1: 1.
Gibson-Graham, J.K. 2006. *A post-capitalist politics.* Minneapolis: University of Minnesota Press.
Grenier, J. 1998. *Working with Indigenous knowledge: A guide for researchers.* Ottawa, Canada: International Development Research Centre (IDRC).
Hoppers, O. C. 2002. *Indigenous knowledge and the integration of knowledge systems: Towards a philosophy of articulation.* Claremont, South Africa: New African Books.
Hart, G. 2002. *Disabling globalisation, places of power in post-apartheid South Africa.* Berkeley: University of California Press and University of Natal Press.
Kurtz, D. 2013. Indigenous methodologies: Traversing Indigenous and Western worldviews in research. *AlterNative: An International Journal of Indigenous Peoples* 9 (3): 217–229. doi:10.1177/117718011,300900,303.
Maffi, L. 2005. Linguistic, cultural, and biological diversity. *Annual Review of Anthropology* 34: 599–617.
Mollel, N.J. 1994. *Utilization of Indigenous knowledge: Toward sustainable farming in agro and pastoral societies in the lowlands of Northern Tanzania.* Arusha, Tanzania Selian Agricultural Research Institute Press.
Morphy, H. 1995. Landscape and the introduction of the ancestral past. In E. Hirsch and M. O'Hanlon (eds.), *The anthropology of landscape: Perspectives on place and space.* Oxford: Clarendon Press, 184–209.
Mshana, R. 2002. Globalisation and intellectual property rights. In H. Odora (eds.), *Indigenous knowledge and the integration of knowledge systems: Towards a philosophy of articulation.* Claremont, South Africa: New African Books, 200–208.
Muin'iskw, J. and Crowfeather, D. 2016. Mi'kmaw spirituality – Talking circles. Retrieved from http://www.muiniskw.org/pgCulture2c.htm.

Oguamanam, C. 2006. *International law and Indigenous knowledge: Intellectual property, plant biodiversity, and traditional medicine.* Toronto: University of Toronto Press.

Ouma, A. 2013. *From rural gift to urban commodity: Traditional medicinal knowledge and socio-spatial transformation in the eastern Lake Victoria region.* Ph.D. diss., Umea University.

Petrov, A., BurnSilver, S., Stuart Chapin III, F., Fondahl, G., Graybill, J., and Keil.K. 2016. Arctic sustainability research: Toward a new agenda. *Polar Geography* 39(3): 165–178. doi:10.1080/1088937X.2016.1217095.

Rössler, M. 2005. World heritage cultural landscapes: A global perspective. In J.R. Brown, N. Mitchell, and M. Beresford (eds.), *The protected landscape approach linking nature, culture and community.* Gland, Switzerland: IUCN, 37–46.

Sandberg, L. 2011. *Fear of violence and gendered power relations: Responses to threat in public space in Sweden.* Ph.D. diss., Umea University.

Silverman, H. 2011. Contested cultural heritage: A selective historiography. In H. Silverman (ed.), *Contested cultural heritage.* New York, NY: Springer, 1–49.

Simmons, D., Bayha, W., Beaulieu, D., Gladu, D., and Manseau, M. 2012. Aboriginal talking circle: Aboriginal perspectives on caribou conservation – Overview by the aboriginal talking circle coordinating team. *Rangifer* 20: 17–19.

Tanzania Gateway. 2010. Tanzania Indigenous knowledge database: Sharing and dissemination of IK information. Retrieved from www.tanzaniagateway.org/ik/.

UNESCO Digital Library. 2011. *UNESCO biodiversity initiative.* Retrieved from https://unesdoc.unesco.org/ark:/48223/pf0000213313.

Whittaker, N. 2015. *Traditional ecological knowledge and sustainable food sourcing: Dayutang village, Hani Rice Terraces.* Independent Study Project (ISP) Collection. Retrieved from https://digitalcollections.sit.edu/isp_collection/2069.

Wiles, J., Rosenberg, M. and Kearns, R. 2005. Narrative analysis as a strategy for understanding interview talk in geographic research. *Area* 37 (1): 89–99.

Index

Note: Page numbers followed by "n" refer to notes.

For Product Safety Concerns and Information please contact our EU representative GPSR@taylorandfrancis.com
Taylor & Francis Verlag GmbH, Kaufingerstraße 24, 80331 München, Germany

www.ingramcontent.com/pod-product-compliance
Ingram Content Group UK Ltd.
Pitfield, Milton Keynes, MK11 3LW, UK
UKHW022010180926
475701UK00003B/287